OPTICAL PROPERTIES OF METALS / INTERMOLECULAR INTERACTIONS

OPTICHESKIE SVOISTVA METALLOV / MEZHMOLEKULYARNOE VZAIMODEISTVIE

ОПТИЧЕСКИЕ СВОЙСТВА МЕТАЛЛОВ / МЕЖМОЛЕКУЛЯРНОЕ ВЗАИМОДЕЙСТВИЕ

The Lebedev Physics Institute Series

Editor: Academician D. V. Skobel'tsyn

Director, P. N. Lebedev Physics Institute, Academy of Sciences of the USSR

Volume 25	Optical Methods of Investigating Solid Bodies
Volume 26	Cosmic Rays
Volume 27	Research in Molecular Spectroscopy
Volume 28	Radio Telescopes
Volume 29	Quantum Field Theory and Hydrodynamics
Volume 30	Physical Optics
Volume 31	Quantum Electronics in Lasers and Masers, Part 1
Volume 32	Plasma Physics
Volume 33	Studies of Nuclear Reactions
Volume 34	Photomesic and Photonuclear Processes
Volume 35	Electronic and Vibrational Spectra of Molecules
Volume 36	Photodisintegration of Nuclei in the Giant Resonance Region
Volume 37	Electrical and Optical Properties of Semiconductors
Volume 38	Wideband Cruciform Radio Telescope Research
Volume 39	Optical Studies in Liquids and Solids
Volume 40	Experimental Physics: Methods and Apparatus
Volume 41	The Nucleon Compton Effect at Low and Medium Energies
Volume 42	Electronics in Experimental Physics
Volume 43	Nonlinear Optics
Volume 44	Nuclear Physics and Interaction of Particles with Matter
Volume 45	Programming and Computer Techniques in Experimental Physics
Volume 46	Cosmic Rays and Nuclear Interactions at High Energies
Volume 47	Radio Astronomy: Instruments and Observations
Volume 48	Surface Properties of Semiconductors and Dynamics of Ionic Crystals
Volume 49	Quantum Electronics and Paramagnetic Resonance
Volume 50	Electroluminescence
Volume 51	Physics of Atomic Collisions
Volume 52	Quantum Electronics in Lasers and Masers, Part 2
Volume 53	Studies in Nuclear Physics
Volume 55	Optical Properties of Metals and Intermolecular Interactions
Volume 56	Physical Processes in Lasers

In preparation

Volume 54	Photomesic and Photonuclear Reactions and Investigation Methods with Synchrotrons
Volume 57	Theory of Interaction of Elementary Particles at High Energies
Volume 58	Investigations on Nonlinear Optics and Hyperacoustics
Volume 59	Luminescence and Nonlinear Optics
Volume 60	Spectroscopy of Laser Crystals
Volume 61	Theory of Plasmas
Volume 62	Stellar Atmospheres and Interplanetary Plasma: Techniques for Radioastronomical Devices
Volume 63	Nuclear Reactions and Interactions of Neutrons with Matter
Volume 64	Primary Cosmic Radiation
Volume 65	Stellarators
Volume 66	Theory of Coherent Acceleration of Particles and Emission of Relativistic Bunches
Volume 67	Physical Investigations of Strong Magnetic Fields
Volume 68	Emissive Recombination in Semiconducting Crystals
Volume 69	Nuclear Reactions and Accelerators of Charged Particles

Proceedings (Trudy) of the P. N. Lebedev Physics Institute

Volume 55

OPTICAL PROPERTIES OF METALS
AND
INTERMOLECULAR INTERACTIONS

Edited by
Academician D. V. Skobel'tsyn
Director, P. N. Lebedev Physics Institute
Academy of Sciences of the USSR, Moscow

Translated from Russian by
G. D. Archard

CONSULTANTS BUREAU
NEW YORK–LONDON
1973

Library of Congress Cataloging in Publication Data

Motulevich, G P
Optical properties of metals.

(Proceedings (Trudy) of the P. N. Lebedev Physics Institute, v. 55)
Translation of Opticheskie svoĭstva metallov.
Includes bibliographies.
1. Metals–Optical properties. 2. Molecular spectra. I. Malyshev, V. I. Mezhmolekuli͡arnoe vzaimodeĭstvie. English. 1973. II. Skobel'ṫsyn, Dmitriĭ Vladimirovich, 1892– ed. III. Title. IV. Title: Intermolecular interactions. V. Series: Akademii͡a nauk SSSR. Fizicheskiĭ institut. Proceedings, v. 55.
QC1.A4114 vol. 55 [QD473] 530'.08s [547'.3] 72-94827
ISBN 978-1-4684-8374-1 ISBN 978-1-4684-8372-7 (eBook)
DOI 10.1007/978-1-4684-8372-7

The original Russian text was published by Nauka Press in Moscow in 1971 for the Academy of Sciences of the USSR as Volume 55 of the Proceedings of the P. N. Lebedev Institute. The present translation is published under an agreement with Mezhdunarodnaya Kniga, the Soviet book export agency.

A Division of Plenum Publishing Corporation
227 West 17th Street, New York, N. Y. 10011
United Kingdom edition published by Consultants Bureau, London
A Division of Plenum Publishing Company, Ltd.
Davis House (4th Floor), 8 Scrubs Lane, Harlesden, London, NW10 6SE, England

PREFACE

The first part of this collection sets out the results of some experimental and theoretical investigations into the optical properties of nontransition metals. The extensive future prospects of metal optics are indicated; the use of metal optics enables a whole series of important electron properties of metals to be determined.

Results obtained by studying intermolecular forces (the hydrogen bond and van der Waals forces) using spectroscopic methods (Raman effect and infrared absorption) are presented in the second part. A method of studying the true absorption of the drop phase of a water cloud is described. Methods of increasing the dispersion of manufactured spectral instruments and constructing various infrared spectrometers are indicated.

The publication is intended for scientific workers, graduates, and students concerned with problems of metal optics, the electron properties of metals, and molecular spectroscopy.

CONTENTS

OPTICAL PROPERTIES OF NONTRANSITION METALS
G. P. Motulevich

EXPERIMENTAL STUDIES OF INTERMOLECULAR FORCES BY SPECTROSCOPIC METHODS AND THE DEVELOPMENT OF SPECTRAL APPARATUS

V. I. Malyshev

Dedicated to the Memory of the Author's Father

OPTICAL PROPERTIES OF NONTRANSITION METALS*

G. P. Motulevich

INTRODUCTION

Metal-optical research started developing vigorously at the beginning of the present century, when Drude, regarding electrons as free particles, established a relationship between the optical constants and the concentration of conduction electrons [1]. The first approximate values of this quantity were obtained from optical measurements at this time.

However, serious disagreements soon developed between Drude's theory and experiment. The absorption of light by a metal at low temperature should vanish, according to Drude, whereas in fact it remains constant and quite large. By way of example (Table 1) we may present some data relating to copper [2, 3].

The difference is particularly great at helium temperatures, at which the experimental values for different metals are hundreds of times greater than the theoretical. This large discrepancy was attributed to the fact that, in optics, one is concerned with the surface layer, the properties of which differ considerably from those of the bulk metal. This point of view retarded the development of optical measurements.

In 1953-1954 Dingle [3] and Ginzburg [4, 5] applied the theory of the anomalous skin effect to optics; this was previously developed by Pippard [6], Reuter and Sondheimer [7], Chambers [8], Holstein [9], and others for radio-frequency measurements. According to this theory, there is a nonlocalized relationship between the current density within the metal and the electric field of a light wave. Allowance for this fact in considering the diffuse reflection of electrons from a metal surface† leads to the finite absorption of light by the metal, even at $T = 0$. These investigations served as an impetus to the development of metal-optical research.

It was later found that there was yet another reason for the effect. This was associated with the quantum character of the interaction between electrons and light. After absorbing a photon, an electron may emit a whole phonon spectrum at $T = 0$, leading to a finite absorption of light at absolute zero. These questions were considered theoretically by Gurzhi [11, 12] and Holstein [13].

Allowance for the anomalous skin effect and the quantum character of the interaction of an electron with a photon leads to agreement between the experimental and theoretical temperature dependence of the light absorption coefficient (Chapter VII).

* Abridged text of Doctoral Dissertation defended on January 22, 1968, at the P. N. Lebedev Physics Institute, Academy of Sciences of the USSR.

† The diffuse reflection of electrons from a metal surface is confirmed experimentally [8, 10].

TABLE 1. Temperature Dependence of the Absorption Coefficient A (in %) for Copper

T, °K	A, experimental	A, Drude's theory
293	1.17	0.31
78	0.86	0.058
4.2	0.39	0.0020

The state of metal optics in 1955 was analyzed in our previous publications [2, 5]. In the review [2] we presented a theory relating the optical constants of a metal to its microcharacteristics; this was based on the isotropic model of almost free electrons, allowing for the anomalous character of the skin effect. We emphasized the importance of metal-optical measurements, particularly in the infrared region, in order to determine the concentration of the conduction electrons and other microcharacteristics of metals. We also analyzed existing experimental investigations, and indicated that, at that time, there were no reliable data regarding the optical constants of metals, especially in the infrared region. There were indeed no reasonably accurate methods of measuring in the infrared region. Methods of preparing samples for investigation suffered from serious defects.

Thus the first problem presented for our consideration lay in developing new methods of measuring the optical constants n and $\varkappa$ ($n - i\varkappa$ being the complex refractive index) giving the required accuracy in the infrared part of the spectrum, and also in developing methods of preparing samples with properties the same as or close to the properties of the bulk metal.

We developed a new polarization method of measuring the optical constants of metals, using the multiple reflection of light from the mirrors under consideration. This method enabled us to measure n and $\varkappa$ in the infrared and visible parts of the spectrum to an accuracy of 1-2% over a wide temperature range, from ordinary to helium temperatures [14-20].

We also developed methods of preparing the samples to be studied, and obtained samples of the required quality [17-24].

All this methodical work was carried out principally at a time at which the complexity and anisotropy of the Fermi surface was being recognized in solid-state theory (see, for example, a review by Lifshits and Kaganov [25]). It accordingly came to be realized that the simple idea of "almost free" electrons, particularly in the case of polyvalent metals, failed to approximate the true properties of matter. This also cast doubt on existing theories associating the optical properties of metals with their microcharacteristics.

The question as to the applicability of the model of almost free electrons to metal optics arose a long time ago. The Fourier component of the potential of the ions inside the metal, corresponding to the interatomic distance, actually amounts to tens of electron volts; this is considerably greater than the energy of the electron at the Fermi surface, which approximately equals 10 eV. It is strange that the electron should behave as almost free at such high potentials.

Thus the second problem lay in theoretically considering the relation between the optical properties of metals and their principal microcharacteristics on the basis of the latest achievements in solid-state physics.

The development of solid-state theory has recently led to some very fruitful ideas associated with the concept of the pseudopotential. These ideas were developed by Harrison [26], Heine [27, 28], Ziman [29-32], and a number of other authors. According to these papers, the screening of the ions by the electrons, and also the possibility of smoothing out the rapid electron oscillations inside the ionic core, justify the use of a pseudopotential and a pseudo-wave function.

The pseudopotential is essentially the difference between the screened potential of the ion and the effective potential associated with the rapid oscillatory motion of the electron inside the ion; it may be smaller than the original potential and constitute a far smoother function than the true potential. The pseudo-wave function corresponds to the true wave function without allowing for the rapid oscillations inside the ionic core. Mathematically the rapid oscillations of the wave function inside the atomic core are associated with the fact that the wave function of a conduction electron must be orthogonal to all the wave functions of the inner shells.

After the aforementioned investigations, it became clear that the valence electron might be regarded as simply "feeling" the weak pseudopotential in many practical problems. This gives a reasonable basis for the model of "almost-free" electrons, and enables us to consider the optical properties of ordinary metals in the infrared and visible parts of the spectrum using the concept of almost-free electrons as well. However, the properties of these almost-free electrons differ from the properties of completely free electrons. The coherent scattering of electrons by the lattice planes has a considerable effect on the optical properties of metals, even in the case of the weak pseudopotential.

We considered the relation between the Fourier components of the pseudopotential and the principal integrated electron characteristics of a metal such as the concentration of conduction electrons N, the total area of the Fermi surface S_F, the average* velocity of the electrons on the Fermi surface $\langle v_F \rangle$, the density of states of the electrons on the Fermi surface $(dY/dE)_F$, and the shape of the Fermi surface. We also established a link between these electron characteristics (and their temperature dependence) and the optical properties of metals in the infrared part of the spectrum, as well as a relation between the optical properties of metals in the visible and near infrared and the Fourier components of the pseudopotential [33-35]. This made it possible to determine the Fourier components from optical measurements. Since the same Fourier components also determine the band structure and other properties depending on electron — ion interaction (for example, the de Haas — van Alphen effect), cyclotron resonance, the phonon spectrum, the absorption of ultrasound in a magnetic field, and so on), we may also use the values obtained in optics for interpreting other data.

Thus, after the foregoing theoretical work, it became possible to use the results of optical measurements for obtaining many basic electron properties of nontransition metals.

It should be noted that the approach to the problem considered in this dissertation may be extended not only to metals in the crystalline state but also to molten metals, alloys (both ordered and disordered), and to amorphous states of metals.

The relationships presented in this treatise apply to ordinary (i.e., nontransition) metals. It is as yet uncertain how far they may be applied to transition metals as well.

Detailed measurements of the optical properties of polyvalent nontransition metals have demonstrated the validity of the picture thus developed.

The third problem lay in carrying out detailed experimental investigations into the optical properties of metals and in determining their main electron characteristics.

* The angular brackets $\langle\ \rangle$ denote an average value over the Fermi surface.

In the experimental work, principal attention was directed at a study of polyvalent metals. Specially studied were metals of the third and fourth groups in the periodic table, namely, In, Al, Pb, and Sn [14, 15, 17-23, 36-38]. The lattice should have a greater effect on the behavior of the conduction and valence electrons of these metals than it does in the case of univalent metals, since the momenta of the valence electrons are large and the original Harrison sphere plotted in momentum space intersects a large number of Bragg planes. Despite the great interest offered by these metals, they have hardly been studied at all hitherto. The only exception is aluminum, which has been thoroughly studied in view of the wide use of sprayed-on aluminum films in a number of scientific and technological fields.

Apart from these polyvalent metals, we studied [24] the univalent metal Au; a complex study of the properties of this metal augmented our considerations of the relation between the optical properties and the band structure of a metal. A great deal of work has been done on the optical properties of noble metals. We succeeded in producing films with characteristics much closer to those of the bulk metal than other authors.

Since we were interested in the properties of both conduction electrons and also the electrons taking part in interband transitions, this determined the spectral range to be studied, viz., the infrared and visible parts of the spectrum. Analysis of this range facilitated the experimental determination of the main electron properties of the metals.

A number of special characteristics of the interaction between electrons, light, and the lattice called for measurements over a wide temperature range. In view of this, we measured the optical constants of a number of metals at 298, 78, and 4.2°K. These measurements enabled us to establish the temperature dependence of the Fourier components of the pseudopotential, the concentration of the conduction electrons, and various other leading electron parameters of the metals.

In addition to the optical investigations (constituting the principal measurements), we also obtained the density and a number of electrical, superconducting, and other characteristics of simultaneously prepared samples. Only by making such complex measurements were we able, firstly, to develop the necessary technology for preparing the samples and, secondly, to obtain the microcharacteristics of the metals required for the construction of the microscopic theory of the latter more completely [39].

The absence of data relating to the quality of the samples studied prevents us from using the results of the majority of previous investigations. The reason is that different methods of sample preparation may give different structures, and this will lead to substantial changes in the properties of the metals determined by the conduction electrons, the valence electrons, and their interaction with the lattice (owing to the changes in band structure and the additional scattering associated with impurities, defects in the lattice structure, and the boundaries of the crystallites). Thus it is essential to monitor the quality of the test layers. This may be done if the optical investigations are supplemented by a study of a number of other properties.

As a result of the work in question we were able to determine such microcharacteristics of the metals studied as the concentration of conduction electrons, the area of the Fermi surface, the average velocity of the electrons on the Fermi surface, the Fourier components of the pseudopotential, the frequency of the collisions between electrons, phonons, and crystal defects, and also the temperature dependence of these.

Comparison between the electron characteristics obtained by optical and other methods (the de Haas — van Alphen effect, cyclotron resonance, the absorption of ultrasound in magnetic fields, radio-frequency measurements of surface impedance, etc.) indicate close agreement between all of these.

Thus, metal optics offer great prospects for the determination of many of the basic electron characteristics of metals. It is to be hoped that our investigations will present a picture of metal-optical phenomena in accordance with the present-day state of solid-state physics.

CHAPTER I

METHOD OF THE KINETIC EQUATION IN METAL OPTICS

§ 1. Kinetic Equation for the Infrared Part of the Spectrum

Experimental investigations into the optical properties of metals determine the real and imaginary parts of the complex refractive index $n-i\varkappa$. By carrying out measurements over wide spectral and temperature ranges we obtain two functions $\varkappa(\omega, T)$ and $n(\omega, T)$. Let us consider how these quantities are associated with the microcharacteristics of the metal. For this purpose we use the method of the kinetic equation [40, 41].

1. We shall describe the state of the electrons by a distribution function $\tilde{f}(\mathbf{p}, \mathbf{r}, t)$ depending on the momentum $\mathbf{p}$, the coordinate $\mathbf{r}$, and the time t. The equilibrium distribution function holding (in the absence of a field) in a homogeneous metal at temperature T is shown in Fig. 1. This is the well-known Fermi function

$$f_0(\mathbf{p}) = f_0(E(\mathbf{p})) = \left[\exp\left(\frac{E-E_F}{kT}\right)+1\right]^{-1}, \tag{I.1}$$

where E is the energy of the electron, and E_F is the energy of an electron on the Fermi surface.

Let us use $f(\mathbf{p}, \mathbf{r}, t)$ to denote the increment to the distribution function associated with the effect of the field of the light wave. In all cases so far encountered in metal physics, we may consider that $|f| \ll |f_0|$. Thus the total distribution function equals

$$\tilde{f}(\mathbf{p}, \mathbf{r}, t) = f_0(\mathbf{p}) + f(\mathbf{p}, \mathbf{r}, t). \tag{I.2}$$

The spin variables are not taken into account in (I.2).

In the linear approximation with respect to f and $\mathscr{E}$ the kinetic equation takes the form

$$\frac{\partial f}{\partial t} + \mathbf{v}\frac{\partial f}{\partial \mathbf{r}} + e\mathscr{E}\frac{\partial f_0}{\partial \mathbf{p}} + \nu f = 0. \tag{I.3}$$

Here, $\mathbf{v}$ is the velocity of the electron; $\mathscr{E}$ is the electric field of the light wave; and ν is the effective collision frequency of the electron.

In this approximation the collision integral is written in the form νf. When only the first term in the expansion of the collision integral in powers of f is considered, this form is justified for high (room) and low (helium) temperatures [11, 41, 42]; it may be used for ex-

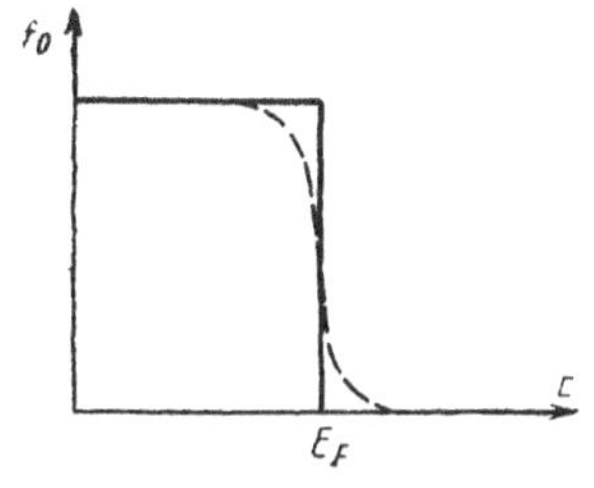

Fig. 1. Dependence of the equilibrium distribution function f_0 on the electron energy E. E_F is the Fermi energy. The broken line relates to $T \neq 0$.

trapolation at intermediate temperatures. We shall therefore use the foregoing expression for the collision integral at all temperatures without restriction.* In the approximation under consideration we may consider (see below) that the effective electron collision frequency is $\nu = \nu_{ep} + \nu_{ee} + \nu_{ed}$, where ν_{ep} is the electron—phonon collision frequency; ν_{ee} is the frequency of interelectron collisions; and ν_{ed} is the frequency of collisions between electrons and impurities or defects.

2. Equation (I.3) is the classical one. However, at high frequencies, for which $\hbar\omega > kT$, the interaction between the electrons and the field of the light wave exhibits a quantum character. An electron absorbing a light quantum changes its energy by an amount $\hbar\omega$ in one jump. The difference between the classical and quantum case is particularly substantial in the limiting case of low temperatures and large values of $\hbar\omega$, i.e., when the following equation is satisfied:

$$\hbar\omega \gg k\theta \gg kT, \tag{I.4}$$

where θ is the Debye temperature. This case usually occurs in metal optics in low-temperature measurements. Here we shall be interested in how this affects the kinetic equation (I.3). Clearly it is more correct to use the quantum kinetic equation. Gurzhi, developing the work of Bogolyubov and Gurov [44] and Klimontovich and Silin [45], obtained a quantum kinetic equation for electrons in a metal situated in the field of an electromagnetic wave [11, 12]. As quantum distribution function, a mixed representation of the density matrix was taken. The equation obtained for this function contains a certain operator constituting a quantum generalization of the ordinary collision integral. We shall not present the complex expressions for this operator here. We simply note that if, in the case of metal-optical problems, the following inequalities are satisfied:

$$\exp(\hbar\omega/kT) \gg 1 \quad \text{and} \quad \exp(\hbar\omega/kT) \gg \exp(\theta/T), \tag{I.5}$$

then instead of the collision operator we may use a certain function of temperature. This enables us to introduce the effective collision frequency between electrons and phonons $\nu_{ep}(T)$, which may be written in the form†

$$\nu_{ep}(T) = \nu_\theta \frac{T}{\theta} \varphi(T). \tag{I.6}$$

On satisfying inequalities (I.5) we have the following for $\varphi(T)$:

$$\varphi(T) = 2\left(\frac{T}{\theta}\right)^4 \int_0^{\theta/T} x^4 \coth\frac{x}{2}\, dx. \tag{I.6a}$$

The form of the function $\varphi(T)$ was calculated by Gurzhi [11, 12]. For

$$\begin{gathered} T \gg \theta \quad \varphi(T) \approx 1, \\ \nu_{ep} \approx \nu_{ep}^{\text{cl}} = \nu_\theta \frac{T}{\theta}, \end{gathered} \tag{I.6b}$$

* In the case of a sharply anomalous skin effect this approximation is substantiated for any temperatures [43].

† More precisely Eqs. (I.6)-(I.6b) contain not the Debye temperature θ but a characteristic temperature θ_R determining the temperature dependence of the static conductivity. Usually these quantities lie close together, $\theta_R \approx \theta$.

for

$$T \ll \theta \quad \varphi(T) \approx \frac{2}{5} \frac{\theta}{T},$$
$$\nu_{ep} \approx \frac{2}{5} \nu_\theta.$$

It is well known that the classical frequency of electron – phonon collisions ν_{ep}^{cl} is proportional to T^5 at low temperatures. The limiting value of ν_{ep} for $T \ll \theta$ may also be obtained from Holstein's paper [13].

Thus, allowing for the quantum nature of the interaction between electrons and light has the effect that, in the presence of photons, the electron – phonon collision frequency depends relatively little on temperature, remaining finite even for T = 0. We studied this effect experimentally. The results of the comparison between theory and experiment are presented in Chapter VII.

In this chapter we shall allow for the fact that the ν_{ep} entering into Eq. (I.3) by virtue of ν differs from the classical value of this quantity.

3. Let us consider one further interaction leading to a change in ν. This is the effect of interelectron interaction on the quantity in question. This problem was considered by Ginzburg and Silin [46], Pitaevskii [47], and Gurzhi [48]. The influence of interelectron interaction on the optical constants in the infrared part of the spectrum is slight, being severely restricted by the Pauli principle. An electron lying in the diffuse Fermi region can only collide with other electrons from the same region. After the collision the two electrons should again fall into the diffuse Fermi region. We may allow for interelectron collisions by introducing an additional frequency ν_{ee} leading to an additional absorption of light. According to [48] we have

$$\nu_{ee}(\omega, T) = \nu_{ee}^{cl}(T)\left[1 + \left(\frac{\hbar\omega}{2\pi kT}\right)^2\right]. \tag{I.7}$$

Here ν_{ee}^{cl} is the corresponding classical frequency of interelectron collisions proportional to T^2. In the near infrared $\hbar\omega \gg 2\pi kT$ and $\nu_{ee}(\omega) \gg \nu_{ee}^{cl}$. Thus the appearance of the interelectron collision frequency is much more probable in the optical than in the static case. It initially appeared that the frequency ν_{ee} would occur in metals in the near infrared part of the spectrum [49]. Later, however, after carefully studying the temperature dependence of the component of ν proportional to ω^2, we became convinced that ν_{ee} could only be observed at helium temperatures in metals not having any interband transitions in the infrared region [18, 22, 23]. This question is considered in more detail by Golovashkin [50, 51].

We should mention one further mechanism leading to the absorption of light, associated with scattering at impurities. By impurities we mean both chemical and physical impurities, including scattering at inhomogeneities, crystallite boundaries, and so forth. Allowance for this circumstance requires the introduction of yet another frequency ν_{ed}, which, according to the Matthiessen rule [41, 52], should be added to ν_{ep} and ν_{ee}.

4. Thus, in general, we have the following expressions for the effective electron collision frequency [39]:

$$\nu = \nu_{ep} + \nu_{ee} + \nu_{ed},$$
$$\nu_{ep} = \nu_\theta \frac{T}{\theta} \varphi(T),$$
$$\nu_{ee} = \nu_{ee}^{cl}\left[1 + \left(\frac{\hbar\omega}{2\pi kT}\right)^2\right]. \tag{I.8}$$

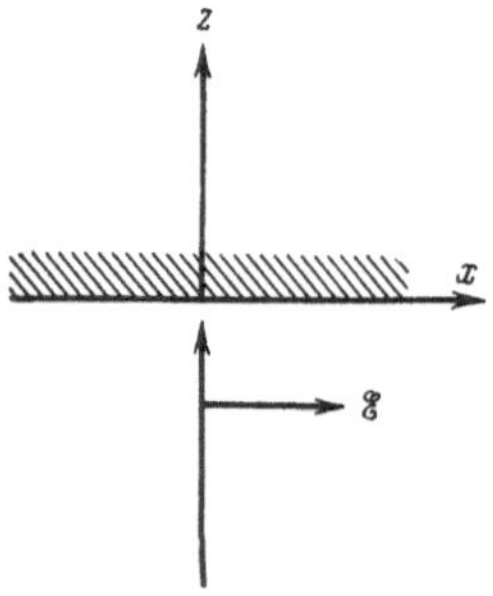

Fig. 2. Normal incidence of light on a metal occupying the half space z > 0 $\mathscr{E} \equiv \mathscr{E}_x$.

Equations (I.8) allow for the fact that the absorption of light by an electron in a metal is only possible when the electron simultaneously interacts with a phonon, another electron, or an impurity. We also derived the frequency of collisions between an electron and the surface of the metal ν_{ez} in [21, 22]. Here we shall consider these effects more consistently by solving the kinetic equation.

5. Equation (I.3) has to be solved simultaneously with the field equations. Let the metal occupy the upper half-space. We direct the z axis perpendicular to the interface (Fig. 2). For simplicity we confine attention to the case in which the direction of the electric field $\mathscr{E}$ in the metal coincides with the direction of the current density $\mathbf{j}$ (isotropic case, polycrystalline aggregates). Let linearly polarized light with the cyclical frequency ω fall on the metal normally to the surface. We take the x axis along the electric field of the light wave, i.e., $\mathscr{E} \equiv \mathscr{E}_x$. The complete system of equations will in this case take the following form (see [2]):

$$v_z \frac{\partial f}{\partial z} + (i\omega + \nu) f(z) = -e\mathscr{E}_x(z) \frac{\partial f_0}{\partial p_x}, \tag{I.3a}$$

$$\frac{d^2\mathscr{E}_x}{dz^2} + \frac{\omega^2}{c^2}\mathscr{E}_x = i\frac{4\pi\omega}{c^2} j_x, \tag{I.9}$$

$$j_x = \frac{2e}{(2\pi\hbar)^3}\int v_x f d^3p, \tag{I.10}$$

where c is the velocity of light.

In (I.3a) we allow for the fact that $\partial f/\partial t = i\omega f$. Since $(\partial f_0/\partial p_x) = (\partial f_0/\partial E)(\partial E/\partial p_x) = v_x(\partial f_0/\partial E)$ (E is the electron energy), while the function $\partial f_0/\partial E$ has a δ-like character, the optical properties are determined by the electrons on the Fermi surface.

The value of the first term in Eq. (I.3a), $v_z \partial f/\partial z$, determines the character of the skin effect.

If this term may be neglected, then we have the so-called normal skin effect, which mainly occurs for molten metals and metals in the amorphous state; it also occurs for certain metals in the crystalline state in the visible part of the spectrum and in the near infrared at high (room) temperatures.

If $v_z \partial f/\partial z \gg (i\omega + \nu) f$, then we have a sharply anomalous skin effect. This case arises in the uhf region for crystalline metals, but in the infrared range only for Group I metals at low (helium) temperatures.

If the term $v_z \partial f/\partial z$ is small, but still has to be regarded as a correction term, then we have the weakly anomalous skin effect. This is the most widespread case for metals in the crystalline state in the infrared part of the spectrum. For polyvalent metals it occurs at all temperatures, but for monovalent metals only at high (room) temperatures.

We shall now briefly consider the characteristics of the general solutions of Eqs. (I.3a), (I.9), and (I.10), giving special attention to the practically important cases of the weakly anomalous and normal skin effects.

§ 2. Anomalous Skin Effect

The general solution of Eqs. (I.3a), (I.9), and (I.10) has been considered by a number of authors [3, 4, 7, 53, 54]; it is also given in our own review [2] and in Sokolov's book [55]. In all these cases the Fermi surface was treated as spherical.

In general there is a nonlocal relationship between the current density and polarization inside the metal and the electric field of the light wave. In this case the complex dielectric constant ε' cannot be introduced. However, even here we may use the surface impedance Z and associate an effective complex dielectric constant $\varepsilon'_{\mathrm{eff}} = (n_{\mathrm{eff}} - i\varkappa_{\mathrm{eff}})^2$ with this. Usually, $1/|\varepsilon'_{\mathrm{eff}}|$ may be neglected in comparison with unity. Then*

$$Z = \frac{4\pi}{c\sqrt{\varepsilon'_{\mathrm{eff}}}}. \tag{I.11}$$

The surface impedance completely characterizes the reflection of electromagnetic waves from the metal surface for both the normal and the anomalous skin effects. The relation between the surface impedance and the boundary conditions was developed in [56] and presented in [2]. Experiment enables us to determine n_{eff} and $\varkappa_{\mathrm{eff}}$ if we use the properties of the reflected light, just as n and $\varkappa$ are determined in the case of the normal skin effect. In future we shall omit the subscript "eff."

Let us turn to the solution of Eqs. (I.3a), (I.9), and (I.10). The solution of this system depends essentially on the character of electron reflection from the metal surface. Experiment shows that reflection is diffuse for both single crystals and polycrystalline aggregates [8, 10]. Without going into the details of the methods which have been specially developed for solving this system of equations, we may simply indicate that, instead of the ordinary Ohm's law, in this case we have the following integrated relationship between the current density and the field:

$$j(z) = \int_0^\infty K\left(\frac{z-\xi}{l}\right)\mathscr{E}(\xi)\,d\xi, \tag{I.12}$$

where l is the free path of the electron and K is a kernel differing substantially from zero for an argument less than or equal to unity. This relationship between the current and field leads to an integrodifferential equation for the field in the metal of the following type:

$$\frac{d^2\mathscr{E}}{dz^2} + \frac{\omega^2}{c^2}\mathscr{E} = \mathrm{const}\int_0^\infty K\left(\frac{z-\xi}{l}\right)\mathscr{E}(\xi)\,d\xi. \tag{I.13}$$

The solution is complex and cumbersome. The final results are only obtained after numerical integration. The fall-off in the field $\mathscr{E}$ with depth is very rapid, although not exponential.

*It is quite easy to allow for the corrections associated with the terms $\sin^2\varphi/\varepsilon'$ (φ is the angle of incidence of the light on the surface of the metal). The corresponding corrections for n_{eff} and $\varkappa_{\mathrm{eff}}$ will be considered below.

In order to calculate the surface impedance Z, it is sufficient to calculate $(\partial \mathscr{E}/\partial z)_0$, where the subscript 0 signifies that the value of the derivative is taken at the surface of the metal. In fact [2, 56],

$$Z = \frac{4\pi}{c}\left[\frac{\mathscr{E}_x}{H_y}\right]_0 = -i\frac{4\pi\omega}{c^2}\left[\frac{\mathscr{E}_x}{\partial \mathscr{E}_x/\partial z}\right]_0. \tag{I.14}$$

The field on the surface of the metal is given $(\mathscr{E}_x)_0$ so that we only have to calculate $(\partial \mathscr{E}_x/\partial z)_0$. However, the calculation of this quantity is also very difficult and cumbersome. Fairly simple results are obtained in the limit of the sharply anomalous skin effect, when $l \gg \delta$ (here, δ is the classical depth of the skin layer). On satisfying this inequality, all the calculations may be completed quite simply. However, the relation fails to hold either in the infrared or in the visible region, and will not be discussed any further. If this relation is not satisfied, then usually we obtain expressions for the real and imaginary parts of the surface impedance Z in the form of expansions in series of small parameters. One of the parameters is $v_F/(\omega\delta)$ (v_F is the velocity of the electrons on the Fermi surface). The smallness of the parameter means that the path traveled by the electron in one period is small compared with the depth of the skin layer. The second parameter is ν/ω or ω/ν. Here ν is the effective electron collision frequency determined by Eq. (I.8). For the intermediate layer, in which $\omega \sim \nu$, only cumbersome, slowly converging series are obtained, in practice quite unsuitable for analyzing experimental data in view of their complexity.

We have already noted that in the infrared region from 1 to 12 μ there is usually a weakly anomalous skin effect. The complete solution of the system of equations (I.3a)-(I.10) is only necessary for Group I metals at low temperatures. In this case, the inequality $\nu/\omega \ll 1$ is valid. For the real and imaginary parts of the surface impedance $Z = R + iX$ we have the relations

$$cX/\pi = 1.34\cdot 10^{11}(1+P^2G)/(\sqrt{N}\,\lambda), \tag{I.15}$$

$$cR/\pi = 0.75\beta + 3.54\cdot 10^{-5}N^{-1/2}\nu P^2\beta D. \tag{I.16}$$

Here $\beta = v_F/c$; λ is the wavelength of the light in μ, and N is the concentration of conduction electrons in cm^{-3},

$$P = 2.59\cdot 10^{-11}\cdot\beta\sqrt{N}\,\lambda, \tag{I.17}$$

$$G = 0.0865 + 0.216q + 0.125q^2, \tag{I.17a}$$

$$D = 0.215 + 0.748q + 0.750q^2 + 0.216q^3, \tag{I.17b}$$

$$q = 2.045\cdot 10^{-5}\nu/(\sqrt{N}\,\beta) \tag{I.17c}$$

(ν is the effective collision frequency in sec^{-1}). For G and D we give the expressions obtained in [2, 3] by the classical kinetic equation method. According to [11], the calculation of G and D by the method of the quantum kinetic equation leads to slightly different coefficients. Considering that these terms only enter as corrections, they may validly be calculated to a lower accuracy, and instead of the complex quantum expressions Eqs. (I.17a) and (I.17b) may be employed.

In analyzing the experimental data in the infrared region it is best to use the foregoing relations (I.15)-(I.17). This question was fully discussed earlier [39]. We shall therefore not now consider the other relationships connecting such quantities as n, $\varkappa$, $n^2 - \varkappa^2$, $\varkappa/n$, $\sqrt{n^2+\varkappa^2}$, etc., with the microcharacteristics of the metal; these are all given in [2].

A detailed scheme for analyzing experimental results in cases involving the anomalous skin effect will be given below.

§ 3. Normal Skin Effect

The normal skin effect has been frequently expounded and is treated in particular in [2]; it corresponds to the neglect of the first term in (I.3a), after which this equation transforms from differential into algebraic. The conditions of electron reflection from the boundary of the metal are then unimportant, and all the optical constants are determined by volume effects.

The normal skin effect may easily be considered for an arbitrary Fermi surface.

Let us take the x, y, and z axes indicated in Fig. 2 along the principal axes of the ellipsoid ε'. It follows from (I.3a) that

$$f(z) = -\frac{e\mathscr{E}_x(z)}{i\omega+\nu}\frac{\partial f_0}{\partial p_x} = -\frac{e\mathscr{E}_x(z)}{i\omega+\nu}\frac{\partial f_0}{\partial E}\frac{\partial E}{\partial p_x}, \qquad \text{(I.18)}$$

where ν is determined by Eq. (I.8). Then from (I.10) we have

$$j_x(z) \equiv \left[\sigma_{xn} + \frac{i\omega(\varepsilon_{xn}-1)}{4\pi}\right]\mathscr{E}_x(z) \equiv \frac{i\omega}{4\pi}(\varepsilon'_{xn}-1)\mathscr{E}_x(z) = \frac{-2e^2\mathscr{E}_x(z)}{(2\pi\hbar)^3}\int\frac{v_x}{i\omega+\nu}\frac{\partial f_0}{\partial E}\frac{\partial E}{\partial p_x}d^3p. \qquad \text{(I.19)}$$

The index n means that the corresponding quantities relate to the normal skin effect; σ and ε are the conductivity and dielectric constant at the frequency ω; ε' is the complex dielectric constant at the same frequency ω.

It follows from solid-state theory that $\partial E/\partial p_x = v_x$. We note that $\partial f_0/\partial E$ is a very sharp function, and transform from a volume integral to a surface integral. Then

$$d^3p = dp_x dp_y dp_z = \frac{\partial E}{v}dS. \qquad \text{(I.20)}$$

Here dS is an element of the isoenergy surface E = const in momentum space, while v is the velocity of the electrons on this surface.

We obtain

$$j_x(z) = \frac{2e^2}{(2\pi\hbar)^3}\frac{\mathscr{E}_x(z)}{i\omega+\nu_x}\oint(v_x^2/v)\,dS_F. \qquad \text{(I.21)}$$

In (I.21) the integral is taken over the Fermi surface and dS_F is an element of the latter. We have taken the quantity $1/(i\omega+\nu)$ out from under the integral sign, i.e., we have used the theorem of the mean, giving ν the subscript x.

It follows from (I.21) that

$$\varepsilon'_{xn} - 1 = -\frac{4\pi e^2}{\omega(\omega - i\nu_x)}\left[\frac{2}{(2\pi\hbar)^3}\oint\frac{v_x^2}{v}dS_F\right]. \qquad \text{(I.22)}$$

Orienting the field of the incident wave along the y axis we have

$$\varepsilon'_{yn} - 1 = -\frac{4\pi e^2}{\omega(\omega - i\nu_y)}\left[\frac{2}{(2\pi\hbar)^3}\oint\frac{v_y^2}{v}dS_F\right]. \qquad \text{(I.22a)}$$

Analogously for ε'_{zn} we shall have

$$\varepsilon'_{zn} - 1 = -\frac{4\pi e^2}{\omega(\omega - i\nu_z)}\left[\frac{2}{(2\pi\hbar)^3}\oint\frac{v_z^2}{v}dS_F\right]. \qquad \text{(I.22b)}$$

If we determine the three complex quantities ε'_{xn}, ε'_{yn}, and ε'_{zn}, experimentally, we may determine the six quantities

$$\left(\frac{N}{m}\right)_i = \frac{2}{(2\pi\hbar)^3}\oint\frac{v_i^2}{v}dS_F \text{ and } \nu_i \quad i = 1, 2, 3. \qquad \text{(I.23)}$$

Here m is the mass of the free electron.

For metals of the cubic system

$$\oint \frac{v_x^2}{v} dS_F = \oint \frac{v_y^2}{v} dS_F = \oint \frac{v_z^2}{v} dS_F = \frac{1}{3} \oint v dS_F, \tag{I.24}$$

$$\nu_x = \nu_y = \nu_z = \nu,$$

so that by measuring the complex dielectric constant we may determine the quantities

$$\frac{N}{m} = \frac{2}{3(2\pi\hbar)^3} \oint v dS_F \text{ and } \nu. \tag{I.25}$$

The case of cubic symmetry is the most important. We shall subsequently confine attention to this case. Using (I.22) and (I.24) we obtain

$$\varepsilon_n - 1 \equiv (n_n - i\varkappa_n)^2 - 1 = -\frac{4\pi e^2}{\omega(\omega - i\nu)} \frac{N}{m}, \tag{I.26}$$

$$\varepsilon_n - 1 \equiv n_n^2 - \varkappa_n^2 - 1 = -\frac{4\pi e^2}{\omega^2 + \nu^2} \frac{N}{m},$$

$$\sigma_n \equiv \frac{n_n \varkappa_n \omega}{2\pi} = \frac{e^2 \nu}{\omega^2 + \nu^2} \frac{N}{m}. \tag{I.26a}$$

Equations (I.26) have exactly the same form as for a free classical electron gas with a conduction electron concentration N, and with a mass m equal to the mass of a free electron.* It is in this sense that we shall subsequently call the quantity N so defined the concentration of the conduction electrons.† The quantity N is defined by Eq. (I.25); it coincides with the concentration N which occurs in the rule of sums. The relation between this quantity and the concentration of valence electrons will be considered later. For the isotropic case

$$\frac{N}{m} = \frac{2}{3(2\pi\hbar)^3} v_F 4\pi p_F^2 = \frac{8\pi}{3} \frac{m_{\text{eff}} v_F^3}{(2\pi\hbar)^3}. \tag{I.27}$$

In this case the quantity N is simply related to the density of states $(dY/dE)_F$ near the Fermi surface

$$N = \frac{2}{3} m v_F^2 (dY/dE)_F. \tag{I.28}$$

Let us consider the conditions necessary for achieving the normal skin effect. For simplicity we confine attention to the isotropic case. Let us substitute the expression for the current density $j(z) = (i\omega/4\pi)(\varepsilon'_n - 1)\,\mathscr{E}(z)$ into (I.9). We obtain

$$\frac{d^2\mathscr{E}}{dz^2} + \frac{\omega^2}{c^2} \varepsilon'_n \mathscr{E} = 0, \tag{I.29}$$

whence

$$\mathscr{E} = \mathscr{E}(0) \exp\left(-i\frac{\omega}{c}\sqrt{\varepsilon'_n}\, z\right), \tag{I.30}$$

where $\mathscr{E}(0)$ is the field at the surface of the metal (i.e., at $z = 0$). The attenuation of the field

*Vonsovskii [57] showed that a relation between the dielectric constant and the concentration of the conduction electrons analogous to (I.26) might also be obtained by a quantum method, using the density matrix.

†Sometimes instead of N and m other authors use N_{val} and m_{eff}. The relation between these is given by the equation $N_{val}/m_{eff} = N/m$. Since we can only determine the ratio N/m by experiment, the two approaches are equivalent. We ourselves shall use N.

with increasing depth is exponential. Using (I.18) and (I.30) we obtain

$$\frac{\partial f}{\partial z}=-i\frac{\omega}{c}\sqrt{\varepsilon_n'}f=-\frac{\omega}{c}(\varkappa_n-in_n)f, \quad (I.31)$$

where

$$\sqrt{\varepsilon_n'}=n_n-i\varkappa_n. \quad (I.32)$$

Equation (I.31) gives the value of the omitted terms. On comparing the real and imaginary parts of the two terms on the left-hand side of Eq. (I.3a), we see that, in order to use the equations relating to the normal skin effect, we must satisfy the following inequalities:

$$n_n v_F/c \ll 1, \quad (I.33)$$

$$\varkappa_n v_F/c \ll \nu/\omega. \quad (I.34)$$

In obtaining these we have replaced v_z by v_F. Considering that $l = v_F/\nu$, inequality (I.34) may be written in the form

$$l \ll \delta = c/(\omega\varkappa_n)=\lambda/(2\pi\varkappa_n), \quad (I.35)$$

where l is the electron free path, δ is the classical depth of the skin layer, and λ is the light wavelength.

The physical meaning of conditions (I.33) and (I.34) is that, on satisfying these, the current and polarization at a given point are determined by the field at the same point. In the infrared region inequality (I.33) is satisfied for all metals, but this is by no means always the case for (I.34).

We have already indicated that the normal skin effect holds, in particular, for metals of the first group in the crystalline state in the visible and near infrared regions. In this case we usually have $\nu/\omega \ll 1$; we may therefore expand all relations linking the optical constants with the microcharacteristics of the metal in series in powers of ν/ω. The corresponding equations are presented in the review [2]. The conditions of the normal skin effect are also satisfied for all metals in the liquid and amorphous states in the visible and near infrared parts of the spectrum. However, in this case ν is of the same order as ω, and we cannot use an expansion in the parameter ν/ω.

In order to determine N and ν it is convenient to use the following equations [21, 22]:

$$(\varepsilon_n'-1)^{-1}=-\frac{\omega(\omega-i\nu)}{4\pi e^2}\frac{m}{N}. \quad (I.36)$$

Separating real and imaginary parts we obtain

$$N=\frac{0.1115\cdot 10^{22}}{\lambda}\frac{1}{\mathrm{Re}(\varepsilon_n'-1)^{-1}}, \quad (I.37)$$

$$\frac{\nu}{N}=\frac{\lambda}{5.918\cdot 10^{5}}\,\mathrm{Im}(\varepsilon_n'-1)^{-1}. \quad (I.38)$$

Here λ is in μ, N in cm^{-3}, and ν in sec^{-1}.

The method of analyzing experimental data under conditions in which the equations of the normal skin effect are applicable will be considered later.

§ 4. Weakly Anomalous Skin Effect

The weakly anomalous skin effect occurs most frequently for metals in the crystalline state; it was considered in detail in our earlier publications [37, 58]. In contrast to the general case in which solutions exist either for $\nu \ll \omega$ or for $\nu \gg \omega$, in the case of the weakly anomalous skin effect we have analytical solutions valid for any relation between ν and ω.

For many metals (particularly those of the polyvalent type) the interaction between the electrons and phonons is substantial, so that ν becomes comparable with ω in the infrared part of the spectrum. However, a large value of ν reduces the electron range l, and this becomes smaller than the depth of the skin layer δ. However, l and δ remain of the same order of magnitude. In this case we have to allow for the reflection of electrons from the surface of the metal; an allowance for this effect may be made by extending consideration to first-order terms in the small parameter $(v_F/c)\sqrt{\varepsilon'}/(1 - i\nu/\omega)$.

As before, we consider that the field $\mathscr{E}$ in the metal coincides in direction with $\mathbf{j}$ (isotropic case or the case of cubic symmetry with four-fold symmetry axes parallel to the x and z axes) (Fig. 2).*

We shall solve the system of equations (I.3a), (I.9), (I.10) by the method of successive approximations. In the zero approximation we neglect the term $v_z\,\partial f/\partial z$ in (I.3a), after which we obtain the foregoing normal skin effect. The solution for the field — see (I.30) — is substituted into the right-hand side of Eq. (I.3a). We obtain the first-approximation equation

$$v_z\frac{\partial f}{\partial z}+(i\omega+\nu)f=-\frac{\partial f_0}{\partial E}\frac{v_x}{v_z}e\mathscr{E}(0)\exp\left[-i\frac{\omega}{c}(n_n-i\varkappa_n)z\right]. \tag{I.39}$$

We remember that, according to (I.26) and (I.25),

$$\varepsilon_n-1=-\frac{4\pi e^2}{\omega(\omega-i\nu)}\frac{N}{m},$$

$$\frac{N}{m}=\frac{2}{3(2\pi\hbar)^3}\oint v\,dS_F.$$

By way of boundary conditions, we use conditions corresponding to the diffuse reflection of electrons from the surface of the metal:

$$\begin{aligned}&\text{for}\quad v_z>0\quad f=0\ \text{for}\ z=0\ \text{and}\ f=0\ \text{for}\ z\to\infty,\\ &\text{for}\quad v_z<0\quad f=0\ \text{for}\ z\to\infty.\end{aligned} \tag{I.40}$$

The solution satisfying the boundary conditions is

$$f=\frac{\dfrac{\partial f_0}{\partial E}e\mathscr{E}(0)\dfrac{v_x}{v_z}}{\dfrac{i\omega+\nu}{v_z}-i\dfrac{\omega}{c}(n_n-i\varkappa_n)}\left\{\exp\left[-i\frac{\omega}{c}(n_n-i\varkappa_n)z\right]-B\exp\left(-\frac{i\omega+\nu}{v_z}z\right)\right\}, \tag{I.41}$$

$B = 1$ for $v_z > 0$ and $B = 0$ for $v_z < 0$. Putting (I.41) into (I.10) we have

$$j_x(z)=\frac{2e^2\mathscr{E}_x(0)}{(2\pi\hbar)^3}\oint\frac{\exp\left[-i\dfrac{\omega}{c}(n_n-i\varkappa_n)z\right]-B\exp\left(-\dfrac{i\omega+\nu}{v_z}z\right)}{\dfrac{i\omega+\nu}{v_z}-i\dfrac{\omega}{c}(n_n-i\varkappa_n)}\frac{v_x^2}{v_z}\frac{dS_F}{v}, \tag{I.42}$$

and there is no local relation between $j_x z$ and $\mathscr{E}_x(z)$. Then we calculate the total current $I_x \equiv \int_0^\infty j_x(z)\,dz$, the expression for which may be schematically written in the form

$$I_x=\frac{2e^2\mathscr{E}_x(0)}{(2\pi\hbar)^3}\left\{\oint(1)-\oint\limits_{v_z>0}'(2)\right\}=\frac{2e^2\mathscr{E}_x(0)}{(2\pi\hbar)^3}\left\{\oint[(1)-(2)]+\oint\limits_{v_z<0}'(2)\right\}.$$

* Kaganov and Slezov [59] considered the case of arbitrary anisotropy, but these authors were only interested in the region $\nu \ll \omega$.

Here, $\oint'_{v_z<0}$ is the integral over the part of the Fermi surface with $v_z < 0$;

$$[(1)-(2)]=\frac{v_x^2}{v_z}\frac{dS_F}{v}\frac{cv_z}{i\omega(n_n-i\varkappa_n)(i\omega+\nu)}.$$

We make use of the central symmetry of the Fermi surface and transform from the integral over the part of the Fermi surface with $v_z < 0$ to the integral over the part of the Fermi surface with $v_z > 0$. We obtain

$$I_x=\frac{2e^2\mathscr{E}_x(0)}{(2\pi\hbar)^3}\frac{c}{i\omega(n_n-i\varkappa_n)(i\omega+\nu)}\left[\oint\frac{v_x^2}{v}dS_F-\frac{i\omega}{c}\frac{(n_n-i\varkappa_n)}{(i\omega+\nu)}\oint'_{v_z>0}\frac{v_x^2v_z}{v}\frac{dS_F}{\left[1+i\frac{\omega v_z}{c}\left(\frac{n_n-i\varkappa_n}{i\omega+\nu}\right)\right]}\right]. \quad \text{(I.43)}$$

We confine attention to terms of the first order in $(v/c)\,(n_n-i\varkappa_n)/(1+i\nu/\omega)$; then

$$I_x=-\frac{2e^2\mathscr{E}_x(0)}{(2\pi\hbar)^3}\frac{c}{\omega(\omega-i\nu)(n_n-i\varkappa_n)}\left(\oint\frac{v_x^2}{v}dS_F\right)\left[1-\frac{1}{c}\left(\frac{n_n-i\varkappa_n}{1-i\nu/\omega}\right)\frac{\oint'_{v_z>0}\frac{v_x^2v_z}{v}dS_F}{\oint v_x^2dS_F/v}\right]. \quad \text{(I.44)}$$

Allowing for the relation between the surface impedance Z and the total current* I_x we shall have

$$(Z)^{-1}=\frac{I_x}{\mathscr{E}_x(0)}=-\frac{2e^2}{(2\pi\hbar)^3}\frac{c}{\omega(\omega-i\nu)(n_n-i\varkappa_n)}\left(\oint\frac{v_x^2}{v}dS_F\right)(1-\alpha), \quad \text{(I.45)}$$

where

$$\alpha=\frac{1}{c}\left(\frac{n_n-i\varkappa_n}{1-i\nu/\omega}\right)\frac{\oint'_{v_z>0}v_x^2v_zdS_F/v}{\oint v_x^2dS_F/v}. \quad \text{(I.46)}$$

Let us denote the surface impedance for the normal skin effect (i.e., neglecting the term $v_z\partial f/\partial z$) by Z_n. For the normal skin effect $\alpha = 0$ and

$$(Z_n)^{-1}=-\frac{2e^2}{(2\pi\hbar)^3}\frac{c}{\omega(\omega-i\nu)(n_n-i\varkappa_n)}\oint\frac{v_x^2}{v}dS_F, \quad \text{(I.47)}$$

where

$$(Z)^{-1}=(Z_n)^{-1}(1-\alpha). \quad \text{(I.48)}$$

* For cubic symmetry and an arbitrary orientation of the crystal, in the zero approximation the direction $\mathbf{j}$ coincides with the direction of $\mathscr{E}_x$. In the first approximation, in the presence of a field $\mathscr{E}$ both j_x and j_y will be nonzero. To the same degree of accuracy, we obtain

$$\frac{I_y}{I_x^0}=\frac{i\omega}{c}\left(\frac{n_n-i\varkappa_n}{i\omega+\nu}\right)\oint'_{v_z>0}\frac{v_xv_yv_z}{v}dS_F\Big/\oint\frac{v_x^2}{v}dS_F.$$

Here I_y is the total current along the y axis, I_x^0 is the total current along the x axis in the zero approximation. For an arbitrary orientation of the crystal $|I_y/I_x^0| < |\alpha|$. The existence of I_y leads to the appearance of nondiagonal terms Z_{xy} and ε'_{xy}. These terms are small and may be neglected. Thus the resultant equations may in practice also be applied to cubic single crystals. In the isotropic case or in the case in which the four-fold symmetry axes are parallel to the x and y axes, $I_y = Z_{xy} = \varepsilon'_{xy} = 0$.

Using the relation between the surface impedance and the effective dielectric constant as in (I.11), we obtain

$$\sqrt{\varepsilon'}=\sqrt{\varepsilon_n'}(1-\alpha) \tag{I.49}$$

or

$$(n-i\varkappa)=(n_n-i\varkappa_n)(1-\alpha),$$

$$\varepsilon'=\varepsilon_n'(1-2\alpha), \tag{I.50}$$

$$(\varepsilon')^{-1}(1-2\alpha)=(\varepsilon_n')^{-1}. \tag{I.51}$$

To an accuracy limited only by higher-order terms, we may replace $(n_n - i\varkappa_n)$ by $(n - i\varkappa)$, in the expression for α, i.e., by measurable quantities, after which

$$2\alpha=\frac{2}{c}\left(\frac{n-i\varkappa}{1-i\nu/\omega}\right)\frac{\oint'_{v_z>0} v_x^2 v_z dS_F/v}{\oint v_x^2 dS_F/v}. \tag{I.52}$$

Equations (I.51) and (I.52), together with Eqs. (I.26) and (I.26a), give a relation between the optical constants of the metal and its microcharacteristics for the weakly anomalous skin effect.

Let us apply Eq. (I.52) to a spherical Fermi surface, for which

$$\oint v_x^2 dS_F/v=(4/3)\,\pi m^2 v_F^3,$$

$$\oint'_{v_z>0} v_x^2 v_z dS_F/v=(\pi/4)\,m^2 v_F^4, \tag{I.53}$$

whence

$$2\alpha=(3/8)(v_F/c)\,\varkappa(1+in/\varkappa)/(\nu/\omega+i). \tag{I.54}$$

Since our earlier measurements showed that, for the majority of metals, the skin effect was weakly anomalous in the infrared region, i.e., that the corresponding corrections associated with the reflection of electrons from the metal surface were small, we are justified in using (I.54) in order to calculate the coefficients in (I.52).

The complex expansion parameter used for the weakly anomalous skin effect is a combination of several dimensionless ratios:

$$\frac{3}{8}\frac{v_F}{c}\frac{n-i\varkappa}{1-i\nu/\omega}=\frac{3}{8}\frac{\{n(v_F/c)+[l_1/(2\pi\delta)](\nu/\omega)\}+i\,[n(v_F/c)(\nu/\omega)-l_1/(2\pi\delta)]}{1+\nu^2/\omega^2}. \tag{I.55}$$

Here $l_1 = 2\pi v_F/\omega$ is the path traversed by the electron in one period of the field. After carrying out an experiment we may always calculate the value of this complex parameter and establish the character of the skin effect.

In concluding this chapter we may note that optical measurements in the infrared part of the spectrum always (i.e., for a skin effect of any type) enable us to determine the quantity

$$\frac{N}{m}=\frac{2}{3(2\pi\hbar)^3}\oint v_F dS_F. \tag{I.56}$$

CHAPTER II

EFFECT OF THE PERIODIC POTENTIAL OF THE LATTICE ON THE OPTICAL PROPERTIES OF METALS

§ 1. Use of the Pseudopotential Concept

The periodic potential of the lattice has a considerable effect on the optical properties of metals both in the visible and in the infrared parts of the spectrum. The influence in the infrared region is associated with the fact [60] that it changes the concentration of the conduction electrons N, the area of the Fermi surface S_F, and the velocity of the electrons on the Fermi surface v_F. The influence of the lattice potential in the visible region is associated with interband transitions determined by this potential [34, 35].

In order to allow for the effect of the lattice on the main electron characteristics of a metal, we use the concept of the pseudopotential mentioned in the Introduction. According to this concept, an electron "feels" not the total potential of the ion, but a weak pseudopotential, i.e., it is almost free. The coherent scattering of electrons by lattice planes has a considerable effect on the optical properties of the metal, even in the case of a weak pseudopotential. We shall now consider these processes.

Let us expand the pseudopotential in a Fourier series

$$V(\mathbf{r}) = \sum_g V_g \exp(2\pi i g r), \tag{II.1}$$

where g is the reciprocal-lattice vector. The summation is carried out over all the vectors of the reciprocal lattice. We shall subsequently omit the sign of the vector in the summation index g. Experimental determinations of V_g carried out both optically [19, 20, 35] and by the de Haas — van Alphen method [61, 62] show that the following inequality is satisfied for metals:

$$|V_g| \ll E_F. \tag{II.2}$$

This enables us to use the weak-coupling approximation, which has been studied in full detail in a number of publications, in particular in [29, 32, 63].

The effects which we shall be considering include the influence of V_g on the concentration of conduction electrons N, the total area of the Fermi surface S_F, and the average velocity of the electrons on the Fermi surface $\langle v_F \rangle$; all these are of the order of unity. Hence, in order to determine the main features of the phenomenon it is sufficient to consider terms of the first order in $|V_g|/E_F$. This yields simple analytical expressions determining the difference between the main electron characteristics N, S_F, and $\langle v_F \rangle$ and the corresponding characteristics of free electrons. In the approximation in question, the effect of the sum of the Fourier components V_g equals the sum of the effects of each component individually. To each periodic potential in coordinate space described by a term in the sum (II.1) there corresponds a particular Bragg plane in momentum space. Thus the total effect of several Bragg planes equals the sum of the effects of each one of these.

§ 2. Effect of One Bragg Plane on the Energy and Velocity of an Electron and on the Shape of the Fermi Surface

1. Let us construct the sphere of free electrons for a concentration equal to the valence concentration, and consider the intersection of this sphere by one Bragg plane with the index g.

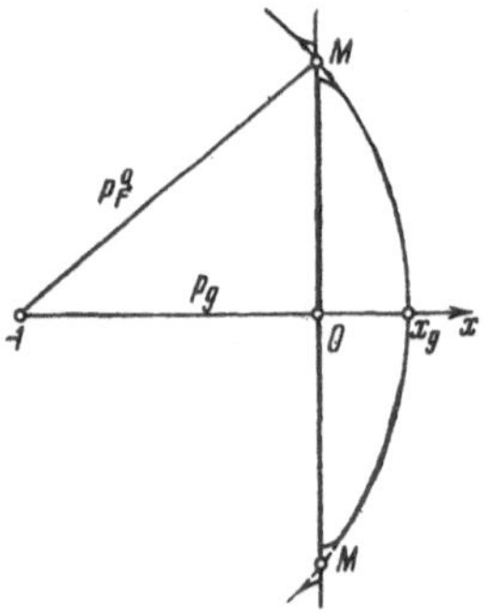

Fig. 3. Intersection of the sphere of free electrons by the Bragg plane MM. Here p_F^0 is the radius of the sphere of free electrons, p_g is the distance from the center Γ of the sphere to the Bragg plane MM; the broken line signifies part of the sphere of free electrons.

We shall use the scheme of expanded zones. The region of intersection of present interest is shown in Fig. 3.

We have already noted that for a weak potential the difference between the motion of the actual electron and a free one is mainly associated with coherent scattering by the lattice planes. In momentum space, coherent scattering only changes the behavior of the electron appreciably when its momentum is close to one of the Bragg reflection planes. For the Fourier component of the potential V_g the equation of the corresponding Bragg plane in p space will have the form

$$2\mathbf{p}_g(\mathbf{p}-\mathbf{p}_g)=0. \quad \text{(II.3)}$$

Here $\mathbf{p}$ is the electron momentum, and $2\mathbf{p}_g = 2\pi\hbar\mathbf{g}$. The distance from the center of the zone Γ to this plane equals $p_g = |\mathbf{p}_g|$. In subsequent calculations we shall require the components of $\mathbf{p}$ and $\mathbf{p}_g$ parallel and perpendicular to the Bragg plane: $p_\parallel$, $p_\perp$, $p_{g\parallel}$, and $p_{g\perp}$. For $\mathbf{p}_g$ we have $p_{g\parallel} = 0, p_{g\perp} = p_g$. In the present approximation the energy of the electron is determined from a secular equation of the second order [29, 32, 63]. Introducing the dimensionless quantities

$$x=(p_\perp-p_g)/p_g, \qquad y=p_\parallel/p_g, \quad \text{(II.4)}$$

$$w=2mE/p_g^2, \qquad \xi=2m|V_g|/p_g^2, \quad \text{(II.5)}$$

where m is the mass of the free electron, we obtain the equation of the isoenergy surface in the form

$$w=1+x^2+y^2+\operatorname{sign}(x)\sqrt{4x^2+\xi^2}, \quad \text{(II.6)}$$

$$\operatorname{sign}(x)=\begin{cases}+1 & \text{for } x>0\\ -1 & \text{for } x<0.\end{cases}$$

On the Bragg plane x = 0 and

$$w=w_0\pm\xi, \quad \text{i.e.,} \quad E=E_0\pm|V_g| \quad \text{(II.7)}$$

(the subscript 0 signifies that the corresponding quantity relates to free electrons). Near the Bragg plane, when $4x^2 \ll \xi^2$, the corrections to the energy will be of the first order with respect to $|V_g|$. A long way from the Bragg plane they will be of the second order.

2. Equation (II.6) enables us to determine the influence of the Bragg plane on the velocity of the electron. The plane in question does not alter the velocity component parallel to itself, i.e., $v_\parallel = \partial E/\partial p_\parallel = v_\parallel^0$; however, it does alter the component perpendicular to the plane, $v_\perp =$

$\partial E/\partial p_\perp$. In the variables of (II.4), (II.5) we shall have

$$v_\parallel = (p_g/m)\, y, \tag{II.8}$$

$$v_\perp = (p_g/m)\left(x + \operatorname{sign} x \frac{2x}{\sqrt{4x^2+\xi^2}}\right), \tag{II.9}$$

$$v^2 = v_\parallel^2 + v_\perp^2 = \frac{p_g^2}{m^2}\left(w - \frac{\xi^2}{4x^2+\xi^2} - \operatorname{sign} x \frac{\xi^2}{\sqrt{4x^2+\xi^2}}\right). \tag{II.10}$$

3. Let us determine the influence of the Bragg plane on the shape of the Fermi surface. We shall suppose that the plane intersects the free-electron sphere, the radius of the latter being given by the relation

$$p_F^0 = 2\pi\hbar\left(\frac{3}{8\pi}\frac{\Delta n}{\Delta\tau}\right)^{1/3}, \tag{II.11}$$

where Δn is the number of valence electrons associated with the volume $\Delta\tau$.

It follows from Eq. (II.6) that, in the case of the action of a single Bragg plane, the Fermi surface is a surface of rotation with a symmetry axis perpendicular to the Bragg plane. The change in the shape of the Fermi surface may readily be seen from Fig. 3. The plane of the figure is perpendicular to the Bragg plane and passes through the center of the zone Γ. The point M lies on the intersection of the Bragg plane with the sphere of free electrons. Clearly this point does not belong to the Fermi surface, since the electron energy which it represents is either smaller than E_F, if we approach it from the left, or greater than E_F, if we approach it from the right. It is easy to show that this change in the Fermi surface leads to a change in the volume enclosed by this surface, determined by quantities of the second order of smallness. This means that, to our present accuracy, $E_F = E_F^0$, and that a long way from the Bragg plane $p_F = p_F^0$. Thus, a long way from the Bragg plane the Fermi surface coincides with the sphere of free electrons; near the plane it deviates sharply from this sphere, acquiring the shape indicated in Fig. 3.

The element of area of the isoenergy surface equals

$$dS = 2\pi p_g^2\, y(x)\sqrt{1+[y'(x)]^2}\, dx, \tag{II.12}$$

where y(x) is given by Eq. (II.6) for w = const, whence

$$dS = 2\pi m p_g v\, dx. \tag{II.13}$$

For the Fermi surface

$$dS_F = 2\pi m p_g v_F\, dx. \tag{II.14}$$

§ 3. Influence of the Fourier Components of the Pseudopotential on N, $S_F \langle v_F \rangle$ and $(dY/dE)_F$

1. Let us apply the equations so derived in order to determine the integrated characteristics of present interest. Let us consider the density of the electron states (dY/dE). It is well known that $(dY/dE)_F \sim \oint dS_F/v$. Earlier we found that $dS_F/v_F = \pi p_g dx$. This expression does not depend on ξ. The limits of integration with respect to x are $-x_g'$ and x_g (Fig. 3), where

$$x_g = p_F/p_g - 1 \geqslant 0,$$
$$x_g' = p_F/p_g + 1 \geqslant 2. \tag{II.15}$$

x_g and x_g' are also independent of ξ. Hence, in the approximation under consideration, linear with respect to $|V_g|/E_F$, the periodic lattice potential does not change the density of states of the electrons on the Fermi surface:

$$(dY/dE)_F=(dY/dE)_F^0. \tag{II.16}$$

2. Let us consider the influence of the periodic potential on the concentration of conduction electrons N. Using (I.56) and (II.14), we obtain

$$N\sim\oint v_F dS_F\sim\int v_F^2 dx.$$

Whence

$$\frac{N_{\text{val}}-N}{N_{\text{val}}}\equiv\frac{v_F^0 S_F^0-\langle v_F\rangle S_F}{v_F^0 S_F^0}, \tag{II.17}$$

where $\langle v_F\rangle$ is the average electron velocity on the Fermi surface.

Equations (II.10) and (II.15) easily enable us to calculate $\langle v_F\rangle S_F$. We note that the integral of the term $(\operatorname{sign} x)\xi^2/\sqrt{4x^2+\xi^2}$ is of the order $\xi^2\ln\xi$ and may be neglected. Thus, as a result of the influence of one Bragg plane we have

$$\left(\frac{N_{\text{val}}-N}{N_{\text{val}}}\right)_g=\frac{\pi}{4}\frac{p_g}{p_F^0}\frac{|V_g|}{E_F^0}\left(\frac{1}{2}+\frac{\varphi_g}{\pi}\right), \tag{II.18}$$

$$\tan^{-1}\frac{2x_g'}{\xi}\simeq\frac{\pi}{2},\quad \tan^{-1}\frac{2x_g}{\xi}=\varphi_g. \tag{II.19}$$

As a result of the action of two Bragg planes we have

$$\frac{N_{\text{val}}-N}{N_{\text{val}}}=\sum_g\frac{\pi}{4}\frac{p_g}{p_F^0}\frac{|V_g|}{E_F^0}\left(\frac{1}{2}+\frac{\varphi_g}{\pi}\right), \tag{II.20}$$

$$\varphi_g=\tan^{-1}\frac{2(p_g/p_F^0)(1-p_g/p_F^0)}{|V_g|/E_F^0}. \tag{II.21}$$

The summation extends over all the Bragg planes intersecting the free electron sphere. We note that on satisfying the condition $x_g\gg\xi/2$, i.e., the condition that the Bragg plane should be a fair way from the tangent plane, we shall have

$$\frac{1}{2}+\varphi_g/\pi\approx 1, \tag{II.22}$$

and Eq. (II.20) simplifies.

3. Let us consider the influence of the periodic potential on the area of the Fermi surface. As a result of the presence of a single Bragg plane the change in the area of the Fermi surface is

$$\delta\oint dS_F\equiv S_F^0-\oint dS_F=2\pi m p_g\int_{-x_g'}^{x_g}dx\,(v_F^0-v_F)=$$

$$=2\pi p_g^2\int_{-x_g'}^{x_g}dx\left[\sqrt{w_F}-\sqrt{w_F-\xi^2/(4x^2+\xi^2)-(\operatorname{sign}x)\xi^2/\sqrt{4x^2+\xi^2}}\right]. \tag{II.23}$$

The term $(\operatorname{sign}x)\xi^2/\sqrt{4x^2+\xi^2}$ gives a contribution proportional to $\xi^2\ln\xi$ as before, and may be neglected. Hence the change associated with one Bragg plane may be approximately written in the following form:

$$\delta \oint dS_F = 2\pi p_g^2 \xi \sqrt{w_F} \int_0^\infty dt \left\{ 1 - \sqrt{1 - (1/w_F)[1/(1+t^2)]} \right\}, \tag{II.24}$$

where

$$t = 2x/\xi. \tag{II.25}$$

In (II.24) we have omitted the term

$$\mathcal{I}' = \int_{x_g}^\infty dx \left[\sqrt{w_F} - \sqrt{w_F - \xi^2/(4x^2+\xi^2)}\right] \approx \xi^2/(2\sqrt{w_F}) \int_x^\infty dx/(4x^2) = \xi^2/(8\sqrt{w_F}x_g). \tag{II.26}$$

Here we have assumed for the calculation that

$$(1/w_F)[\xi^2/(4x_g^2+\xi^2)] \ll 1, \quad 4x_g^2 \gg \xi^2. \tag{II.26a}$$

As we shall subsequently see, these inequalities are satisfied in the case of present interest, viz., that of ter- and tetravalent nontransition metals. Returning to Eq. (II.24), after integrating by parts we obtain

$$\delta \oint dS_F = 2\pi p_g^2 \xi / \sqrt{w_F} \int_0^{\pi/2} d\varphi (\sin^2\varphi) / \sqrt{1 - (\cos^2\varphi)/w_F}. \tag{II.27}$$

Here

$$t = \tan\varphi. \tag{II.27a}$$

We consider then $\sin^2\varphi = 1 + w_F[1 - (1/w_F)\cos^2\varphi - 1]$. Finally we obtain

$$\delta \oint dS_F = 2\pi p_g^2 \xi \left\{ \left(\frac{1}{\sqrt{w_F}} - \sqrt{w_F} \right) K\left(\frac{1}{\sqrt{w_F}}\right) + \sqrt{w_F} E\left(\frac{1}{\sqrt{w_F}}\right) \right\}. \tag{II.28}$$

Here

$$K(z) = \int_0^{\pi/2} \frac{d\varphi}{\sqrt{1 - z^2 \sin^2\varphi}}, \quad E(z) = \int_0^{\pi/2} \sqrt{1 - \sin^2\varphi}\, d\varphi \tag{II.28a}$$

are complete elliptical integrals of the first and second kinds (Jahnke and Emde [64]).

On integrating between 0 and $\pi/2$ we may replace $\cos\varphi$ by $\sin\varphi$.

Thus on allowing for the effects of all the Bragg planes we have

$$\frac{S_F^0 - S_F}{S_F^0} \cong \sum \frac{1}{2} \frac{|V_g|}{E_F^0} f\left(\frac{p_g}{p_F^0}\right), \tag{II.29}$$

where

$$f(z) = \frac{1}{z} E(z) - \left(\frac{1}{z} - z\right) K(z). \tag{II.29a}$$

Let us consider the case in which ξ/x_g is not small. We calculate the integral $\mathcal{I}'$ which was omitted above. Using (II.25) and (II.27a), making the substitution $\varphi \to (\pi/2) - \varphi$, $\sin\varphi \to \cos\varphi$, $\cos\varphi \to \sin\varphi$, $d\varphi \to -d\varphi$, and considering that

$$\cos^2\varphi = 1 + w_F[1 - (1/w_F)\sin^2\varphi - 1],$$

we obtain

$$\mathcal{Y}' = -\pi p_g^2 \xi \sqrt{w_F} \left\{ -\frac{2x_g}{\xi} \left[1 - \sqrt{1 - \frac{1}{w_F} \frac{\xi^2}{(4x_g^2 + \xi^2)}} \right] + \right.$$

$$\left. + E\left(\frac{1}{\sqrt{w_F}},\ \frac{\pi}{2} - \tan^{-1}\frac{2x_g}{\xi}\right) - \left(1 - \frac{1}{w_F}\right) F\left(\frac{1}{\sqrt{w_F}},\ \frac{\pi}{2} - \tan^{-1}\frac{2x_g}{\xi}\right) \right\}. \tag{II.30}$$

Here

$$E(z, \psi) = \int_0^{\psi} d\varphi \sqrt{1 - z^2 \sin^2\varphi}; \qquad F(z, \psi) = \int_0^{\psi} \frac{d\varphi}{\sqrt{1 - z^2 \sin^2\varphi}} \tag{II.30a}$$

(see [64]).

Finally, for the case of a value of ξ / x_g which cannot be regarded as small, on allowing for the effects of all the Bragg planes we shall have

$$\frac{S_F^0 - S_F}{S_F^0} = \sum_g \frac{1}{2} \frac{|V_g|}{E_F^0} \left[f\left(\frac{p_g}{p_F^0}\right) - \frac{1}{2} \tilde{f}\left(\frac{p_g}{p_F^0},\ \varphi_g\right) \right], \tag{II.31}$$

$$f(z) = \frac{1}{z} E(z) - \left(\frac{1}{z} - z\right) K(z), \tag{II.31a}$$

$$\tilde{f}(z, \varphi) = \frac{1}{z} E\left(z, \frac{\pi}{2} - \varphi\right) - \left(\frac{1}{z} - z\right) F\left(z, \frac{\pi}{2} - \varphi\right) - \frac{1}{z}\left(1 - \sqrt{1 - z^2\cos^2\varphi}\right)\tan\varphi, \tag{II.31b}$$

$$E(z) \equiv E\left(z, \frac{\pi}{2}\right), \qquad K(z) \equiv F\left(z, \frac{\pi}{2}\right). \tag{II.31c}$$

The functions $F(z, \psi)$ and $E(z, \psi)$ are determined by (II.30a). These are the complete elliptic integrals of the first and second kinds, respectively [64]. The value of φ_g is given by (II.21).

4. After determining the effect of the periodic potential on N and S_F, we may calculate the average velocity of the electrons on the Fermi surface $\langle v_F \rangle$ by using the relation

$$\frac{\langle v_F \rangle}{v_F^0} = \frac{N}{N_{val}} \frac{S_F^0}{S_F}. \tag{II.32}$$

5. We may now show how to make an approximate calculation of the area of the Fermi surface and the average velocity on the latter, knowing the ratio N/N_{val}. It follows from (II.16) that

$$S_F^0 \frac{1}{v_F^0} = S_F \left\langle \frac{1}{v_F} \right\rangle.$$

Considering that $\left\langle \frac{1}{v_F} \right\rangle \approx \frac{1}{\langle v_F \rangle}$, we obtain

$$\frac{N}{N_{val}} = \frac{S_F \langle v_F \rangle}{S_F^0 v_F^0} \approx \left(\frac{S_F}{S_F^0}\right)^2 \approx \left(\frac{v_F}{v_F^0}\right)^2. \tag{II.33}$$

Whence

$$S_F / S_F^0 \approx \langle v_F \rangle / v_F^0 \approx \sqrt{N / N_{val}}. \tag{II.33a}$$

In Chapter VII we shall compare Eq. (II.33a) with the results of our calculations based on (II.31) and (II.32).

6. Using Eqs. (II.16), (II.20), (II.21), and (II.29) or (II.31), we may calculate the influence of the periodic lattice potential on $(dY/dE)_F$, N, and S_F. Equations (II.16), (II.29), and (II.31) are also applicable to noncubic crystals. Equation (II.20) also gives the change in $\oint v_F dS_F$ for any crystal lattice. For noncubic symmetry the latter quantity is associated with a tensor N_{ik}

$$\sum_{l=1}^{3} \frac{N_{ll}}{m} = \frac{2}{(2\pi\hbar)^3} \oint v_F dS_F.$$

For polycrystalline samples we may consider that the average value of N is given by the integral of $v_F dS_F$.

The inaccuracy in the expressions derived for $(N_{val} - N)/N_{val}$ and $(S_F^0 - S_F)/S_F^0$ is associated with the use of the secular equation of the second order instead of the secular equation of higher order. The error usually amounts to about 10% (Chapter VII).

7. We shall subsequently apply the equations thus obtained to determine the influence of the pseudopotential on the foregoing integrated characteristics (N and S_F) for various crystal structures.

Starting from the theory of weakly coupled electrons, we may express the Fourier component of the pseudopotential in the following form [31, 32, 65, 66]:

$$V(\mathbf{K}) = F(\mathbf{K}) U(\mathbf{K}). \tag{II.34}$$

Here **K** is the wave vector, U(**K**) is the Fourier component of the self-consistent atomic potential determined by the Fourier component of the potential of an individual ion and the dielectric constant of the metal (electron screening being taken into account); F(**K**) is the structure factor, depending solely on the position of the ions [32, 63],

$$F(\mathbf{K}) = \frac{1}{N_c} \sum_{l} \exp(-i\mathbf{K}\mathbf{l}), \tag{II.35}$$

where N_c is the total number of unit cells per unit volume; $\mathbf{l}$ is the lattice vector corresponding to the point l. The same structure factor enters as a factor into the structure amplitude defining the intensity of x-ray diffraction maxima for lattices containing only one type of atom. If the lattice is simple, then

$$F(\mathbf{K}) = \begin{cases} 1 & \text{for} \quad \mathbf{K} = 2\pi\mathbf{g} \\ 0 & \text{for} \quad \mathbf{K} \neq 2\pi\mathbf{g}. \end{cases} \tag{II.36}$$

If the lattice is complex, with a basis incorporating s atoms, then we must allow for the effects of interference between various sublattices. Let the positions of the atoms composing the basis of the complex lattice be given by the radius vector $\mathbf{r}_n$

$$\mathbf{r}_n = \sum_{i=1}^{3} x_{ni} \mathbf{a}_i, \tag{II.37}$$

where $\mathbf{a}_i$ are the base vectors of the lattice.

The Bragg plane of current interest we shall define by the subscripts $(n_1 n_2 n_3)$. Then

$$F_{n_1 n_2 n_3} = \sum \exp\left[2\pi i \left(n_1 x_{n_1} + n_2 x_{n_2} + n_3 x_{n_3}\right)\right]. \tag{II.38}$$

The summation extends over all the atoms forming the basis.

The effect of the different sublattices is that only some of the Fourier components of the pseudopotential become nonzero. In our approximation the quantity $F_{n_1n_2n_3}$ is proportional to $V_{n_1n_2n_3}$.

§ 4. Face-Centered Cubic Lattice

The face-centered cubic lattice is of greatest interest to us, since a whole series of nontransition metals have this type of lattice; it may conveniently be considered as a simple cubic lattice with a basis: (0 0 0), (0 ½ ½), (½ 0 ½), (½ ½ 0).

Using Eq. (II.38) we find that the structure factors for this lattice equal

$$F_{n_1n_2n_3}=1+\cos\pi(n_2+n_3)+\cos\pi(n_3+n_1)+\cos\pi(n_1+n_2). \quad \text{(II.39)}$$

Table 2 shows the moduli of the structure factor for various Bragg planes and the number of physically equivalent planes as well. We see from Eq. (II.39) and Table 2 that for a structure of this type $|F_{n_1n_2n_3}| = 4$ if all three numbers n_1, n_2, and n_3 are of the same parity. In other cases, $|F_{n_1n_2n_3}| = 0$. Since the structure factors of all nonzero Fourier components of the pseudopotential are equal to each other, the values of $|V_g|$ are determined by the second factor of U. The effectiveness of a particular family of Bragg planes $\{n_1n_2n_3\}$ is the greater, the greater their number.

The distance from the center of the zone to the corresponding Bragg plane in momentum space is $p_g \equiv p_{n_1n_2n_3}$

$$p_{n_1n_2n_3}\Big/\left(\frac{2\pi\hbar}{a}\right)=\frac{1}{2}\sqrt{n_1^2+n_2^2+n_3^2}. \quad \text{(II.40)}$$

These distances are also shown in Table 2 for different Bragg planes.

TABLE 2. Structure Factors and Distances to the Center of Various Bragg Planes for Certain Types of Structures of the Cubic System

$\{n_1n_2n_3\}$	No. of planes	$p_{n_1n_2n_3}/(2\pi\hbar/a)$	$\|F_{n_1n_2n_3}\|$		
			F.c.	B.c.	diamond
100	6	0.500	0	0	0
110	12	0.707	0	2	0
111	8	0.866	4	0	$4\sqrt{2}$
200	6	1.000	4	2	0
210	24	1.118	0	0	0
211	24	1.225	0	2	0
220	12	1.414	4	2	8
221	24	1.500	0	0	0
300	6	1.500	0	0	0
310	24	1.581	0	2	0
311	24	1.658	4	0	$4\sqrt{2}$
222	8	1.732	4	2	0
320	24	1.803	0	0	0
321	48	1.871	0	2	0
400	6	2.000	4	2	8

Note: $\{n_1n_2n_3\}$ are the indices of physically equivalent Bragg planes; $p_{n_1n_2n_3}$ is the distance of the Bragg plane from the center of the zone in momentum space; a is the lattice constant; $|F_{n_1n_2n_3}|$ is the modulus of the structure factor.

TABLE 3. Dependence of the Radius of the Sphere of Free Electrons on the Concentration of Valence Electrons for Certain Types of Structures of the Cubic System

No. of valence electrons per atom	$p_F^0/(2\pi\hbar)/a)$ face-centered	body-centered	diamond	No. of valence electrons per atom	$p_F^0/(2\pi\hbar)/a)$ face-centered	body-centered	diamond
0.5	0.620	0.492	0.782	3	1.127	0.895	1.420
0.75	0.710	0.564	0.895	4	1.241	0.985	1.563
1	0.782	0.620	0.985	5	1.337	1.061	1.684
1.25	0.842	0.668	1.061	6	1.420	1.127	1.790
1.5	0.895	0.710	1.127	7	1.495	1.187	1.884
1.75	0.942	0.748	1.187	8	1.563	1.241	1.970
2	0.985	0.782	1.241	9	1.626	1.290	2.048
2.5	1.061	0.842	1.337	10	1.684	1.337	2.122

Note: p_F^0 is the radius of the sphere of free electrons for the specified concentration of valence electrons.

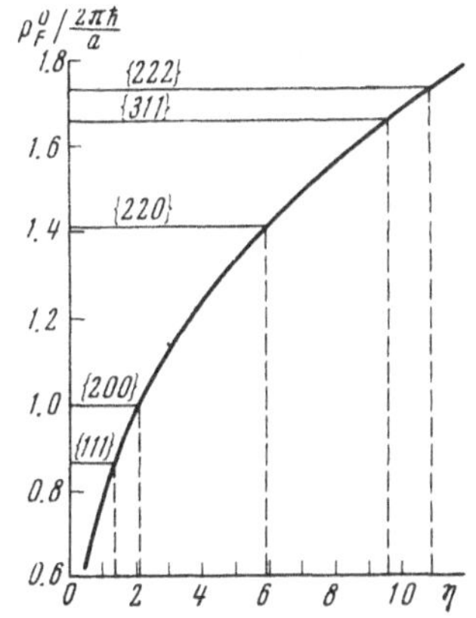

Fig. 4. Radius of the sphere of free electrons p_F^0 as a function of the number of valence electrons per atom η for an fcc lattice. Horizontal lines indicate distance from the center of the sphere Γ to the corresponding Bragg plane.

The unit cell of the lattice under consideration contains four atoms. Hence, for a volume $\Delta\tau = a^3$ there are 4η valence electrons, where η is the valence of the metal. Using (II.11) we may calculate the radius of the sphere of free electrons as a function of the concentration of the valence electrons. The results of the calculation are presented in Table 3 and Fig. 4. We see from the figure that in the case of the face-centered lattice the closest Bragg planes {111} only start intersecting the sphere of free electrons for $\eta = 1.4$. These planes can only have a substantial influence on the optical properties for a concentration of valence electrons greater than or equal to 1.4 electrons per atom.* For ter-, tetra-, and pentavalent metals only the Bragg planes {111} and {200} can have a marked effect on the optical properties. For hexavalent metals the effect of the {220} planes must also be taken into account. The {311} planes are only important for $\eta > 9.5$, etc.

*In such metals as gold, silver, and copper, the influence of the Fourier component of the pseudopotential V_{111} on their optical properties is considerable, although for these metals $\eta = 1$. This circumstance is associated with the high value of the ratio $|V_{111}|/E_F^0 \approx 0.3$.

Let us give some more detailed attention to ter- and tetravalent metals with a face-centered cubic lattice. For these metals the V_{111} and V_{200} Fourier components of the pseudopotential will have a marked effect on the optical properties.

First let us consider the tervalent metals. We see from Tables 2 and 3 and Fig. 4 that

$$p_{111}/p_F^0 = 0.769; \quad p_{200}/p_F^0 = 0.888. \tag{II.41}$$

Using Eqs. (II.20) and (II.21) and remembering that we have eight {111} and six {200} planes, we obtain

$$\frac{N_{\text{val}} - N}{N_{\text{val}}} = 4.84 \frac{|V_{111}|}{E_F^0}\left(\frac{1}{2} + \frac{\varphi_{111}}{\pi}\right) + 4.18 \frac{|V_{200}|}{E_F^0}\left(\frac{1}{2} + \frac{\varphi_{200}}{\pi}\right), \tag{II.42}$$

$$\varphi_{111} = \tan^{-1}\frac{0.355}{|V_{111}|/E_F^0}, \quad \varphi_{200} = \tan^{-1}\frac{0.199}{|V_{200}|/E_F^0}. \tag{II.43}$$

Relations (II.41) and (II.43) easily enable us to calculate the effect of the foregoing components of the pseudopotential on the total area of the Fermi surface from Eqs. (II.31).

Now let us consider tetravalent metals. We see from Tables 2 and 3 and Fig. 4 that

$$p_{111}/p_F^0 = 0.698; \quad p_{200}/p_F^0 = 0.806. \tag{II.44}$$

As a result of this,

$$\frac{N_{\text{val}} - N}{N_{\text{val}}} = 4.39 \frac{|V_{111}|}{E_F^0}\left(\frac{1}{2} + \frac{\varphi_{111}}{\pi}\right) + 3.80 \frac{|V_{200}|}{E_F^0}\left(\frac{1}{2} + \frac{\varphi_{200}}{\pi}\right), \tag{II.45}$$

$$\varphi_{111} = \tan^{-1}\frac{0.422}{|V_{111}|/E_F^0}, \quad \varphi_{200} = \tan^{-1}\frac{0.313}{|V_{200}|/E_F^0} \tag{II.46}$$

Equations (II.44) and (II.46) easily enable us to calculate $(S_F^0 - S_F)/S_F^0$ from Eqs. (II.31).

Later in Chapter VII we shall compare the results calculated from the formulas presented in this section with our own experimental results for aluminum, indium, and lead. In the case of lead we shall also compare the calculated values of $(S_F^0 - S_F)/S_F^0$ obtained from Eqs. (II.31) with the results of a computer calculation based on the fourth-order secular equation [62].

§ 5. Body-Centered Cubic Lattice

A body-centered cubic lattice may be considered as a simple cubic lattice with a basis (0 0 0), ($\frac{1}{2}$ $\frac{1}{2}$ $\frac{1}{2}$). Using (II.38) we find in this case that the structure factors are equal to

$$F_{n_1 n_2 n_3} = 1 + \cos\pi(n_1 + n_2 + n_3). \tag{II.47}$$

The modulus of the structure factor is shown for various Bragg planes in Table 2. It follows from (II.47) that if the sum $(n_1 + n_2 + n_3)$ equals an odd number, then $F_{n_1 n_2 n_3} = 0$; if it equals an even number, then $F_{n_1 n_2 n_3} = 2$. The structure factors of all the nonzero Fourier components of the pseudopotential are equal. Hence, for this lattice, in the same way as for the fcc lattice, the values of $|V_g|$ are determined by the second factor of U. The effectiveness of a specified family of Bragg planes $\{n_1 n_2 n_3\}$ depends on the number of planes as well as the value of $|V_{n_1 n_2 n_3}|$. The latter is also presented in Table 2. The distance from the center of the zone Γ to the corresponding Bragg plane is given in Eq. (II.40). These distances are given in Table 2 as well.

The unit cell of the lattice under consideration contains two atoms. Hence the volume $\Delta\tau = a^3$ is associated with 2η valence electrons. The radius of the sphere of free electrons is given in relation to the concentration of valence electrons in Table 3 and Fig. 5. We see from the figure that the set of Bragg planes {110} nearest to the center of the zone Γ only

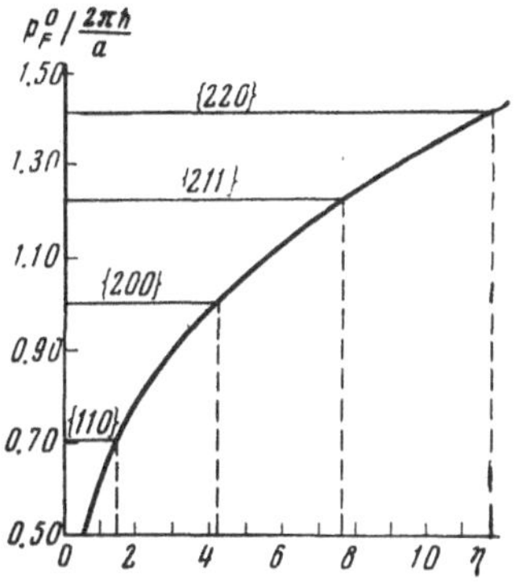

Fig. 5. Radius of the sphere of free electrons p_F^0 as a function of the number of valence electrons per atom η for a bcc lattice. Notation as in Fig. 4.

starts having a substantial influence on the optical properties of the metal for $\eta \geq 1.5$ electrons per atom. For $\eta \geq 4.2$ electrons per atom, the {200} Bragg planes also have to be taken into account. The next set of Bragg planes {211} only start having an effect for $\eta \geq 7.7$ electrons per atom, and so on.

Let us take the examples of penta-, hexa-, and heptavalent metals. For these metals the V_{110} and V_{200} Fourier components of potential are important. In using Eqs. (II.20) and (II.21) we allow for the fact that there are twelve {110} and six {200} Bragg planes.

Let us first consider pentavalent metals. We see from Tables 2 and 3 and Fig. 5 that

$$p_{111}/p_F^0 = 0.666; \qquad p_{200}/p_F^0 = 0.9425, \tag{II.48}$$

This gives

$$\frac{N_{\text{val}} - N}{N_{\text{val}}} = 6.27 \frac{|V_{110}|}{E_F^0}\left(\frac{1}{2} + \frac{\varphi_{110}}{\pi}\right) + 4.44 \frac{|V_{200}|}{E_F^0}\left(\frac{1}{2} + \frac{\varphi_{200}}{\pi}\right), \tag{II.49}$$

$$\varphi_{110} = \tan^{-1}\frac{0.445}{|V_{110}|/E_F^0}, \qquad \varphi_{200} = \tan^{-1}\frac{0.108}{|V_{200}|/E_F^0}. \tag{II.50}$$

For hexavalent metals an analogous calculation gives

$$p_{110}/p_F^0 = 0.627, \qquad p_{200}/p_F^0 = 0.887, \tag{II.51}$$

$$\frac{N_{\text{val}} - N}{N_{\text{val}}} = 5.91 \frac{|V_{110}|}{E_F^0}\left(\frac{1}{2} + \frac{\varphi_{110}}{\pi}\right) + 4.18 \frac{|V_{200}|}{E_F^0}\left(\frac{1}{2} + \frac{\varphi_{200}}{\pi}\right), \tag{II.52}$$

$$\varphi_{110} = \tan^{-1}\frac{0.468}{|V_{110}|/E_F^0}, \qquad \varphi_{200} = \tan^{-1}\frac{0.200}{|V_{200}|/E_F^0}. \tag{II.53}$$

For heptavalent metals we shall have

$$p_{110}/p_F^0 = 0.596, \qquad p_{200}/p_F^0 = 0.842, \tag{II.54}$$

$$\frac{N_{\text{val}} - N}{N_{\text{val}}} = 5.61 \frac{|V_{110}|}{E_F^0}\left(\frac{1}{2} + \frac{\varphi_{110}}{\pi}\right) + 3.97 \frac{|V_{200}|}{E_F^0}\left(\frac{1}{2} + \frac{\varphi_{200}}{\pi}\right), \tag{II.55}$$

$$\varphi_{110} = \tan^{-1}\frac{0.482}{|V_{110}|/E_F^0}, \qquad \varphi_{200} = \tan^{-1}\frac{0.266}{|V_{200}|/E_F^0}. \tag{II.56}$$

These formulas enable us very easily to calculate the value of $(S_F^0 - S_F)/S_F^0$ for all the cases under consideration.

§ 6. Cubic Lattice of the Diamond Type

We shall consider the cubic lattice of the diamond type as a simple cubic lattice with a basis

$$(0\ 0\ 0),\ (0\ {}^1/_2\ {}^1/_2),\ ({}^1/_2\ 0\ {}^1/_2),\ ({}^1/_2\ {}^1/_2\ 0),\ ({}^1/_4\ {}^1/_4\ {}^1/_4),\ ({}^3/_4\ {}^1/_4\ {}^3/_4),\ ({}^3/_4,\ {}^1/_4,\ {}^3/_4),\ ({}^3/_4\ {}^3/_4\ {}^1/_4).$$

Using (II.38) and making some simple transformations, we find that the structure factor is in this case equal to

$$F_{n_1n_2n_3}=[1+\cos\pi(n_2+n_3)+\cos\pi(n_3+n_1)+\cos\pi(n_1+n_2)]\times$$
$$\times\left[1+\cos\frac{\pi}{2}(n_1+n_2+n_3)+i\sin\frac{\pi}{2}(n_1+n_2+n_3)\right].$$

Whence it follows that

$$|F_{n_1n_2n_3}|=2\left|\cos\frac{\pi}{4}(n_1+n_2+n_3)\right|\cdot|1+\cos\pi(n_2+n_3)+\cos\pi(n_3+n_1)+\cos\pi(n_1+n_2)|. \quad \text{(II.57)}$$

The structure-factor moduli for various Bragg planes are shown in Table 2. It follows from Eq. (II.57) that the second factor, coinciding with the corresponding expressions for the fcc lattice, is nonzero when all three numbers n_1, n_2, and n_3 are of the same parity. If all the n_i are odd then $|F_{n_1n_2n_3}| = 4\sqrt{2}$. If all the n_i are even, then the first factor is only nonzero when the sum $(n_1 + n_2 + n_3)$ is a multiple of 4. In this case, $|F_{n_1n_2n_3}| = 8$.

The unit cell of this lattice contains eight atoms. Hence a volume $\Delta\tau = a^3$ is associated with 8η valence electrons (η is the valence of the metal). The radius of the sphere of free electrons is given as a function of the concentration of valence electrons in Table 3 and Fig. 6. We see from the figure that the system of Bragg planes closest to the center of the zone Γ, the $\{111\}$ system, starts intersecting the sphere of free electrons for $\eta = 0.7$ electrons per atom. For $\eta = 3$ electrons per atom, the sphere of free electrons starts intersecting the $\{220\}$ set of planes; the next set of planes $\{311\}$ is intersected at $\eta = 4.8$ electrons per atom; and so on.

Let us take the example of tetravalent metals. For these metals we must allow for the V_{111} and V_{220} Fourier components of pseudopotential. It is reasonable to expect that the V_{220} component will affect the optical properties more than the V_{111}, first because there are twelve of the former and only eight of the latter (Table 3), and second because $|F_{220}|/|F_{111}| = \sqrt{2}$. We

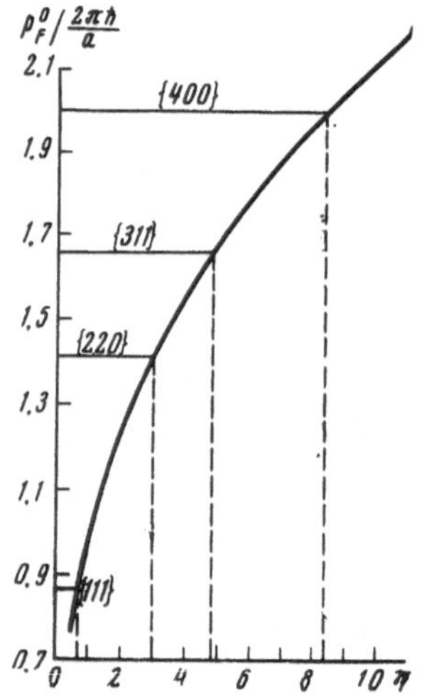

Fig. 6. Radius of the sphere of free electrons p_F^0 as a function of the number of valence electrons per atom η for a cubic lattice of the diamond type. Notation as in Fig. 4.

see from Tables 2 and 3 and Fig. 6 that

$$p_{111}/p_F^0 = 0.554; \qquad p_{220}/p_F^0 = 0.905. \tag{II.58}$$

Using Eqs. (II.20) and (II.21) we obtain

$$\frac{N_{\text{val}} - N}{N_{\text{val}}} = 3.48 \frac{|V_{111}|}{E_F^0}\left(\frac{1}{2} + \frac{\varphi_{111}}{\pi}\right) + 8.53 \frac{|V_{220}|}{E_F^0}\left(\frac{1}{2} + \frac{\varphi_{220}}{\pi}\right), \tag{II.59}$$

$$\varphi_{111} = \tan^{-1}\frac{0.494}{|V_{111}|/E_F^0}, \qquad \varphi_{220} = \tan^{-1}\frac{0.172}{|V_{220}|/E_F^0}. \tag{II.60}$$

The set of {220} planes separates out a closed volume in momentum space just sufficient for four electrons per atom. Thus, for a certain value of the Fourier component of the pseudopotential V_{220} all the electrons are enclosed in this zone. We may approximately estimate the value of the ratio $|V_{220}|/E_F^0$ for which $N \approx 0$ even if we put $V_{111} = 0$. It follows from Eqs. (II.59) and (II.60) that this happens for $|V_{220}|/E_F^0 \approx 0.15$. Actually, N vanishes even earlier, i.e., for smaller values of $|V_{220}|$, since in reality $|V_{111}| \neq 0$.

§ 7. Some Metals with a Tetragonal Lattice

We have already indicated that the equations derived in § 3 are also suitable for noncubic crystals. In this section we shall apply the results given in § 3 to certain metals with tetragonal lattices. We denote the main translational vectors for this lattice by a_x, a_y, and a_z. These vectors lie at right angles to one another. Hence, the coordinate axes in the direct and reciprocal lattices coincide. However, $a_z \neq a_x = a_y$, so that Eq. (II.40) has to be slightly altered. In this case we have

$$p_{n_1n_2n_3} \Big/ \left(\frac{2\pi\hbar}{a_x}\right) = \frac{1}{2}\sqrt{n_1^2 + n_2^2 + n_3^2 (a_x/a_z)^2}. \tag{II.61}$$

Equation (II.11) may be expressed in the following form for the tetragonal lattice.

The volume

$$\Delta\tau = a_x^3 (a_z/a_x) \tag{II.62}$$

contains Δn electrons, so that

$$p_F^0 \Big/ \left(\frac{2\pi\hbar}{a_x}\right) = \left(\frac{3}{8\pi}\Delta n\right)^{1/3} \left(\frac{a_x}{a_z}\right)^{1/3}. \tag{II.63}$$

Considering the last two equations, we may calculate the effect of the pseudopotential components on the quantity $\langle v_F \rangle S_F$. Let us suppose that for polycrystalline samples we have the relation

$$\frac{N_{\text{val}} - N}{N_{\text{val}}} \simeq \frac{v_F^0 S_F^0 - \langle v_F \rangle S_F}{v_F^0 S_F^0}. \tag{II.64}$$

We shall now give special consideration to indium and tin.

A. Indium

Indium is a tervalent metal with a face-centered tetragonal lattice [67] and an axial ratio $a_z/a_x = 1.07$. Using (II.63) and (II.61) we obtain

$$p_F^0 \Big/ \left(\frac{2\pi\hbar}{a_x}\right) = 1.100, \tag{II.65}$$

$$p_{n_1n_2n_3} \Big/ \left(\frac{2\pi\hbar}{a_x}\right) = \frac{1}{2}\sqrt{n_1^2 + n_2^2 + 0.862 n_3^2}. \tag{II.66}$$

The structure factors for indium coincide with the corresponding structure factors for an fcc lattice. The sphere of free electrons will be intersected by eight {111} Bragg planes, four (200) and (020) planes, and two (002) planes. Hence, in order to calculate N and S_F, we must allow for three components of pseudopotential, the V_{111}, V_{200}, and V_{002}. It follows from (II.65) and (II.66) that

$$p_{111}/p_F^0=0.769, \quad p_{002}/p_F^0=0.844, \quad p_{200}/p_F^0=0.909. \tag{II.67}$$

Using (II.20), (II.21), and (II.64), we obtain

$$\frac{N_{val}-N}{N_{val}}=4.83\frac{|V_{111}|}{E_F^0}\left(\frac{1}{2}+\frac{\varphi_{111}}{\pi}\right)+1.33\frac{|V_{002}|}{E_F^0}\left(\frac{1}{2}+\frac{\varphi_{002}}{\pi}\right)+2.85\frac{|V_{200}|}{E_F^0}\left(\frac{1}{2}+\frac{\varphi_{200}}{\pi}\right); \tag{II.68}$$

$$\varphi_{111}=\tan^{-1}\frac{0.355}{|V_{111}|/E_F^0}; \quad \varphi_{002}=\tan^{-1}\frac{0.263}{|V_{002}|/E_F^0}, \quad \varphi_{200}=\tan^{-1}\frac{0.165}{|V_{200}|/E_F^0}. \tag{II.69}$$

B. Tin

Tin is a tetravalent metal with a complex tetragonal lattice [68], which may be regarded as a simple tetragonal lattice with a basis: (0 0 0), (½ ½ ½), (½ 0 ¼), (0 ½ ¾). Using (II.38), we find that the structure factor in this case is equal to

$$F_{n_1n_2n_3}=\left[1+\cos\pi(n_1+n_2+n_3)+\cos\frac{\pi}{2}(2n_1+n_3)+\cos\frac{\pi}{2}(2n_2-n_3)\right]+$$
$$+i\left[\sin\pi(n_1+n_2+n_3)+\sin\frac{\pi}{2}(2n_1+n_3)+\sin\frac{\pi}{2}(2n_2-n_3)\right]. \tag{II.70}$$

For tin $a_x/a_z = 1.83$, so that

$$p_F^0\Big/\left(\frac{2\pi\hbar}{a_x}\right)=1.52, \tag{II.71}$$

$$p_{n_1n_2n_3}\Big/\left(\frac{2\pi\hbar}{a_x}\right)=\frac{1}{2}\sqrt{n_1^2+n_2^2+3.55n_3^2}. \tag{II.72}$$

According to (II.70) and (II.72), the sphere of free electrons will be intersected by the Bragg planes shown in Table 4. We see from the table that for tin the following Fourier components of the pseudopotential are important: V_{200}, V_{101}, V_{220}, and V_{211}.

Using (II.71) and the results of Table 4 we obtain:

$$p_{200}/p_F^0=0.658; \quad p_{101}/p_F^0=0.684; \quad p_{220}/p_F^0=0.928; \quad p_{211}/p_F^0=0.947. \tag{II.73}$$

This gives

$$\frac{N_{val}-N}{N_{val}}=2.07\frac{|V_{200}|}{E_F^0}\left(\frac{1}{2}+\frac{\varphi_{200}}{\pi}\right)+4.30\frac{|V_{101}|}{E_F^0}\left(\frac{1}{2}+\frac{\varphi_{101}}{\pi}\right)+$$
$$+2.91\frac{|V_{220}|}{E_F^0}\left(\frac{1}{2}+\frac{\varphi_{220}}{\pi}\right)+11.9\frac{|V_{211}|}{E_F^0}\left(\frac{1}{2}+\frac{\varphi_{211}}{\pi}\right); \tag{II.74}$$

TABLE 4. Structure Factors and Other Characteristics of the Bragg Planes Intersecting the Sphere of Free Electrons for Tin

$(n_1n_2n_3)$	No. of planes	$\frac{p_{n_1n_2n_3}}{2\pi\hbar/a_x}$	$\lvert F_{n_1n_2n_3}\rvert$	$(n_1n_2n_3)$	No. of planes	$\frac{p_{n_1n_2n_3}}{2\pi\hbar/a_x}$	$\lvert F_{n_1n_2n_3}\rvert$
(200) and (020)	4	1.00	4	(220)	4	1.41	4
(101) and (011)	8	1.04	$2\sqrt{2}$	(211) and (121)	16	1.44	$2\sqrt{2}$

$$\varphi_{200}=\tan^{-1}\frac{0.450}{|V_{200}|/E_F^0},\qquad \varphi_{101}=\tan^{-1}\frac{0.432}{|V_{101}|/E_F^0},$$

$$\varphi_{220}=\tan^{-1}\frac{0.134}{|V_{220}|/E_F^0},\qquad \varphi_{211}=\tan^{-1}\frac{0.100}{|V_{211}|/E_F^0}. \tag{II.75}$$

Since p_{101} is close to p_{200} and p_{211} is close to p_{220} we may expect that $V_{101} = V_{011}$ will be close to $V_{200} = V_{020}$ and $V_{211} = V_{121}$ close to V_{220}. This will simplify Eqs. (II.74) and (II.75). Let us put $V_1 \approx V_{101} \approx V_{200}$ and $V_2 \approx V_{220} \approx V_{211}$. We introduce the mean distances to the Bragg planes p_1 and p_2 determined by the relations

$$p_1=\frac{(n_{200}+n_{020})\,p_{200}+(n_{101}+n_{011})\,p_{101}}{n_{200}+n_{020}+n_{101}+n_{011}},$$

$$p_2=\frac{n_{220}p_{220}+(n_{211}+n_{121})\,p_{211}}{n_{220}+n_{211}+n_{121}}.$$

Here $n_{n_1n_2n_3}$ is the number of corresponding Bragg planes. Then

$$p_1/p_F^0=0.675,\qquad p_2/p_F^0=0.943, \tag{II.73a}$$

$$\frac{N_{\text{val}}-N}{N_{\text{val}}}\approx 6.36\frac{|V_1|}{E_F^0}\left(\frac{1}{2}+\frac{\varphi_1}{\pi}\right)+14.8\frac{|V_2|}{E_F^0}\left(\frac{1}{2}+\frac{\varphi_2}{\pi}\right), \tag{II.74b}$$

$$\varphi_1=\tan^{-1}\frac{0.439}{|V_1|/E_F^0},\qquad \varphi_2=\tan^{-1}\frac{0.108}{|V_2|/E_F^0}. \tag{II.75a}$$

§ 8. Interband Transitions Associated with the Bragg Energy Splitting, Without Allowing for Relaxation Processes

The next two sections are devoted to establishing the possibility of determining the Fourier components of the pseudopotential V_g from optical measurements [19, 20, 34, 35, 69].

The values of V_g were only determined experimentally comparatively recently. This was first done on the basis of measurements of the de Haas—van Alphen effect [61, 62], and a little later on the basis of cyclotron resonance [70]. A fairly complex computing technique was required. The Fourier components of the pseudopotential thus obtained only relate to helium temperatures. However, optical measurements enable us to find V_g by a much more direct method over a wide temperature range.

In order to determine V_g from optical data, we have to consider the interband transitions associated with the corresponding Bragg planes. In polyvalent metals these arise mainly in the visible and near-infrared parts of the spectrum. We shall shortly show that for these transitions there is a sharp peak in the combined interband density of states at frequencies $\hbar\omega \approx 2|V_g|$, which leads to the development of maxima in the interband conductivity $\tilde{\sigma}$, and also in the imaginary part of the complex interband dielectric constant $\tilde{\varepsilon}_2$. Thus the maxima in the dispersion of $\tilde{\sigma}$ (or $\tilde{\varepsilon}_2$) may be used to determine the Fourier components of the pseudopotential.*

Let us consider interband transitions associated with the Bragg splitting of the energy. Figures 7 and 8 show the intersection of the sphere of free electrons by one Bragg plane. The broken line indicates the band of higher number in the scheme of bands indicated. It is most convenient to use this scheme in the present section. We shall be interested in interband transi-

* The idea of weakly bound electrons for calculating the interband conductivity and interband dielectric constant was first used by Sergeiev and Tchernikovsky in 1934 [71]. However, only the development of the theory of the pseudopotential made it possible to establish the specific role of the transitions near the Bragg planes.

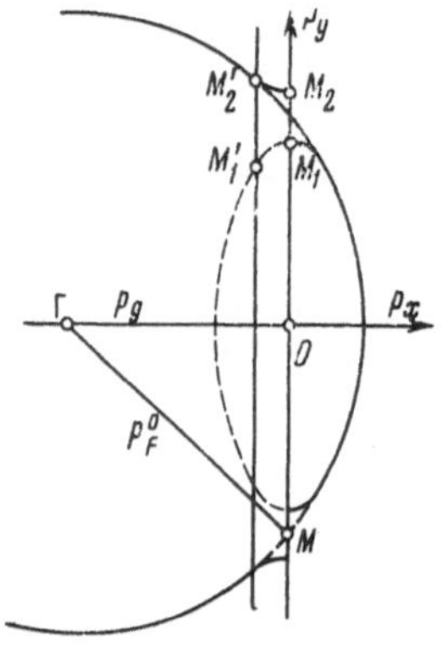

Fig. 7. Intersection of the sphere of free electrons by one Bragg plane. The Bragg plane coincides with the plane $p_y p_z$; $\Gamma M = p_F^0$, $\Gamma O = p_g$; the broken curve shows the second band in the band scheme; the plane $M_1' M_2'$ is parallel to the Bragg plane.

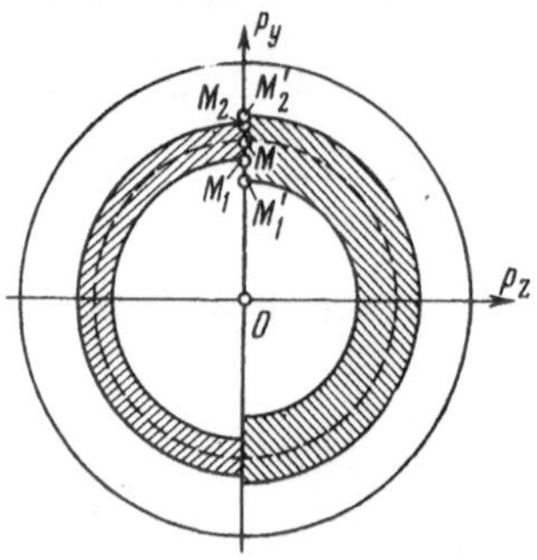

Fig. 8. Intersection of the Bragg plane and a plane parallel to it with the Fermi surface. On the left is the M_1M_2 ring on the Bragg plane; on the right is the $M_1' M_2'$ ring on the plane parallel to the Bragg plane. Broken line is the trace of the intersection of the Bragg plane with the sphere of free electrons.

tions between electron states close to the Bragg plane in question. As in the previous sections, we use the approximation of weakly bound electrons. In order to describe the electron states near the Bragg plane the wave function of the electron may be regarded as the linear combination of two plane waves [26]. The wave functions of the electron normalized to unit volume in the lower and upper band will then, respectively, equal

$$\Psi_1 = a_{11} \exp\left(\frac{i\mathbf{pr}}{\hbar}\right) + a_{12} \exp\left[\frac{i(\mathbf{p}-2\mathbf{p}_g)\mathbf{r}}{\hbar}\right],$$
$$\Psi_2 = a_{21} \exp\left(\frac{i\mathbf{pr}}{\hbar}\right) + a_{22} \exp\left[\frac{i(\mathbf{p}-2\mathbf{p}_g)\mathbf{r}}{\hbar}\right], \tag{II.76}$$

$$a_{11} = \frac{1}{\sqrt{2}} (1 + X^2 - X\sqrt{1+X^2})^{-1/2},$$
$$a_{12} = a_{11}(X - \sqrt{1+X^2}), \tag{II.77}$$
$$a_{21} = \frac{1}{\sqrt{2}} (1 + X^2 + X\sqrt{1+X^2})^{-1/2},$$
$$a_{22} = a_{21}(X + \sqrt{1+X^2});$$

$$X = p_g(p_g - p_\perp)/(m\,|\,V_g\,|\,). \tag{II.78}$$

The energy of the electron in the lower and upper bands is given by Eq. (II.6). Let us consider the vertical transitions (transitions with conservation of quasi-momentum). The energy difference for such transitions equals

$$\Delta E = 2\,|\,V_g\,|\,\sqrt{1+X^2}. \tag{II.79}$$

We shall use the following notation:

$$2 \mid V_g \mid = \hbar\omega_g. \tag{II.80}$$

It follows from (II.78) and (II.79) that for all points of a plane parallel to the Bragg plane X = const and ΔE = const. Under the action of light with a frequency ω for which $\hbar\omega = \Delta E$, there may be a transition from the lower band into the upper band for all electrons, the state of which is represented by points on the ring $M_1' M_2'$ (Figs. 7 and 8). For these electrons the law of energy conservation is in fact satisfied, and furthermore the lower state is occupied, since its energy falls below that of the Fermi level by an amount much greater than kT, while the upper state is free, since its energy exceeds that of the Fermi level by an amount much greater than kT. The minimum frequency of the interband transitions near the Bragg plane in this approximation is ω_g; it corresponds to the ring on the Bragg plane M_1M_2.

Thus, in metals (in contrast to semiconductors), the basic structure of the bands of interband conductivity is associated with the Bragg energy splitting. In semiconductors this structure is due to energy splitting near critical points constituting points of high symmetry.* In metals the points of high symmetry may also give slight maxima of interband conductivity, but these are much smaller than the maxima associated with the rings on the Bragg planes, since the number of electrons taking part in the transitions near the points in question (those of high symmetry) is much smaller than the number taking part in transitions near the rings.

We may note two other circumstances: first, the points situated on the M_1M_2 ring (Fig. 8) are not points of high symmetry; second, in the transitions under consideration both the lower and the higher states have an energy differing from the Fermi energy by an amount much greater than kT.

Let us determine the combined interband density of states $dY/d\omega$. Here dY is the number of states for the interband-transition frequency range $(\omega, \omega + d\omega)$. Clearly,

$$\frac{1}{\hbar}\left(\frac{dY}{d\omega}\right) = \int 2\frac{d^3p}{(2\pi\hbar)^3}\,\delta(\Delta E - \hbar\omega). \tag{II.81}$$

Let us calculate $d^3p = S_X dp_\perp$. Here, S_X is the area of the ring $M_1'M_2'$. It is easy to see (Fig. 7) that

$$S_X = 2\pi m\hbar\omega_g\sqrt{1+X^2}. \tag{II.82}$$

It follows from (II.78) that

$$dp_\perp = -\frac{m\hbar\omega_g}{2p_g}\,dX. \tag{II.83}$$

Using (II.82) and (II.83) we obtain

$$d^3p = \pi m^2\frac{\hbar^2\omega_g^2}{p_g}\sqrt{1+X^2}\,dX. \tag{II.84}$$

Further,

$$\delta(\Delta E - \hbar\omega) = \delta(\hbar\omega_g\sqrt{1+X^2} - \hbar\omega). \tag{II.85}$$

* The influence of critical points on the spectral density of the lattice vibrations was considered by van Hove [72] and Phillips [73]. Later analogous considerations were applied by Phillips [74, 75], and Brust et al. [76] to electrons in metals. In theoretical review articles by Kohn [77] and Phillips [75], the role of the critical points was again emphasized. Ehrenreich, Phillip, et al. [78] and Shklyarevskii et al. [79] attempted to compare the experimental maxima in $\tilde{\sigma}(\omega)$ with the foregoing singular points.

Hence,

$$\left(\frac{dY}{d\omega}\right)_g = \frac{m^2}{4\pi^2\hbar p_g}\frac{\omega^2}{\sqrt{\omega^2-\omega_g}} \quad \text{for} \quad \omega > \omega_g,$$
$$\left(\frac{dY}{d\omega}\right)_g = 0 \quad \text{for} \quad \omega < \omega_g. \tag{II.86}$$

At the point $\omega = \omega_g$, $dY/d\omega$ has a radical singularity passing to infinity. An analogous result was obtained by Harrison [69], with the one difference that he made no allowance for the dependence of S_X on X and used an expansion of $\sqrt{1+X^2}$ in X^2.

The singularity in the combined interband density of states leads to the appearance of maxima in the relationships for $\tilde{\sigma}(\omega)$ and $\mathrm{Im}\,\tilde{\varepsilon}'$.

Let us find the contribution to the interband conductivity $\tilde{\sigma}_g$ of the transitions near the Bragg plane under consideration. We shall determine $\tilde{\sigma}_g$ from the electromagnetic-field losses*

$$\tilde{\sigma}_g = \frac{w_g \hbar\omega}{\overline{\mathscr{E}^2}}. \tag{II.87}$$

Here, w_g is the probability of the absorption of a light quantum of frequency ω in unit time; $\overline{\mathscr{E}}$ is the electric field of the light wave. The dash denotes time averaging.

The time-dependent Hamiltonian of the interaction between the electrons and the electromagnetic field $H_i(t)$ we take in the form

$$H_i(t) = i\frac{\hbar e}{mc}\nabla\mathbf{A}(t), \tag{II.88}$$

$$\mathbf{A}(t) = \frac{1}{2}[\mathbf{A}\exp(-i\omega t) + \mathbf{A}^*\exp(i\omega t)], \tag{II.88a}$$

where $\mathbf{A}(t)$ is the vector potential of the electromagnetic field. We neglect terms proportional to A^2, which means neglecting two-photon processes by comparison with one-photon processes.

Using (II.87) and (II.88) we obtain† [80] the expression

$$\tilde{\sigma}_g = \frac{1}{8\pi^2}\frac{e^2}{m^2c^2\hbar}\frac{\omega}{\overline{\mathscr{E}^2}}\int |\,\Psi_1\,|\,\nabla\mathbf{A}\,|\,{}_2\Psi\,|^2\delta(\Delta E - \hbar\omega)\,d^3p. \tag{II.89}$$

Equation (II.89) does not allow for relaxation processes. From (II.76) and (II.88a) (neglecting the dependence of $\mathbf{A}$ on $\mathbf{r}$), we obtain

$$|\langle\Psi_1\,|\,\nabla\mathbf{A}\,|\,\Psi_2\rangle|^2 = \frac{1}{\hbar^2}\frac{1}{1+X^2}(\mathbf{A}\mathbf{p}_g)^2. \tag{II.90}$$

Let us allow for the relation between $\mathscr{E}$ and $\mathbf{A}$

$$\mathscr{E} = (-1/c)(\partial\mathbf{A}/\partial t) = (i\omega/c)\,\mathbf{A}.$$

It follows from (II.89), (II.90), and (II.81)-(II.86) that

$$\tilde{\sigma}_g = \frac{e^2 p_g \cos^2\theta_g}{4\pi\hbar^2}\frac{\omega_g^2}{\omega\sqrt{\omega^2-\omega_g^2}}. \tag{II.91}$$

* For simplicity we confine attention to cubic crystals.

† An analogous expression may be obtained by using the results of [81].

Here θ_g is the angle between* the vectors $\mathscr{E}$ and $\mathbf{p}_g$. Let us average $\tilde{\sigma}_g$ over all directions of the electromagnetic field, on the grounds that we are concerned with a polycrystalline metal. Then $\overline{\cos^2\theta} = {}^1/_3$. Let us consider all the physically equivalent planes $\{g\}$. These make the following contribution to the conductivity:

$$\tilde{\sigma}_{\{g\}} = \frac{e^2 p_g n_g}{12\pi\hbar^2} \frac{\omega_g^2}{\omega\sqrt{\omega^2 - \omega_g^2}}. \quad \text{(II.92)}$$

Here n_g is the number of physically equivalent Bragg planes. This equation is also valid for cubic single crystals. In what follows we shall omit the curly brackets. In the presence of physically nonequivalent planes we shall use the additivity of $\tilde{\sigma}$. This gives

$$\tilde{\sigma} = \frac{e^2}{12\pi\hbar^2} \frac{1}{\omega} \sum_g \frac{p_g n_g \omega_g^2}{\sqrt{\omega^2 - \omega_g^2}}, \quad \text{(II.93)}$$

where the summation extends over all the physically nonequivalent Bragg planes.

It follows from Eq. (II.92) that the interband conductivity $\tilde{\sigma}_g$ develops very sharply for $\omega = \omega_g$ and has an infinitely large maximum. For $\omega > \omega_g$, the value of $\tilde{\sigma}_g$ will be finite. The asymptotic value of $\tilde{\sigma}_g$ for $\omega \gg \omega_g$ equals

$$\tilde{\sigma}_g = \frac{1}{12\pi} \frac{e^2}{\hbar^2} p_g n_g \left(\frac{\omega_g}{\omega}\right)^2. \quad \text{(II.94)}$$

§ 9. Interband Transitions Associated with the Bragg Energy Splitting, Allowing for Relaxation

The infinitely large value of $\tilde{\sigma}_g$ for $\omega = \omega_g$ obtained in the previous section is mainly associated with the following two circumstances. First, we have used an approximation in which the wave function of the electron equals the sum of only two plane waves. This means that each time we are considering the action of only one Fourier component of the pseudopotential. Allowance for the simultaneous action of several components has the effect that the energy difference between the upper and lower bands will not be constant at all points of the ring $M_1^\prime M_2^\prime$. As a result of this the maximum of $\tilde{\sigma}_g$ will become finite. We may expect that the relative width of this maximum will be of the order of $|V_g|/E_F^0$. Second, we have neglected relaxation processes, which lead to the blurring of the energy levels. These processes may be associated both with interelectron interaction and also with interaction between the electrons and phonons.

The second factor is evidently more important than the first. Experiments which we carried out for a number of metals (Chapter VI) showed that the blurring (spreading) of the energy levels was considerable, much greater than that which might be expected from the simultaneous action of several planes. Furthermore, this blurring increased with rising temperature.

A proper account of relaxation processes can only be taken if we use the corresponding kinetic equation. We shall allow for the finite lifetime of the excited state, and in Eq. (II.89) replace the δ function by the Lorentz function φ. Remembering that

$$\delta(\xi) = \lim_{\gamma \to 0} \varphi(\xi),$$

*Equation (II.91) means that the light causes interband transitions associated with the Bragg plane g only if the projection of the electric field of the light wave on the normal to the Bragg plane under consideration differs from zero.

where

$$\varphi(\xi) = \frac{\gamma}{\pi(\xi^2 + \gamma^2)}, \tag{II.95}$$

we use $\varphi(\Delta E - \hbar\omega)$, regarding γ as a finite constant quantity. This may only be done to a first approximation. The exact expression for γ requires the solution of the corresponding kinetic equation, as indicated above. Thus,

$$\tilde{\sigma}_g = \frac{1}{8\pi^2} \frac{e^2}{m^2 c^2 \hbar^2} \frac{\omega}{\mathscr{E}^2} \int |\langle \Psi_1 | \nabla \mathbf{A} | \Psi_2 \rangle|^2 \varphi(\Delta E - \hbar\omega)\, d^3 p. \tag{II.96}$$

Subsequently it will be convenient to use the following dimensionless quantities:

$$\omega' = \omega/\omega_g, \quad \gamma' = \gamma/(\hbar\omega_g). \tag{II.97}$$

In this notation,

$$\xi/(\hbar\omega_g) = \sqrt{1 + X^2} - \omega'. \tag{II.98}$$

Consequently,

$$\tilde{\sigma}_g = \frac{e^2}{12\pi^2\hbar^2} p_g n_g \mathscr{I}, \tag{II.99}$$

$$\mathscr{I} = \frac{\gamma'}{\omega'} \int_0^\infty \frac{dX}{\sqrt{X^2+1}\left[\left(\sqrt{X^2+1} - \omega'\right)^2 + \gamma'^2\right]}. \tag{II.100}$$

The upper limit is of the order of $2(p_g^2/p_F^2)(E_F/|V|_g) \gg 1$, and may be taken as infinity. Existing tables [82] enable us to calculate the indefinite integral. We shall not give the result here as it is very cumbersome.

The value of the integral $\mathscr{I}$ depends on ω', γ'. Let us consider the function $\mathscr{I}(\omega')$, treating γ' as a parameter. The general expression for $\mathscr{I}$ cannot be given in a convenient form. Calculations show that the function $\mathscr{I}(\omega')$ has a finite maximum at about $\omega' \approx 1$. Let us introduce the coefficient $t = \omega_{max}/\omega_g$ characterizing the displacement of the maximum of $\mathscr{I}$ and $\tilde{\sigma}_g$ from the value of $\omega' = 1$ (ω_{max} is the frequency corresponding to the maximum of $\mathscr{I}$). The dependence of t on γ' is given in Fig. 9.

We see from the figure that the displacement is not very great. This result constitutes a good basis for determining the Fourier component of the pseudopotential from optical measurements. If in fact we determine the frequency ω_{max} corresponding to the maximum of $\tilde{\sigma}_g(\omega)$,

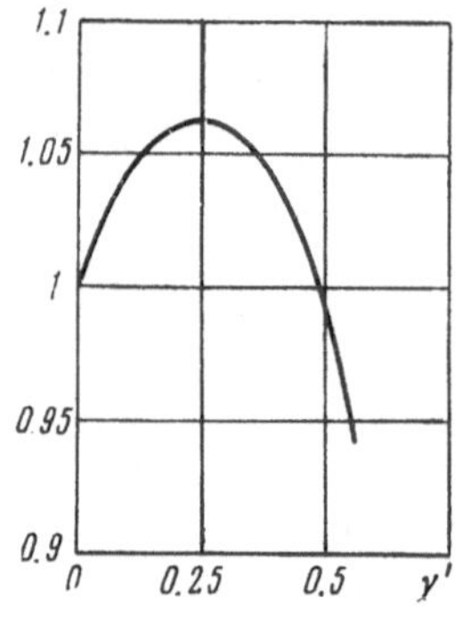

Fig. 9. Dependence of $t = \omega_{max}/\omega_g$ on $\gamma' = \gamma/\hbar\omega_g$.

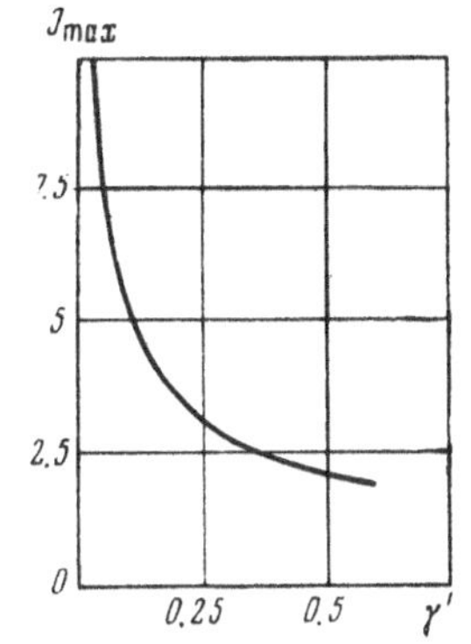

Fig. 10. Dependence of the maximum value of the function $\mathcal{I}(\omega')$ on $\gamma' = \gamma/\hbar\omega_g$.

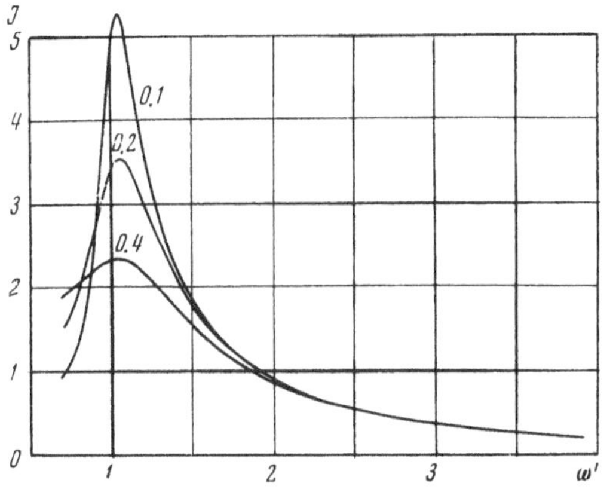

Fig. 11. Shape of the $\mathcal{I}(\omega')$ curve for three values of the parameter γ'.

we may determine the value of $|V_g|$ by using the equation

$$2\,|\,V_g\,| \equiv \hbar\omega_g = \frac{\hbar\omega_{max}}{t}, \qquad \text{(II.101)}$$

in which t is almost equal to unity. The method of determining $|V_g|$ from optical measurements will be set out in more detail in Chapter VI.

Figure 10 shows the maximum values of the function $\mathcal{I}(\omega')$ for various γ'. These results, together with Eq. (II.99), readily enable us to obtain the maximum values of $(\tilde{\sigma}_g)_{max}$. It should be noted that $(\tilde{\sigma}_g)_{max}$ only depends on the product $p_g n_g$ and the relative width of the energy levels γ'. As we should expect, the value of $(\tilde{\sigma}_g)_{max}$ increases on reducing γ'.

The shape of the $\tilde{\sigma}_g(\omega)$ curve is determined by the $\mathcal{I}(\omega')$ relation. This is shown in Fig. 11 for $\gamma' = 0.1$, 0.2, and 0.4. We see from the figure that the corresponding curves are asymmetrical. The width of the curves on the large ω side is greater. It is reasonable to expect that the curves presented will give a fairly good description of the experimental results for values of ω' close to unity. For $\omega' \ll 1$ these equations are not valid, since for such frequencies replacing the δ function by the Lorentz function constitutes too coarse an approximation. For $\omega' \gg 1$ the asymptotic equation (II.94) holds independently of relaxation processes.

Subsequently, when analyzing experimental data, it is important to be able to determine γ' from optical experiments. This parameter may best be found by using the long-wave part of the $\mathcal{I}(\omega')$ curve. Let us consider a point of the $\mathcal{I}(\omega')$ curve with an abscissa of $(\omega'_{max} - \gamma')$. Let us denote the ordinate of this point by $\mathcal{I}$ The results of a calculation of $\mathcal{I}$ for various values of the parameter γ' are presented in Fig. 12.

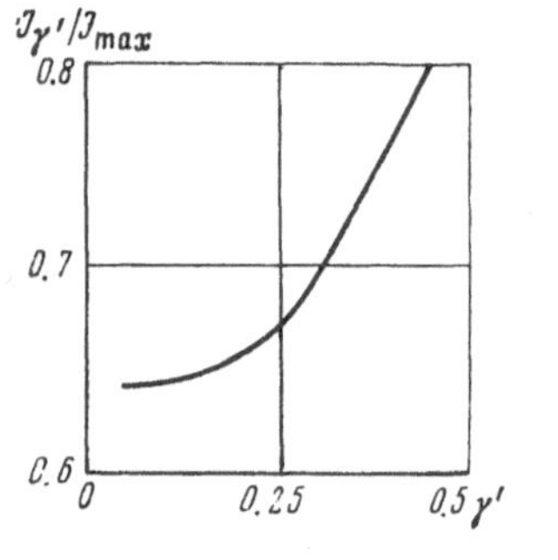

Fig. 12. Ratio $\mathcal{J}_{\gamma'}/\mathcal{J}_{max}$ as a function of the parameter γ'.

We see from Fig. 12 that for $\gamma' < 0.4$ the value of the ratio $\mathcal{J}_{\gamma'}/\mathcal{J}_{max} \approx 0.7$. This means that we may approximately find γ' from the experimental $\tilde{\sigma}(\omega)$ relationship by determining the abscissa of the point having an ordinate approximately 70% of the maximum ordinate. This value of γ' may then be refined by using Fig. 12.

A comparison of the results with experiment is presented in Chapter VII. Here we shall simply note that experiment confirms the conclusions of these two sections. We thus have a basis for determining the Fourier components of the pseudopotential from optical measurements.

A detailed scheme for analyzing the results of optical measurements in order to determine the Fourier components of the pseudopotential will be presented in Chapter VI.

§ 10. Temperature Dependence of the Conduction-Electron Concentration

Earlier, in Section 3, we established a relationship between N, S_F, $\langle v_F \rangle$, and V_g. In obtaining the basic equations of this section we considered the lattice potential without considering the lattice vibrations. Here we shall show that allowance for the lattice vibrations leads to a temperature dependence of V_g, which, in turn, leads to the temperature dependence of N, S_F, and $\langle v_E \rangle$ [33].

The N(T) dependence is associated with the fact that the Fourier component of the pseudopotential is proportional to the structure factor determined by Eq. (II.35). The thermal vibrations of the lattice have the effect that the structure factor contains a temperature factor $\exp[-W(T)]$, called the Debye – Waller factor. An analogous factor determines the intensity of the diffraction maxima in x-ray diffraction [83, 84]. The procedure for obtaining this factor is quite standard [32]. The radius vector corresponding to the point l may be expressed in the form

$$\mathbf{R}_l = \mathbf{l} + \mathbf{u}_l = \mathbf{l} + \sum_{\mathbf{q}} \frac{1}{\sqrt{2}} [\mathbf{u}_{\mathbf{q}} \exp(i\mathbf{q}\mathbf{l}) + \mathbf{u}^*_{\mathbf{q}} \exp(-i\mathbf{q}\mathbf{l})]. \quad \text{(II.102)}$$

Here $\mathbf{u}_q$ is the vector amplitude of the lattice vibrations, including the time factor $\exp(-i\Omega_q t)$; $\mathbf{q}$ is the wave vector of the lattice vibrations.

We have to calculate the value of F(**K**) determined by Eq. (II.35). Using (II.102) we shall have

$$\exp(-i\mathbf{K}\mathbf{R}_l) = \exp(-i\mathbf{K}\mathbf{l}) \prod_{\mathbf{q}} \exp\left\{-\frac{i}{\sqrt{2}} [(\mathbf{K}\mathbf{u}_{\mathbf{q}}) e^{i\mathbf{q}\mathbf{l}} + (\mathbf{K}\mathbf{u}^*_{\mathbf{q}}) e^{-i\mathbf{q}\mathbf{l}}]\right\}. \quad \text{(II.103)}$$

We shall in future omit the vector sign on the index q.

We consider the fact that $\mathbf{u}_q$ are small and expand the corresponding exponentials in series. Then

$$\prod_q \exp\{\ \} = \prod_q \left\{ 1 - \frac{i}{\sqrt{2}} [(\mathbf{K}\mathbf{u}_q) e^{i\mathbf{q}\mathbf{l}} + (\mathbf{K}\mathbf{u}_q^*) e^{-i\mathbf{q}\mathbf{l}}] - \frac{1}{4} [(\mathbf{K}\mathbf{u}_q) e^{i\mathbf{q}\mathbf{l}} + (\mathbf{K}\mathbf{u}_q^*) e^{-i\mathbf{q}\mathbf{l}}]^2 + \cdots \right\}. \quad (\text{II.104})$$

This expression contains various products of the form $\mathbf{u}_q \exp(i\mathbf{q}\mathbf{l})$. Each factor corresponds to the generation or annihilation of a phonon. The probability of multiple-phonon processes falls rapidly as their order increases. If we are simply interested in phononless processes, we obtain

$$\prod_q \exp\{\ \} = \prod_q \left(1 - \frac{1}{2} |\mathbf{K}\mathbf{u}_q|^2 \right). \quad (\text{II.105})$$

The terms under the product sign may be collected back into an exponential. We thus obtain

$$\prod_q \left(1 - \frac{1}{2} |\mathbf{K}\mathbf{u}_q|^2 \right) = \exp\left(-\frac{1}{2} \sum_q |\mathbf{K}\mathbf{u}_q|^2 \right) = \exp(-W_\mathbf{K}), \quad (\text{II.106})$$

$$W_\mathbf{K} = \frac{1}{2} \sum_q |\mathbf{K}\mathbf{u}_q|^2. \quad (\text{II.107})$$

With increasing temperature u_q becomes greater, and this leads to an increase in $W_\mathbf{K}$. Using (II.35), (II.103), and (II.106), we obtain

$$F(\mathbf{K}) = \frac{1}{N_c} \sum_l \exp(-i\mathbf{K}\mathbf{l}) \exp(-W_\mathbf{K}) = \exp(-W_\mathbf{K}) F_g \delta_{\mathbf{K}\mathbf{K}_g}. \quad (\text{II.108})$$

For the Fourier components of the pseudopotential we shall have

$$V(\mathbf{K}) = U(\mathbf{K}) F(\mathbf{K}) = U(\mathbf{g}) F_\mathbf{g} \exp(-W_\mathbf{g}) \delta_{\mathbf{K}\mathbf{K}_g}, \quad \mathbf{K}_g = 2\pi\mathbf{g}. \quad (\text{II.109})$$

We have found that the Fourier components depend on the temperature. To our present degree of accuracy we may put the resultant functions $V_g(T)$ into Eqs. (II.20), (II.31), and (II.32) and determine the temperature dependence of N, S_F, and $\langle v_F \rangle$.

With increasing temperature, the Fourier components of the pseudopotential diminish, so that the difference $N_{val} - N(T)$ becomes smaller, i.e., N(T) increases. For very high temperatures, $N \to N_{val}$.

In studying the optical properties of polyvalent metals we observed a slight increase in N on raising the temperature from that of liquid helium to room temperature for all the metals examined [17, 19, 20, 22, 23, 36]. A discussion of the experimental results, and a comparison with theory, will be presented in Chapter VII.

In order to compare the observed changes in N(T) with theory, we must find $W_g(T)$. An exact determination of this quantity requires summation over the whole spectrum of elastic vibrations. The calculation is very cumbersome. However, we may estimate the order of magnitude of W_g and its temperature variation by using the Debye model of a solid. In this model the solid is regarded as a continuous medium, i.e., the Brillouin zone is replaced by the Debye sphere, the radius of this being determined from the condition

$$N_a = \frac{1}{8\pi^3} \frac{4}{3} \pi q_D^3, \quad (\text{II.110})$$

where N_a is the number of atoms in unit volume (we shall suppose that one atom belongs to each unit cell). For simplicity we suppose that the velocities of all three branches of the acoustic spectrum are identical. In this model all three modes are degenerate for each value of q. The polarization vectors may be taken arbitrarily, only their orthogonality being taken into account. We choose one with $\mathbf{u}_q \| \mathbf{K}$ and two others with $\mathbf{u}_q \perp \mathbf{K}$. We then obtain

$$|\mathbf{K}\mathbf{u}_q|^2 = K^2 |\mathbf{u}_q|^2 \text{ and } W_{\mathrm{K}} = \frac{K^2}{2} \sum_q |\mathbf{u}_q|^2. \tag{II.111}$$

Equation (II.111) is satisfied best of all for cubic crystals consisting of atoms of one type only. We shall be chiefly considering this case in what follows. Allowing for the relation between the amplitude of the vibrations and their mean energy, we obtain

$$|\mathbf{u}_q|^2 = \frac{\left(\bar{n}_q + \frac{1}{2}\right)\hbar\Omega_q}{N_a M \Omega_q^2}, \tag{II.112}$$

$$\bar{n}_q = \frac{1}{\exp(\hbar\Omega_q/kT) - 1}. \tag{II.113}$$

Whence

$$W_{\mathrm{K}}(T) = \frac{1}{2}\frac{K^2\hbar}{N_a M} \sum_q \left[\frac{1}{\exp(\hbar\Omega_q/kT) - 1} + \frac{1}{2}\right]\frac{1}{\Omega_q} = \frac{3}{2} K^2 \frac{\hbar}{M} \int_0^{\Omega_q} \left[\frac{1}{\exp(\hbar\Omega_q/kT) - 1} + \frac{1}{2}\right]\frac{\Omega\, d\Omega}{\Omega_D^3}. \tag{II.114}$$

Here, $\Omega_D = sq_D$ is the Debye frequency, M is the mass of the atom, s is the velocity of sound. Let us change the variables: $z = \hbar\Omega/kT$, where k is Boltzmann's constant. Then

$$W_{\mathrm{K}}(T) = \frac{3}{2} K^2 \frac{\hbar^2}{Mk\theta}\left(\frac{T}{\theta}\right)^2 \int_0^{\theta/T} \left(\frac{1}{\exp z - 1} + \frac{1}{2}\right) z\, dz, \tag{II.115}$$

where $k\theta = \hbar\Omega_D$, and θ is the Debye temperature. For $T \gg \theta$ we may expand exp z in series. We obtain

$$W_{\mathrm{K}}(T) \simeq \frac{3}{2} K^2 \frac{\hbar^2}{Mk\theta} \frac{T}{\theta} \tag{II.116}$$

For $T \ll \theta$, the upper limit $\approx \infty$,

$$\int_0^\infty \frac{z\, dz}{\exp z - 1} = \frac{\pi^2}{3}, \quad W_{\mathrm{K}}(0) = \frac{3}{2} K^2 \frac{\hbar^2}{Mk\theta}. \tag{II.117}$$

Thus, as $T \to 0$, W tends to a constant finite value. This fact is associated with the zero vibrations of the lattice. We note that $W(0) = W(\theta)/4$. Let us write

$$\varphi(x) = \frac{1}{x}\int_0^x \frac{z\, dz}{\exp z - 1} + \frac{x}{4}. \tag{II.118}$$

The function $\varphi(x)$ differs from the well-known Debye function by an amount $x/4$. The values of the Debye function have been tabulated in [83, 85]. Allowing for (II.115)-(II.118), we obtain

$$W(T) = W(0)\; 4\varphi(x)/x, \qquad x = \theta/T. \tag{II.119}$$

Figure 13 gives the values of the ratio $W(T)/W(0)$ in relation to T/θ. On increasing T, this ratio first changes very slowly and then passes into a linear dependence.

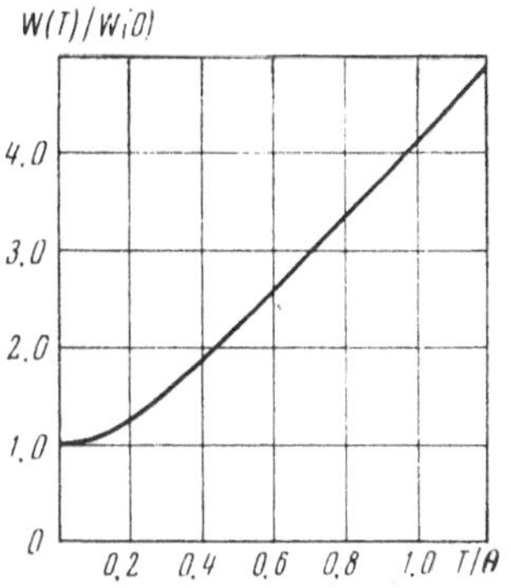

Fig. 13. Temperature dependence of the power index of the Debye — Waller factor.

Using the notation of Section 2, we obtain $K\hbar = 2\pi g\hbar = 2p_g$, whence

$$W(0) = \frac{3}{2}\frac{p_g^2}{M}\frac{1}{k\theta}. \tag{II.120}$$

Let us estimate the change in N on passing from liquid-nitrogen to room temperature.* The different terms under the summation sign in (II.20) have different temperature factors, since they have different values of p_g.

In order to estimate the quantity of present interest we introduce a certain mean value of $\overline{W(T)}$. We remember that for the temperatures in question W is small and so put $\exp(-W) \simeq 1 - W$. On these assumptions,

$$\left[\frac{N(T_K)}{N(T_N)} - 1\right] \simeq \left[\frac{N_{val}}{N(T_N)} - 1\right]\left[\overline{W(T_K)} - \overline{W(T_N)}\right], \tag{II.121}$$

where T_K represents room temperature and T_N nitrogen temperature.

A comparison of the results obtained in this section with experiment will be presented in Chapter VII.

CHAPTER III

MEASURING METHODS AND EXPERIMENTAL APPARATUS

§ 1. Methods of Measuring the Optical Constants of Metals

A review of the methods used for determining the optical constants of metals up to 1955, mainly intended for measurements in the visible and near-ultraviolet parts of the spectrum, was presented in [2]. Recently new methods suitable for the infrared region have been successfully developed. In this section we shall consider the most promising of these, namely, polarization methods and methods based on measuring the reflection coefficient of light for normal incidence, with subsequent use of dispersion relations of the Kramers — Krönig type. In both cases reflected light is used. Only reflected light will give the constants characteristic of the bulk metal. Transmitted light may be used with very thin films, the properties of which differ from those of the bulk metal in that the Fourier components of the pseudopotential are differ-

* The temperature range chosen coincides with the temperature range of the corresponding x-ray experiments, the results of which are compared with the optical experiments.

ent in the two cases. The methods indicated may be used over a wide spectral range, from the ultraviolet to the infrared. The accuracy and reliability are greater for the polarization methods.

1. Polarization methods involve the determination of the ellipticity of the light. Linearly polarized light falling on the boundary of the metal at an angle of $\varphi \neq 0$ is elliptically polarized after reflection from the metal. By measuring the phase shift Δ between the p and s components of the reflected light and the azimuth ρ ($\tan \rho$ determines the ratio of the p and s amplitudes of the reflected light) we obtain two relations, from which n and $\varkappa$ may be determined [2, 86].

If

$$1/|\varepsilon'| \ll 1, \tag{III.1}$$

where $\varepsilon' = (n - i\varkappa)^2$ is the complex dielectric constant, then

$$n = \frac{\sin\varphi \tan\varphi \cos 2\rho}{1 - \sin 2\rho \cos \Delta}, \tag{III.2}$$

$$\varkappa = \frac{\sin\varphi \tan\varphi \sin 2\rho \sin \Delta}{1 - \sin 2\rho \cos \Delta}. \tag{III.3}$$

The minus sign in the denominator allows for the fact that, for normal incidence, the p and s components of the reflected light have a phase shift $\Delta = 0$.

In the infrared region, inequality (III.1) is always satisfied for metals and we may use Eqs. (III.2) and (III.3). In the visible part of the spectrum $1/|\varepsilon'|$ may reach several percent. This means that we must allow for the dependence of the surface impedance on the angle of incidence of the light. Allowance for first-order terms in $1/|\varepsilon'|$ gives the following relation [23] between the complex refractive index, which in the present case we shall denote by $\tilde{n} - i\tilde{\varkappa}$, and Δ and ρ:

$$\tilde{n} = n\left(1 + \frac{1}{2}\,\frac{\sin^2\varphi}{n^2 + \varkappa^2}\right), \tag{III.4}$$

$$\tilde{\varkappa} = \varkappa\left(1 - \frac{1}{2}\,\frac{\sin^2\varphi}{n^2 + \varkappa^2}\right), \tag{III.5}$$

where n and $\varkappa$ are given by Eqs. (III.2) and (III.3). In what follows we shall always denote the optical constants by n and $\varkappa$ without the ~ sign over them.

The resultant values of the optical constants enable us to determine the absorption capacity A and reflection capacity R of the metal for normal light incidence. We already know [2, 86] that

$$A = \frac{4n}{(n+1)^2 + \varkappa^2}, \quad R = \frac{(n-1)^2 + \varkappa^2}{(n+1)^2 + \varkappa^2}, \quad R + A = 1. \tag{III.6}$$

The values of Δ and ρ in the visible and ultraviolet parts of the spectrum may easily be measured by using compensators and polarizers [2]. For the infrared regions there are no compensators available, and Δ must be determined by some other method. We first established the phase shift $\Delta = (\pi/2)(2i - 1)$ (here i is a whole number). The use of multiple light reflection greatly increases the accuracy of the measurement [14, 15]. This may be seen quite clearly from the following relation:

$$\frac{r_p^{(m)}}{r_s^{(m)}} = \frac{A_p}{A_s}(\tan\rho)^m \exp(im\Delta). \tag{III.7}$$

[In Eq. (III.7), i is the imaginary unit; in other cases in this section, i represents a whole num-

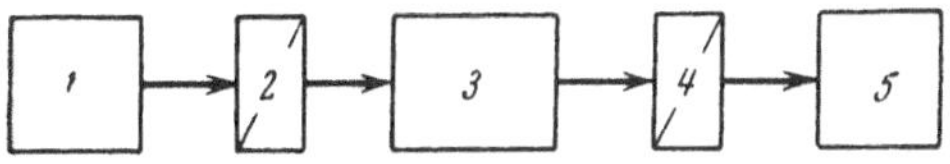

Fig. 14. Block diagram of the polarization method. 1) Illuminator; 2) polarizer; 3) mirrors under consideration; 4) analyzer; 5) radiation receiver.

ber.] Here, $r_p^{(m)}$ and $r_s^{(m)}$ are components of the electric vector of the reflected light wave after an m-fold reflection from the metal surface (mirror); A_p and A_s are the components of the incident light wave; Δ is the phase shift after a single reflection; ρ is the azimuth after a single reflection.

Thus the phase shift increases by a factor of m, i.e., $\Delta_m = m\Delta$, and the resultant azimuth is determined from the equation $\tan\rho_m = (\tan\rho)^m$. The use of multiple reflection enables us to work with smaller angles of incidence of the light; this facilitates working with longer waves.

In order to increase the accuracy of the measurements we used a null modulation method. A block diagram of the method is presented in Fig. 14. The illuminating system creates a parallel beam of unpolarized light. This light passes through the polarizer and becomes linearly polarized; it then passes into the system of test mirrors and experiences multiple reflection. The elliptically polarized light obtained as a result of the reflections passes through a second polarizer (analyzer) and falls on a radiation receiver. In order to modulate the light, one of the polarizers is rotated. The modulation frequency equals twice the frequency of rotation of the polarizer (in the apparatus used in our experiments the light modulation frequency was 9 cps). The radiation receiver thus accepts a pulsating signal comprising steady and alternating parts. As we shall later see, for $\Delta_m = (\pi/2)(2i-1)$ and $\tan\rho_m \tan\alpha = \pm 1$ (α is the azimuth of the stationary polarizer) the alternating part of the signal vanishes, and this event is recorded by a device at the output of the electronic amplifier coupled to the receiver. The measurements are carried out in monochromatic light. The light is made monochromatic by means of a monochromator placed either in the unit designated as the illuminating system or else in the unit designated as the radiation receiver. The position of the monochromator is not immaterial; it should be placed in the unit which is associated with the stationary polarizer or analyzer so that the polarization of the light due to the monochromator may not have any deleterious effect.

Let us consider the intensity of the light transmitted through the polarizer — mirror — analyzer system:

$$I = \frac{1}{2} I_0 \left[\left(\frac{r_p}{A_p}\right)^{2m} \cos^2\alpha_P \cos^2\alpha_A + \left(\frac{r_s}{A_s}\right)^{2m} \sin^2\alpha_P \sin^2\alpha_A + \right.$$
$$\left. + \frac{1}{2}\left(\frac{r_p}{A_p}\right)^m \left(\frac{r_s}{A_s}\right)^m \sin 2\alpha_P \sin 2\alpha_A \cos\Delta_m \right]. \qquad \text{(III.8)}$$

Here α_P and α_A are the azimuths of the polarizer and analyzer. We see that α_P and α_A enter symmetrically into Eq. (III.8). This means that for the resultant intensity it is immaterial whether we rotate the polarizer or analyzer. In order to be specific, we shall suppose that the polarizer is rotated at a frequency Ω, i.e., $\alpha_P = \Omega t$, where t is the time. Then

$$I = A + B\cos(2\Omega t + \gamma). \qquad \text{(III.9)}$$

The quantities A, B, and γ are independent of time. We see from (III.8) and (III.9) that the alternating signal vanishes on satisfying the following conditions, equivalent to the conditions indicated above:

$$\cos\Delta_m = 0, \quad \tan^2\alpha_A = (r_p/r_s)^{2m} \qquad \text{(III.10)}$$

We devised several versions of the method in question, enabling us to measure the optical constants of metals to an accuracy of 1-2%.

An analogous polarization method was developed and used by Shklyarevskii et al. [87, 88]. The apparatus used by these authors recorded the phase shift $\Delta = \pi i$ (i is a whole number). The linearly polarized light remained linearly polarized. For modulating the light, a special interrupter was used, this lying in front of the polarizer so as to eliminate the signal arising from the radiation of the test mirrors themselves. The angle α_A (or α_P) was made equal to 45°. It then follows from (III.8) that the signal vanishes under the following conditions:

$$\cos \Delta_m = -1, \quad \tan \alpha_P = (r_p/r_s)^m \tag{III.11}$$

or

$$\cos \Delta_m = +1, \quad \tan \alpha_P = -(r_p/r_s)^m. \tag{III.11a}$$

Considering conditions (III.10) and (III.11), we see that our own method has two advantages over that of Shklyarevskii. First, in our method the minimum phase shift of 90° occurs for smaller angles of incidence of the light on the test mirrors than in the method of Shklyarevskii, in which it equals 180°. This enables us to work at longer wavelengths with a smaller number of reflections. Second, in our own method less stringent demands are made on the quality of the polarizers. Equation (III.10) is in fact equivalent to obtaining circular polarization, and Eq. (III.11) to linear polarization. On detecting the signal with the rotating polarizer, we shall have a zero signal even in the presence of a small proportion of unpolarized light passed by the polarizer, whereas on using an additional modulator the presence of this small proportion of unpolarized light will prevent us from obtaining a true zero signal (only a more or less deep minimum can be recorded).

In both methods, the measurements are based on obtaining a specific polarization of the light reflected from the metal. On using a rotating polarizer, the approach to circular polarization is detected by reference to the vanishing of the alternating part of the signal. On using an auxiliary interrupter, the approach to linear polarization of the light is established by reference to the vanishing of the light flux. Two methods may be used to obtain the desired polarization.

First Version. After setting the light wavelength λ, we vary the angle of incidence of the light φ and the azimuth of one of the polarizers α until the signal vanishes, which gives a pair of values φ and α_1. Then we simply vary α and obtain three values α_2, α_3, and α_4 for which the signal also vanishes. Then $2\rho_m = (\alpha_2 - \alpha_1) = (\alpha_3 - \alpha_4)$. The vanishing of the signal means that Δ_m has taken the specified value. This may correspond either to circular or to linear polarization of the light. In order to be specific, let us consider the case in which Δ_m corresponds to circular polarization. (Analogous considerations apply to linear polarization.)

For specified light wavelength λ circular polarization occurs for m values of $\varphi = \varphi_i$, $i = 1, 2, \ldots, m$. Here m is the number of reflections from the test mirrors. The value of φ_i corresponds to the resultant phase shift $\Delta_m^i = (\pi/2)(2i-1)$. For each φ_i we find ρ_m^i. The optical constants may be calculated by using m groups of three values φ_i, Δ_m^i, and ρ_m^i. The coincidence of the values of n and $\varkappa$ serves as a good check on the correct adjustment of the apparatus.

Second Version. After setting the angle of incidence φ, we vary the light wavelength λ and the azimuth of one of the polarizers α until the signal vanishes. We measure λ and four values of α, which enables us to find ρ_m. For each angle φ the specified polarization arises for m different values of $\lambda = \lambda_i$, $i = 1, 2, \ldots, m$. Considering, for example, the case of circular polarization of the light, we find that the wavelength λ_i corresponds to a phase shift $\Delta_m^i = (\pi/2)(2i-1)$. Naturally each λ_i has its own value of ρ_m^i. We thus have m values of λ_i for which we know the phase shifts Δ_m^i and azimuths ρ_m^i, so that the optical constants may be calculated.

As already mentioned, we usually used a rotating polarizer. In certain cases, when using the second measuring method, a special interrupter was employed for light modulation. This further increased the number of wavelengths for which the optical constants could be determined with a specified angle of incidence of the light on the test mirrors.

Yet another version of the polarization method is based on measuring the intensity of the reflected light for several specially chosen positions of the polarizer. If, in accordance with the Beattie method [89], we measure the intensities $I(\pi/4, 0)$, $I(\pi/4, \pi/2)$, and $I(\pi/4, \pi/4)$, we find from (III.8)

$$(\tan\rho)^m = \sqrt{I(\pi/4,0)/I(\pi/4, \pi/2)}, \tag{III.12}$$

$$\cos\Delta_m = [2I(\pi/4, \pi/4)/I(\pi/4, \pi/2) - 1 - (\tan\rho)^{2m}]/2(\tan\rho)^m. \tag{III.12a}$$

Here the arguments for I are the azimuths of the polarizer and analyzer. Other orientations of the polarizers may also be used. Thus, in work carried out by Noskov and Charikov [90], the accuracy of the method was increased by using the relation

$$(\tan\rho)^m = \sqrt{I(0, 0)/I(\pi/2, \pi/2)}\,(A_p/A_s)^m. \tag{III.13}$$

This method was widely used by Noskov et al. when studying the optical constants of transition metals.

In comparing our own method of measuring the optical constants with the foregoing procedure, we may mention three failings of the latter: (1) It is not a null method, although null methods give the best accuracy; (2) it is more sensitive to the quality of the polarizers, which can never be ideal in the infrared part of the spectrum (this also reduces the accuracy of the measurements); (3) more stringent demands are made upon the accuracy of adjustment of the optical system.

2. The second group of methods which has also become widely accepted required the measurement of only one reflection coefficient for normal light incidence. These measurements should be carried out over a fairly wide spectral range. The method has been developed by a number of authors [91-95], details being presented in a review by Stern [96]. The essence of the method is briefly as follows.

The reflection coefficient $|r(\omega)|^2$ is measured experimentally for normal incidence over a wide spectral range $\omega_1 \le \omega \le \omega_2$. Using the Fresnel formulas for normal light incidence, we have

$$r = \frac{n - i\varkappa - 1}{n - i\varkappa + 1} = |r| \exp(i\theta). \tag{III.14}$$

Then the complex function $\ln r = \ln|r| + i\theta$ is considered, dispersion relationships of the Kramers – Krönig type relating the real and imaginary parts of the complex dielectric constant being obtained for this [97]. The following equations are accordingly obtained for the phase θ:

$$\theta(\omega) = \frac{2\omega}{\pi} \int_0^\infty \frac{\ln|r(\omega')|}{\omega'^2 - \omega^2}\, d\omega' \tag{III.15}$$

or

$$\theta(\omega) = \frac{1}{\pi} \int_0^\infty \frac{d\ln|r(\omega')|}{d\omega'} \ln\left|\frac{\omega' + \omega}{\omega' - \omega}\right| d\omega'. \tag{III.15a}$$

Since the measurements are usually carried out over a limited range $\omega_1 \le \omega \le \omega_2$, extrapolation is conducted for $|r(\omega)|$ in the long- and short-wave regions $\omega < \omega_1$ and $\omega > \omega_2$. This extrapolation is indeterminate, particularly in the short-wavelengths direction. The extrapolation law is chosen so as to make n and $\varkappa$ agree with the directly measured values of these constants in some particular part of the spectrum.

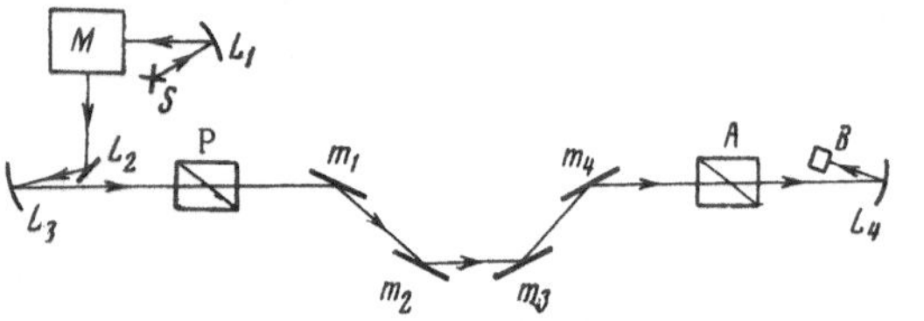

Fig. 15. General arrangement of apparatus 1.

This method has been used by various authors in order to obtain the optical constants of a number of metals and semiconductors at room temperature [78, 94, 95, 98-101]. One weak point of the method is the necessity of carrying out an extrapolation for $|r(\omega)|$, which greatly reduces the accuracy and reliability of the results obtained.

We shall not at this time consider other methods of measuring the optical constants of metals (for example, interference methods), these being considerably less promising for measurements over a wide spectral range. These methods have been considered in a number of review articles [2, 55, 102, 103].

Thus polarization methods give the greatest accuracy in determining the optical constants of metals at the present time, this particularly applying to the infrared part of the spectrum.

§ 2. First Apparatus

For measurements at room temperature, we constructed a four-mirror installation [14], facilitating measurements of the optical constants of metals over the spectral range 0.7-12 μ to an accuracy of 1-2%. The principal arrangement of the apparatus is illustrated in Fig. 15. The light passes from the source S and is focused by a spherical mirror L_1 on the entrance slit of the monochromator M. After passing through the exit slit of the monochromator, the light beam of the desired wavelength is made parallel by means of two mirrors, a plane one L_2 and a spherical one L_3. The use of the additional plane mirror L_2 enables the light to be directed at any desired angle to the normal of mirror L_3, this condition being necessary in order to minimize aberrations. Passing through the polarizer P, the linearly polarized light falls on a system of test mirrors $m_1m_2m_3m_4$. The angle of incidence of the light is exactly the same on all four test mirrors, any deviations from this value not exceeding 2', and is measured to an accuracy of 1' by means of a goniometer. Then the light passes through the rotating analyzer A and is focused on the radiation receiver B by a spherical mirror L_4.

The mechanical system connecting the test mirrors enables all the mirrors to be turned with a single lever, so as to vary the angle of incidence of the light on their surfaces smoothly while remaining exactly the same for each one. The direction of propagation of the light beam after passing the mirror system m_1-m_4 coincides with its previous direction, this being achieved by using servo couplings.* The planes of the mirrors pass through vertical axes of rotation, which we shall denote by the same letters m_1-m_4. The axes m_1 and m_4 are stationary in space. The axes m_2 and m_3 are able to rotate around the axes m_1 and m_4, respectively. On rotating mirror m_1 through a certain angle $(+\theta)$, the mirror m_4 rotates through an angle $(-\theta)$, i.e., in the opposite direction. At the same time as the rotation of mirror m_1, the axis m_2 moves through an angle $(+2\theta)$ around the m_1 axis. Analogously the axis m_3 moves through (-2θ) around the m_4 axis. In addition to these rotations, the mirror m_2 rotates around the m_2 axis by an angle $(-\theta)$, and the mirror m_3 around the m_3 axis by an angle $(+\theta)$.

The system of mirrors is adjusted in the following manner. Mirror m_1 is set so that its plane passes through the rotation axis m_1. A narrow parallel light beam is passed to the rota-

* The basic possibility of using four mirrors for analogous measurements in the visible part of the spectrum was first discussed in [104].

tion axis m_1 in the m_1-m_4 direction. The mirror m_1 is set so that this beam may pass through the rotation axis m_2. This condition must be satisfied for all permissible rotations of the lever which moves the whole mirror system. This serves to check that the plane of the mirror m_1 really passes through the rotation axis m_1. Then the mirror m_2 is set so that, first, its plane may pass through the m_2 axis and, second, that the adjusting beam may pass through the axis m_3. The correct setting of the mirror m_2 is checked in the same way as that of m_1. Then mirrors m_3 and m_4 are adjusted analogously. After setting the mirror m_3, the adjusting beam passes through the rotation axis m_4. The mirror m_4 is set so that the direction of propagation of the adjusting beam after passing through the mirror system may coincide with its original direction, i.e., with the line m_1-m_4.

An additional test for the correctness of the adjustment is agreement between the optical constants obtained for the same light wavelength λ and the four angles of incidence of the light φ_i ($i = 1, 2, 3, 4$) indicated in the previous section.

Apparatus 1 enabled us to make exact measurement of the optical constants n and $\varkappa$ fairly quickly. Thus, for example, detailed $n(\lambda)$ and $\varkappa(\lambda)$ relationships may be obtained for the range of wavelengths 0.7-12 μ in a few hours. However, this method is unsuitable for low-temperature measurements. For this purpose we designed another apparatus.

§ 3. Second Apparatus

Apparatus 2 was specially designed for measurements at low temperatures [16, 17, 19, 20, 51], but also served for room-temperature measurements. The principal arrangement of this apparatus is illustrated in Fig. 16.

The illuminating system S consists of a light source, an intermediate diaphragm, and two spherical mirrors. The first spherical mirror gives an image of the source on the intermediate diaphragm, which lies at the focus of the second spherical mirror. This arrangement ensures small angles between the axis of the beam and the normals to the spherical mirrors, yielding a good, parallel beam. The beam travels horizontally after leaving the illuminating system. Then the plane mirror L_1 turns the beam downward. The light passes through the polarizer P and the system of test mirrors L, where it experiences two or four reflections; it is reflected back from the plane mirror L_0, again experiences two or four reflections from the test mirrors L, passes through the analyzer A, and by means of the plane mirror L_2 returns again to the horizontal plane. The horizontal light beam is focused by the mirror system L_3 on the slit of the monochromator. After passing the monochromator, the beam is focused on the radiation receiver B by the parabolic mirror L_4. The mirror system L_3 comprises one plane and one spherical mirror. This arrangement ensures a narrow angle between the beam axis and the normal to the spherical mirror. In Fig. 16 the illuminating system S and monochromator M with the mirror system L_3 are shown turned through a right angle.

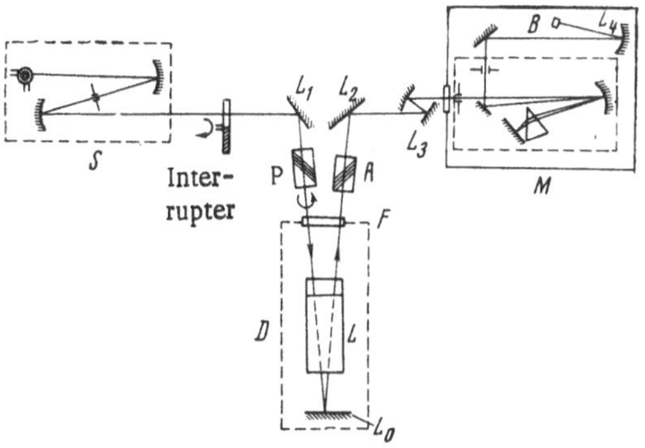

Fig. 16. Basic arrangement of apparatus 2. D is the low-temperature part of the apparatus with a rocksalt window F.

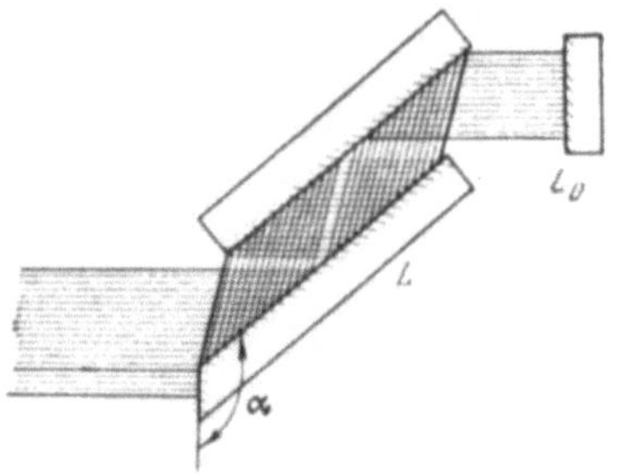

Fig. 17. System of test mirrors. L are the test mirrors, L_0 is the mirror reflecting the beam back again; $\alpha = 180° - \varphi$, where φ is the angle of incidence of the light.

The apparatus may be used in both the foregoing forms. In the first version the interrupter is replaced by a rotating polarizer P. In the second version an additional interrupter is incorporated. The whole system is fixed to a heavy base.

The system of mirrors indicated by L and L_0 in Fig. 16 is shown separately in Fig. 17. The test mirrors L constitute two glass plates coated with a reflecting metal film and fixed to a special guide. The mirrors are made parallel to an accuracy of 1' or better by means of a set of special screws.

In order to determine the angle of incidence of the light on the test mirrors, an extra facet on one of these is polished at a specified angle α to the working surface. Some of the incident light is reflected from the auxiliary face of the mirror and returns toward the source. The mirror L is set so that the auxiliary facet may be perpendicular to the incident beam. In this case the angle of incidence $\varphi = \pi - \alpha$.

In order to minimize the heat capacity of the system, we decided to dispense with any additional movements of the test mirrors after cooling. Thus, in order to obtain the optical characteristics over a wide spectral range, we used mirrors of different lengths, with different angles α. We mentioned earlier that the total number of reflections of the light beam from the test mirrors equalled four or eight. The mirror L_0 first returned the incident beam toward the source, and second, separated the forward and reverse beams in a plane perpendicular to the plane of incidence by a small angle of the order of 4°. The projection of the mirrors L and L_0 in Fig. 17 is perpendicular to their projection in Fig. 16.

In apparatus 2 the mirrors L_1 and L_0 may in principle introduce a slight contribution to the azimuth and phase shift of the test signal. The effect of the mirror L_0, on which the light falls at an angle of about 2°, is negligibly small. The effect of the mirror L_1 is rather greater. We allowed for this effect, although the correction in the infrared amounted to less than 1%.

The low-temperature part of the apparatus [17, 51] is shown in Figs. 18 and 19. The guide with the mirror system L-L_0 is suspended from the seal by means of stainless steel and Plexiglas plates. The cross pieces of the whole system are fairly rigid. The flow of heat along these plates is such that on using a 3- to 4-liter dewar one may work with liquid helium for 5-6 h. On working with liquid nitrogen, the level of the latter remains practically constant for the whole period of the experiment. The guide and mirrors are placed in a sheath so as to prevent direct contact between the test mirrors and the liquid helium or nitrogen. The level of helium or nitrogen in the dewar lies well above the mirrors during the measurements. The test mirrors therefore lie in helium or nitrogen vapor at a temperature practically coinciding with that of the corresponding liquids. The temperature of the mirrors is specially monitored during the optical-constant measurements; at 4.2°K the temperature of the mirrors differs by less than 0.02°K at the beginning and end of the experiments.

The window in the seal is made of rocksalt and is fixed by means of rubber gaskets; it is covered up while the helium is being poured in, and warm nitrogen is blown across it during the experiments. These precautions enable the same salt window to be used four or five times.

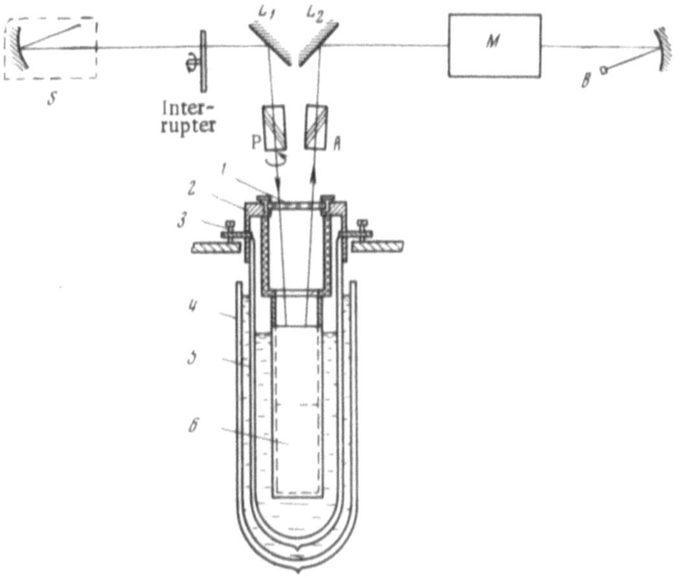

Fig. 18. Low-temperature part of apparatus 2. 1) Rocksalt window; 2) seal; 3) adjusting screws; 4) nitrogen dewar; 5) helium dewar; 6) guide with test mirrors.

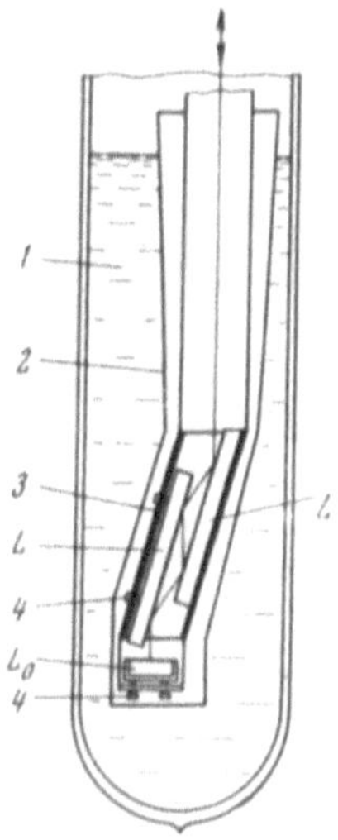

Fig. 19. Arrangement of test mirrors for measuring optical constants. 1) Helium; 2) sheath; 3) guide; 4) adjusting screws; L = test mirrors; L_0 = mirror returning the beam in the reverse direction.

The vertical path of the rays in the low-temperature part of the apparatus avoids the use of complicated and specialized optical dewars. The whole operation is carried out with ordinary glass dewars some 120 mm in diameter.

The low-temperature section is adjusted on the common base of the whole apparatus by means of three screws, after which the position of this section relative to the others may always be restored to a high accuracy. Special experiments confirmed this. The reproducible adjustment of the whole system is also confirmed by the agreement established between optical-constant measurements carried out at room temperature before adding the helium and after reheating.

In [51], Golovashkin made a special study of the systematic errors which might be encountered in working with this apparatus; he showed that in our own version these sources of errors were either entirely removed or else reduced to very small dimensions. In the infrared region for $\lambda > 1.5\ \mu$, allowance for all possible errors gives a correction of under 0.5%. In the near-infrared the corrections are rather greater and reach several percent. These corrections were calculated from the known optical constants of auxiliary mirrors and also measured directly. The experimental determination of the errors is described in the next section. The experimental and calculated results coincided.

§ 4. Measurements in the Visible Part of the Spectrum

Measurements in the visible part of the spectrum were carried out both at room and at low temperatures. For this purpose we used a slightly modified version of apparatus 2. In this we first changed the monochromator and radiation receiver; second, we put a compensator after mirror L_2; and, third, changed the polarizers (Fig. 20). We used four or eight reflections from the test mirrors [19, 20].

For a particular wavelength λ and a specified angle of incidence of the light on the test mirrors φ, the phase shift between the p and s light components was brought to an odd multiple of $\pi/2$ by means of the compensator, after which the analyzer A was rotated to equate the amplitudes of the p and s components. This was recorded by reference to the vanishing of the alternating component of the signal on using the rotating polarizer P. The following equation was thus satisfied:

$$\Delta_m + \Delta_K = \frac{\pi}{2}(2i - 1). \qquad \text{(III.16)}$$

where Δ_m is the phase shift obtained after an m-fold reflection from the test mirrors; Δ_K is the phase shift created by the compensator; and i is a whole number. The azimuth ρ_m was determined in the same way as before. After measuring Δ_m, ρ_m, and φ, the optical constants were calculated from Eqs. (III.2)-(III.5). The measurements were made in the spectral range 0.4-2.6 μ. The $n(\lambda)$ and $\varkappa(\lambda)$ relationships for this range were obtained using one pair of mirrors. However, in order to increase the reliability of the results several series of measurements were carried out. Each series was measured with freshly prepared mirrors.

The use of a compensator enabled us to measure the quantities Δ' and $\tan\rho'$ associated with the effect of all parts of the apparatus apart from the test mirrors on the polarization of the light. For this purpose the foregoing measurements were repeated without the test mirrors. We obtained relationships for the corrections $\Delta'(\lambda)$ and $\tan\rho'(\lambda)$ over the spectral range 0.4-2.6 μ. As might well be expected, the corrections were small. The maximum corrections for Δ with four light reflections from the test mirrors equalled 5-6% and for eight reflections about 2%. The corrections for $\tan\rho$ in both cases were under 1% between 1 and 2.6 μ and 2-6%

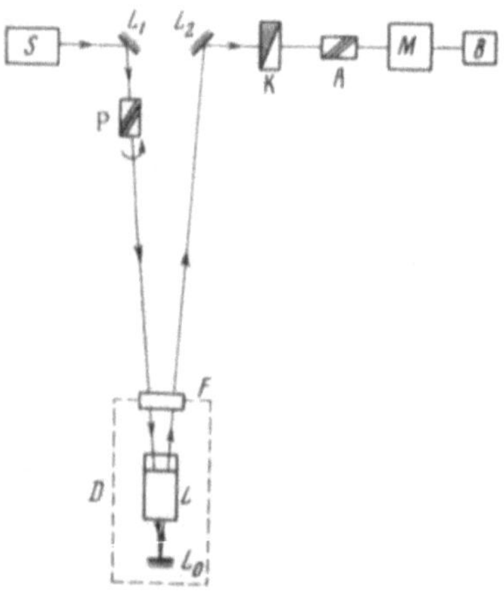

Fig. 20. Basic arrangement of the apparatus for measuring the visible part of the spectrum. S) Illuminating system; L_1, L_2, L_0) plane mirrors; L) test mirrors; P) rotating polarizer; K) compensator; M) monochromator; B) radiation receiver; F) window in the seal; D) low-temperature section.

over the rest of the spectral range. In a narrow spectral region corresponding to the minimum reflection from aluminum mirrors (0.8-0.85 μ), the errors reached 9%. The values obtained for $\Delta'(\lambda)$ and $\tan\rho'(\lambda)$ were allowed for in determining the optical constants.

It should be noted that, on placing the analyzer after the mirror L_2, the effect of the latter on Δ' and $\tan\rho'$ became the same as that of mirror L_1. Hence, in the arrangement including the compensator, the corrections were twice as great as in that of Fig. 16. For measurements in the infrared region the compensator was not employed, and the corresponding corrections were much smaller than those indicated.

§ 5. Sequence of Measurements

The sequence of our measurements of the optical constants of metals was as follows. First we used apparatus 1 at room temperature. The number of series was taken as great as was necessary to determine the optical constants reliably (usually five or six series). In each series freshly prepared mirrors were used. Then low-temperature measurements were made on freshly prepared mirrors in apparatus 2. First the optical constants were again measured at room temperature then (using the same mirrors) at liquid helium or nitrogen temperature, then again at room temperature.* It should be noted that the initial adjustment was excellently maintained during the subsequent measurements. For all working temperatures in this apparatus, the test mirrors were placed in a helium atmosphere. The number of measurements was again determined by the number required to produce reliable results (five to eleven series).

The correctness of the adjustments was confirmed for both forms of the apparatus; the absence of errors was verified by the agreement obtained between the optical constants determined with the two types of instruments at room temperature.

The agreement between the results obtained from the two forms of apparatus also showed that the optical constants of the metals were unaffected by the additional "bias lighting" of the metals with short-wave radiation. In apparatus 1 light fell on the mirrors after passing through a monochromator. In apparatus 2, the spectral range of the light falling on the mirrors was much greater, only being limited by the selenium polarizer, i.e., the incident light had $\lambda \geq 0.7\,\mu$. The experimental results showed that this difference was quite unimportant. We further checked the negligible effect of bias lighting on the optical constants of the metals by incorporating filters to cut off the short-wave radiation. The results obtained with the filters were the same as those obtained without them. It would appear that bias lighting is only important for semiconductors, in which the concentration of conduction electrons is much smaller than in metals.

§ 6. Characteristics of Individual Parts of Systems 1 and 2

As light source we usually employed a dc carbon arc. In our null modulation method, the instability of the arc was not particularly important. In certain special cases we used glow lamps and Silit rods.

In order to monochromatize the light in the infrared region, we used monochromators incorporating a rocksalt prism. The spectral width of the working slits was about 0.04 μ in the range $\lambda \simeq 1\,\mu$, 0.08μ in the range $\lambda \simeq 8\,\mu$, and 0.3 μ in the range $\lambda \approx$ 10-12 μ. With these

*In studying the optical properties of lead, the sequence of measurements in apparatus 2 differed from the above. The measurements were first made at room temperature and then at helium or nitrogen temperature only.

instruments measurements were carried out in the range 0.7-12 μ. On working in the visible and near-infrared regions, we also used a monochromator with a glass prism. The spectral width of the working slits was 0.01 μ in the range $\lambda \approx 0.6-0.7\,\mu$, 0.02 μ in the ranges $\lambda \approx 0.45-0.6$ and 0.7-0.8μ, and 0.05-0.1 μ in the range $\lambda > 0.8\,\mu$. The values of the optical constants obtained by the two instruments in the overlapping spectral ranges agreed with each other, which indicated both the correctness of the monochromator calibration and also the absence of any additional sources of error.

As light receiver, we used the following: in the infrared part of the spectrum (0.7-12 μ) a germanium bolometer containing antimony, developed and made by Shubin; in the visible and near-infrared a photomultiplier (0.4-1.2 μ) and a photoresistance (0.8-2.6 μ). The radiation receivers were chosen so that their ranges of sensitivity overlapped. The resultant values of the optical constants in the overlapping spectral ranges coincided.

The light receiver was connected to the input of a narrow-band amplifier with a maximum amplification factor of 6000. The passband of the amplifier was about 1 cps. Special attention was given to the quality of the input elements of the amplifier, and to careful screening of the circuit. Since the bolometer employed was a high-resistance one, the transition from the bolometer to other receivers necessitated no changes in the input section of the amplifier. The threshold power for the bolometer was $4 \cdot 10^{-10}$ W; for the photomultiplier it was many times smaller. The amplifier has two outputs: one to an oscillograph, and one to a galvanometer or automatic recorder.

As polarizing elements we used selenium polarizers for the spectral range 0.7-12 μ and polaroids for the range 0.4-0.74 μ.

The selenium polarizers constituted a pile of films mounted in a guide; they were placed at nearly the Brewster angle to the incident light. Special attention was paid to the quality of the polarizing elements, since poor polarizers might introduce serious measuring errors (see Conn and Eaton [105]). The method of making selenium polarizers developed by Elliott, Ambrose, and Temple [106] is now widely employed. We ourselves used the slightly different method developed by Shubin. After vacuum-condensing the selenium on a collodion film, the selenium was attached to a metal frame. After the glue had dried, the collodion film was removed with tweezers, and the pure selenium film was left stretched on the frame. As in the earlier case [106], amorphous selenium was used for the polarizers. The pile comprised eight films. The working diameter of the polarizer was 24 mm. The selenium films were placed at an angle of 67.5° to the incident light. When these were used in our own particular instruments, with the apertures mentioned above, for the spectral range 0.7-12 μ, the intensity of the unwanted perpendicular component was under $2 \cdot 10^{-3}$ of that of the parallel component. This corresponds to a degree of polarization greater than 99.8%. The transmission of the polarizers was 40%.

In the spectral range mentioned, polaroids had a ratio of the intensities of the useful and parasitic components of the light equal to 7000 for the region 0.5-0.72 μ, and over 1000 at the ends of the working range; their transmission was 45% in the main part of the spectral range and fell to 20% for $\lambda = 0.45\,\mu$. In the overlap range, the results obtained with the selenium polarizers and the polaroids coincided.

When using the rotating polarizer, the rotation frequency was 4.5 cps. The rotation was effected by means of a geared synchronous motor.

The auxiliary interrupter used for measurements by the second method constituted a half-disc mirror rotating at 9 cps. This rotation was also effected by means of a synchronous motor and gear.

For measurements in the visible and near-infrared regions we used a quartz compensator of the Soleil — Babin type. The phase difference between the p and s components was meaured to an accuracy of 2' with this compensator.

§ 7. Preparation of the Mirror Layers to Be Studied

Special attention was paid to the preparation of the mirror layers (films). We were interested in the optical constants of the bulk metal, and the methods of sample preparation therefore had to satisfy a number of requirements in relation to the nature of the layers. First, since light penetrates into metal to the depth of the skin layer (about $2 \cdot 10^{-6}$ cm), it is essential that the properties of this layer should coincide with or be very similar to the properties of the bulk metal. Second, the surface of the metal should be plane and specular. Third, the thickness of the metal layer should be sufficiently great to permit neglecting the reflection from the second surface.

We decided to evaporate the metal in vacuum and condense it on a polished glass substrate. The material of the evaporator and the mode of evaporation were chosen for each particular metal so as to satisfy the foregoing requirements. Both in determining the mode of evaporation and in preparing the samples for evaporation, the quality of the resultant layer was continuously monitored. We paid special attention to the density, conductivity, temperature coefficient of conductivity, and residual resistance of the metal layers. We also determined various other characteristics of these materials. All the measured static characteristics of the metals studied are presented in Chapter IV.

The metal was evaporated at a pressure of $(2\text{-}7) \cdot 10^{-6}$ mm Hg. We used an oil diffusion pump with water and nitrogen traps. The rate of pumping was about 500 liters/sec. The volume of the bell was over 40 liters. For these parameters the vacuum under the bell remained practically constant during the evaporation.

As evaporators we used the following: for silver, indium, lead, and tin, tantalum boats; for gold, tungsten boats; for aluminum, a spiral of tungsten wire. Before loading the metal into the evaporator, the latter was cleaned and degreased by heating in vacuum for a long period at a temperature slightly greater than that employed for evaporating the metal. After loading the metal into the evaporator the metal was also degreased by vacuum heating. Before preparation of the samples, the first portions of evaporated metal were deposited on a special slide. Then the slide was removed and the evaporated metal was deposited on the substrates. In preparing the samples only a part of the metal was evaporated. The rest contained considerably greater proportions of impurities and was never used. The rate of evaporation was specially chosen for each metal, varying from 100 to 500 Å/sec. The evaporator temperature was also chosen separately for each metal. The distance from the evaporator to the substrate was 250-300 mm, which ensured reasonable uniformity of the layers. The substrates were kept at room temperature.

The substrates on which the metal was condensed were glass plates, one side of which was carefully polished. The deviation from flatness was no greater than 0.3 μ for samples under 100 mm long and 0.5-0.6 μ for longer samples. The plates for apparatus 1 had dimensions 78×20 mm and a thickness of 10 mm. The plates for apparatus 2 had a length from 43 to 180 mm, a width of 30 mm, and a thickness of 10 mm. These were carefully washed in nitric acid, in a mixture of potassium dichromate with sulfuric acid, in ordinary running water, and finally in distilled water; then they were dried or wiped with a special towel. Sometimes before this washing the plates were prewashed in alkali. In the vacuum chamber the plates were further cleaned by a gas discharge at $10^{-1}\text{-}10^{-2}$ mm Hg.

The purity of the original metals was no worse than 99.99%. In this case the purity of the resultant layers was determined not by the chemical impurities but by the dimensions of the crystals and other structural defects.

The test layers were 0.3-1.6 μ thick. Special investigations showed that for samples of this thickness the thickness had no effect on the optical properties [23, 51].

We must now give some special consideration to the means of preparing the specular indium layers. Indium is a metal with a low melting point (157°C) and a comparatively low temperature of evaporation (about 1000°C for a vapor pressure of 10^{-2} mm Hg) [107]; it easily evaporates, but condenses poorly on a glass substrate. In order to obtain good mirror layers we first [21] used an intermediate lead layer 0.03 μ thick. Evaporation of the lead and indium was conducted without any intermediate filling of the vacuum chamber with air. The indium mirrors so made had a slight blue tint; their characteristics differed from those of the bulk metal, but not very much. Another method was used by Shklyarevskii and Yarovaya [79], who deposited the metal on a cooled substrate. The characteristics of these films were not given, but it is well known that the method in question gives a finely dispersed structure [107]. Recently, in the Optical Laboratory of the Physical Institute, Academy of Sciences, Shubin [108] developed a new method of producing indium mirrors. This method requires neither intermediate layers nor a cooled substrate; it is based on the fact that before deposition the glass plates are irradiated with vacuum ultraviolet radiation, after which indium condenses excellently on the substrates. The layers so obtained have properties very similar to those of the bulk metal, and very good mirror properties. In order to effect the necessary irradiation, a special tungsten spiral is introduced into the vacuum and heated to about 2500-3000°C. During the evaporation of the indium the spiral is disconnected; otherwise the indium is excited and, condensing on the substrate, yields layers with properties differing very greatly from those of the bulk metal. The characteristics of the indium layers used in our experiments are given in Chapter IV.

After preparation, the gold, aluminum, and lead samples were annealed in vacuo. The gold was annealed at 400°C, the aluminum at 300°C, and the tin at 150-200°C. The annealing period lasted for several hours. The effect of annealing on the different metals varied. Annealing had no effect on tin. On aluminum the effect was only slight. On gold annealing had a very considerable influence. This result agrees with the conclusions of [109], in which it was also found that annealing had little effect on metals with low melting points.

At the same time as the deposition of samples for optical investigations, we also prepared samples for measuring the density, conductivity, and Hall effect. These samples were also deposited on specially polished glass substrates through appropriate stencils. The substrates for measuring the conductivity and Hall effect had sealed-in Kovar leads. Simultaneous preparation of all the samples ensured an identical nature of the layers used for determining the other characteristics.

§ 8. Measurement of the Static Characteristics of the Metal Layers under Examination

1. The thickness of the deposited layers was measured by an interference method. For these measurements we used a layer forming a step on the carefully polished flat surface of a glass substrate. The height of the step equalled the thickness of the layer. The boundary of the step was made sharp so as to make measurement easier. In order to eliminate errors associated with the phase difference arising on reflection from the metal and the glass, the step and the whole substrate were further coated with an auxiliary metal layer (Fig. 21). Usually (but not always) the auxiliary metal was the same as that being studied. Sometimes, before de-

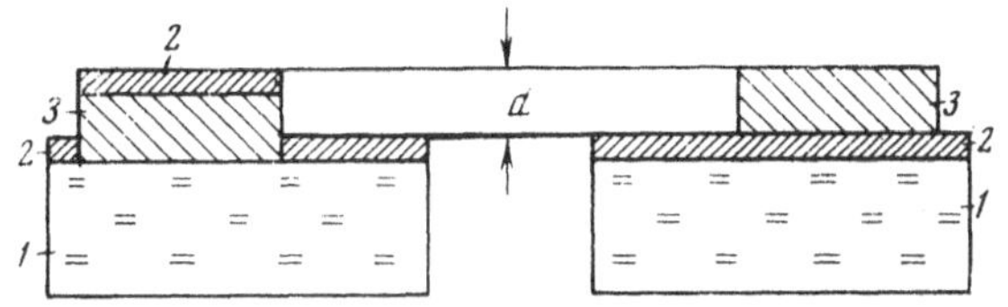

Fig. 21. Cross section of the sample for measuring thickness: d) thickness of working layer; 1) glass substrate; 2) auxiliary metal layer; 3) working layer.

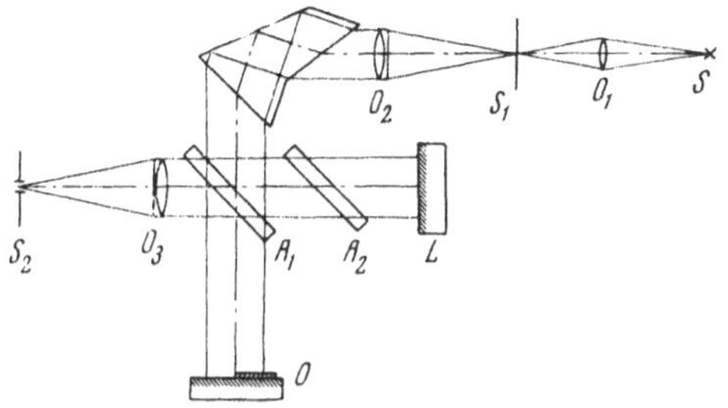

Fig. 22. Optical arrangement of the interference comparator used for measuring thickness. S) Source; O_1, O_2, O_3) objectives; S_1, S_2) entrance and exit slits; P) constant-deviation prism; A_1) semitransparent plate; A_2) compensating plate; L) standard mirror; O) sample.

positing the step, the whole substrate was coated with a preliminary layer of the same metal. The two methods were equivalent.

The thickness was measured with an interference comparator, the optical system being illustrated in Fig. 22; it constituted a combination of a monochromator with a Michelson interferometer. The test plate was placed in one of the arms of the interferometer. Bands of equal thickness were used for the measurements. Two systems of bands were obtained, these being shifted, one with respect to the other. The shift in the bands determined the height of the step. By using as light source a succession of the six lines obtained in a helium discharge tube (the 0.706, 0.668, 0.588, 0.502, 0.492, and 0.447 μ lines), we may measure the thickness of layers 0.3-2 μ thick to an accuracy of about 2%. If we use only a single glow lamp as light source, the errors in determining the thickness amount to $\pm$ 0.03 μ. Thus the thickness of layers of the order of 1 μ thick may be measured with a continuous light source to an accuracy of about 3%.

2. For measuring the density we deposited a layer of specified area through a special stencil, the dimensions being about 4 cm^2 (determined to an accuracy of better than 1%). By weighing on a microbalance we found the weight of the substrate with and without the test layer. The weighing accuracy was 0.01 mg, giving a relative error of a single measurement equal to 0.2-0.3% for tin, lead, and gold; 1.5% for indium; and about 2% for aluminum. The thickness of the layer was measured by the method described above. The density was determined repeatedly for all layers for which the optical measurements were made. The results given below constitute averages for a large number of samples. The relative error was 2-4%.

3. For measuring the conductivity we deposited a strip of the test metal on a glass substrate with metal contacts, using a special stencil with a known ratio of the length of the cut to its width. The thickness of the layer was either measured directly on this substrate or regarded as being equal to the thickness of the layer deposited for determining the density. In the latter case the two substrates were placed close together during deposition, it being first established that the thickness gradient at the location of the two substrates was less than the thickness measuring error. The resultant strips had the following resistances at room temperature: for tin, lead, and indium, about 70-100; for aluminum, about 10-30; for gold, about 10-13 Ω. The resistance was measured at room temperature and at liquid nitrogen and helium temperatures. Apart from these temperatures, some of the metals were also measured at others. Thus for tin and lead we used the temperature of liquid hydrogen and other neighboring temperatures (down to 7°K); for tin and indium we used temperatures below 4.2°K, right down to the temperature of the superconducting transition; for gold we used temperatures between 2 and 4.2°K.

In the resistance measurements we used the bridge and potentiometric circuits. Special measures were taken to reduce the working current. Usually the working current was 20-50 μA. Wherever necessary still lower currents were used. A low working current is essential, first, in the neighborhood of the critical temperatures and, second, in the region of hydrogen temperatures, where the resistances of tin and lead are very temperature dependent. The accuracy of the resistance measurements was higher than 1%.

From the measured quantities (the resistance and thickness of the layer and the ratio of its length and width) we found the specific conductivity and its temperature dependence. The error amounted to 3-4%.

4. The residual resistance R_{res} was determined for tin and indium from the jump taking place in the resistance when the metal passed into the superconducting state. For lead the residual resistance was smaller than that obtained in the manner indicated. In this case, in order to determine the residual resistance we used the experimental temperature dependence of the resistance of the lead layers under test; this enabled us to determine the Debye temperature of these layers and so to extrapolate the resistance value to $T = 0$. In this extrapolation we assumed that the Matthiessen law was obeyed. For gold the residual resistance coincided with the resistance at $T = 4.2$°K (further reducing the temperature made no difference to the resistance of the test gold layers).

5. Temperatures below 4.2°K were obtained by pumping out helium vapor; they were measured with a standard vacuum meter by reference to the vapor pressure. The accuracy of measuring the absolute temperature was about 0.01°. The relative measurements were made to an accuracy of 0.002°.

Temperatures above 4.2°K (up to 20°K) were obtained by heating the helium vapor; they were measured with a carbon thermometer to an accuracy of 0.04-0.08°. Relative measurements were made in this case to an accuracy of 0.005°.

6. The superconducting transition temperature T_c was determined from the jump in resistance in the absence of a magnetic field. The value of T_c was taken as the temperature at which the resistance fell to half its value slightly above the critical point. By studying the temperature dependence of the resistance in the neighborhood of T_c we were able to estimate the range ΔT over which the transition into the superconducting state took place.

7. The Hall effect was measured for gold, lead, and tin. We determined the Hall constant

$$R_X = E_x/(jH). \quad \text{(III.17)}$$

Here E_X is the Hall field in μV/cm; j is the current density in A/cm^2; and H is the magnetic field in kOe. Thus R_X has the units μΩ - cm/kOe.

The test layers were 60 mm long, 4 mm wide, and of the same thickness as the optical test samples. The potential leads for measuring the Hall emf were placed 30 mm from the current leads. The potential leads were placed almost on the same equipotential, the residual small voltage drop being compensated. The measurements were carried out in a steady magnetic field, the sign of the magnetic field then being reversed and a quantity equal to twice the Hall emf determined. Special attention was paid to the dependence of the Hall emf on the magnetic field and on the current density through the sample. Magnetic fields up to 6 kOe were used when studying the Hall effect in gold, and 8 kOe in tin and lead. The current density was 170-360 in gold and 100-7000 A/cm^2 in tin and lead. The Hall emf was measured with a photoelectric amplifier (dc) having a sensitivity of 10^{-9} V/mm. The whole apparatus was checked by determining the Hall emf of 99.99% pure copper foil 0.1 mm thick. The resultant value coincided with the value published in the literature [29] to an accuracy of 5%.

The magnitude of the Hall effect in gold is fairly large. Measurements were easily made without any special precautionary measures. The error for the Hall constant of gold was 5%. The magnitude of the effect in tin and lead, however, was rather small (20-30 times smaller than in gold), and special measures were required for the careful thermostating of the whole apparatus. The tin samples were immersed in silicone oil. The accuracy of the Hall constant determination was 10 and 30% in lead and tin, respectively. The accuracy was determined from the scatter in the values obtained for different samples. In measuring the Hall constant for any particular sample the accuracy was better than this. This great sensitivity to the conditions of sample preparation only characterized the Hall effect in tin and lead, in which the resultant effect was determined by the difference between the emf values associated with the electrons and holes, respectively. The scatter in the values of the other characteristics for different samples was considerably smaller.

8. The characteristic temperature θ_R was determined by comparing the experimental temperature dependence of the resistance with the theoretical dependence based on the use of a single parameter θ_R. According to Grüneisen [52, 110], the reduced resistance of an ideal metal equals

$$r \equiv R(T)/R(\theta_R) = 1.056\,(T/\theta_R)\,F(\theta_R/T). \qquad \text{(III.18)}$$

The function $F(\theta_R/T)$ is tabulated in [52, 110]. Here θ_R is the characteristic temperature. The index R means that the constant is obtained from resistance measurements. This temperature does not coincide with the Debye temperature θ_D obtained from specific heat measurements, although θ_R and θ_D are usually quite close.

Using (III.18), we easily obtain the following:

$$\frac{R(T_2) - R_{res}}{R(T_1) - R_{res}} = \frac{T_2}{T_1}\,\frac{F(\theta_R/T_2)}{F(\theta_R/T_1)}. \qquad \text{(III.19)}$$

Here $R(T_2)$ and $R(T_1)$ are the resistances of the test metal at T_2 and T_1, while R_{res} is the residual resistance. After measuring the three quantities $R(T_1)$, $R(T_2)$, and R_{res}, we may use (III.19) to determine the characteristic temperature θ_R relating to the temperature range T_1-T_2. We shall use this method later in order to determine the θ_R of a number of metals.

CHAPTER IV

EXPERIMENTAL RESULTS

In the complex study of the optical and other properties of metals here described, we devoted our chief attention to metals of the polyvalent type. We studied Group III metals such as indium [19, 21] and aluminum [36, 38] and also Group IV metals such as lead [14, 20, 22] and tin [14, 17, 18, 23]. In addition to this we studied the monovalent metal gold [24].

TABLE 5. Errors Committed in Measuring the Static Characteristics of the Metal Layers Studied

Characteristic	Error %					
	In	Al		Pb	Sn	Au
		unannealed layer	annealed layer			
ρ, g/cm³	4	3	6	3	5	4
σ_K, cgse	4	3	—	3	5	1.5
σ_H, cgse	—	—	—	4	5	—
σ_N/σ_K	1	5	—	1	1	1
R_n/R_K	—	—	—	1	—	—
R_{res}/R_K	18	—	—	1	1	1
T_c, °K	3	—	—	3	0.5	—
θ_R, °K	3	—	—	2	2	3
R_X, μΩ-cm/kOe	—	—	—	10	30	4

For all these metals we determined both the optical constants and also certain static characteristics. The results of the measurements are presented in the tables of this section. In order to increase the accuracy of the measurements, we made many series of measurements in all cases, i.e., for all metals and all temperatures.* Each series of measurements was carried out with freshly prepared samples. The final results were obtained by averaging over all the series. The accuracy of the measurements was determined from the scatter of the values for different series.

In the tables here presented we give the following static characteristics of the test layers: d = thickness of layer in μ; ρ = density in g/cm³; σ_K, σ_N, and σ_H = static conductivity at room, nitrogen, and hydrogen temperatures in cgse units; R_{res}/R_K is the ratio of the residual resistance to the resistance at room temperature. For lead we give the ratio R_n/R_K where R_n is the resistance of the layer in the normal state at a temperature close to T_c. We also indicate T_c, the temperature of the superconducting transition, and $(\Delta T)_c$, the temperature range of the transition into the superconducting state, in °K. At the same time, for θ_R we show the temperature range in which this quantity was determined. We also give R_X, the Hall constant, in μΩ-cm/kOe. For comparison, we also give the values of these characteristics relative to those of the bulk metal taken from the references indicated in the last column of all the tables. The errors committed in measuring the static characteristics are indicated in Table 5.

The tables of optical constants give the dependence of n and $\varkappa$ on λ in the infrared and visible parts of the spectrum for room (T_K), nitrogen (T_N), and helium (T_{He}) temperatures. Here $n - i\varkappa$ is the complex refractive index and λ is the wavelength in μ.

In determining the optical constants, we allowed for the dependence of the surface impedance on the angle of incidence of the light by using Eqs. (III.2)-(III.5). The corresponding corrections were about 2% for $\lambda \approx 0.5$ μ, 0.5% for 1.5 μ, and negligibly small for $\lambda \geq 2$ μ. The thicknesses of the test layers were much greater than the depth of the skin layer and the

*In contrast to the other cases, for annealed aluminum layers we made one control series of measurements, since previous experiments indicated that the qualitative effect of present interest, namely the effect of annealing on N, was very slight for this metal.

mean free path of the electrons in the metal. As regards optical measurements, layers of such thicknesses were equivalent to an infinitely thick layer.

The dependences of n and $\varkappa$ on λ and T are also depicted in the corresponding figures for all the metals. We shall see from the figures that the qualitative picture of these relationships is exactly the same for all the metals studied.

In the long-wave region the resultant experimental functions $n(\lambda)$ and $\varkappa(\lambda)$ are monotonically increasing functions. The n(T) dependence is considerable over the whole range. On reducing the temperature, as might be expected, n falls. This fall occurs not only on passing from room to nitrogen temperature, but also on passing from nitrogen to helium temperature. The $\varkappa(T)$ dependence is greatest at the long-wave end of this range, where $\varkappa$ increases with falling temperature. On passing from room to nitrogen temperature, $\varkappa$ changes much more than on passing from nitrogen to helium temperature. On the whole, the temperature dependence of $\varkappa$ is much smaller than that of n.

In the short-wave region the function $n(\lambda)$ passes through maxima. The height and width of these depend on the temperature. With falling temperature the maxima become higher and narrower. In the function $\varkappa(\lambda)$ singularities appear in the same spectral range as for $n(\lambda)$, but they are much more weakly expressed.

For all the metal studied we used Eqs. (III.6) to obtain the absorption and reflection coefficients A and R for normal light incidence. The dependences of these quantities on λ and T are shown in the corresponding figures.

§ 1. Indium

A. Static Characteristics of Indium Layers

Of all the Group III metals, indium was the one which we studied in most detail [19, 21]. Let us first consider the characteristics of the layers employed. We mentioned earlier that the indium layers were prepared in two ways. Originally an intermediate layer of lead was deposited between the indium and the glass — we shall call this the first method. Later, the layer was prepared by the new Shubin method without any intermediate layer — we shall call this the second method. The results of our determination of the static characteristics of these layers are presented in Table 6.

We see from the table that the new Shubin method of preparing indium layers (second method) gave a better quality than the first method. Shklyarevskii and Yarovaya [79, 118] studied layers obtained by condensing the indium on a cooled substrate. Unfortunately, the

TABLE 6. Static Characteristic of Indium Layers

Characteristic	Method of preparation 1	Method of preparation 2	Bulk metal	Published data (bulk metal)
d, μ	0.6	0.3–0.4	—	—
ρ, g/cm^3	5.9	6.60	7.30	[111]
σ_K, cgse	$0.64 \cdot 10^{17}$	$0.95 \cdot 10^{17}$	$0.98 \cdot 10^{17}$	[112]
σ_N/σ_K	3.24	4.52	5.02	[112]
R_{res}/R_K	$12.3 \cdot 10^{-2}$	$2.8 \cdot 10^{-2}$	$\sim 3 \cdot 10^{-3}$	[113]
$\theta_{R\,(293,78)}$, °K	not determ.	190	190	[112]
T_c, °K	not determ.	3.40	3.407	[114, 115]
$(\Delta T)_c$, °K	not determ.	0.004–0.06	0.01	[113, 116, 117]

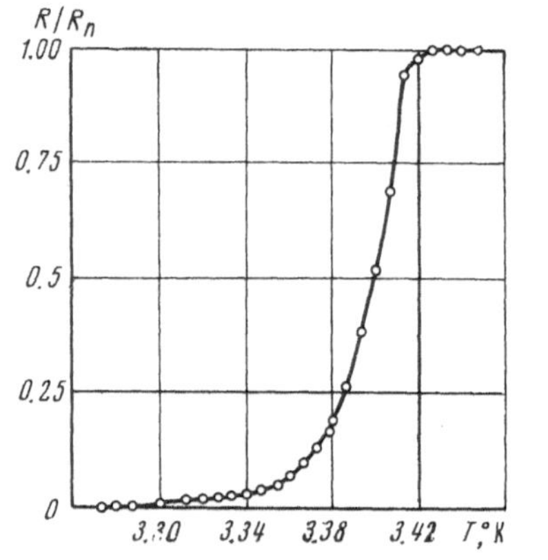

Fig. 23. Dependence of the resistance R of the test layer of indium on temperature T, close to the T_c of one of the samples. R_n is the resistance of the indium layer in the normal state; $(\Delta T)_c = 0.06°K$.

characteristics of these layers were not given. However, condensation on a cooled substrate is often found to give layers with properties differing considerably from those of the bulk metal.

We shall subsequently only consider the properties of layers obtained by the second method.

Let us compare the characteristics of the metal layers studied with those of the bulk metal. We see that the density of our layers is similar to the bulk value. The conductivity at room temperature, the characteristic temperature, and the temperature of the superconducting transition practically coincide with those of the bulk metal. The residual resistance of the samples was low, indicating a relatively large size (as compared with the mean free path) of the crystals forming the polycrystalline test samples. If we subtract the residual resistance from the sample resistance at room and nitrogen temperatures, the ratio σ_N/σ_K will be the same for our samples as for the bulk material.

The temperature spread of the superconducting transition in the test samples was very small. Figure 23 shows the temperature dependence of the resistance of the test layer close to T_c. The curve relates to a sample in which $(\Delta T)_c$ reaches its maximum value. We see from the figure that the transition into the superconducting state is very sharp, which indicates an excellent homogeneity of the layers.

We thus succeeded in producing indium layers with characteristics similar to those of the bulk metal; we may therefore reasonably expect that the measured optical constants of the layers will also agree with those of the bulk material.

It should further be mentioned that the test indium layers also had a high quality of mirror surface.

B. Optical Constants of Indium

In this section we shall give some detailed consideration to the optical constants n and $\varkappa$ measured for the layers obtained by method 2 [19].

The optical properties of indium were studied in the infrared and visible parts of the spectrum at helium and room temperatures. The measurements in the spectral range 1-10 μ at room temperature were carried out in both types of apparatus. The results of the measurements coincided to an accuracy of better than 0.5%. The measurements at helium temperature were carried out in apparatus 2.

The measurements in the spectral range 0.55-2.6 μ (both at room and helium temperatures) were carried out in a modified form of apparatus 2, using a compensator (Chapter III,

TABLE 7. Optical Constants of Indium

λ, μ	295° K		4,2° K		λ, μ	295° K		4,2° K	
	n	$\varkappa$	n	$\varkappa$		n	$\varkappa$	n	$\varkappa$
10.0	24.8	51.9	13.5	62.3	1.60	2.33	11.8	1.48	11.9
8.0	18.4	45.3	7.5	50.6	1.55	—	—	1.45	11.45
6.0	12.4	37.2	4.2	38.7	1.50	2.19	11.0	1.39	10.9
5.0	9.77	32.2	3.1	32.7	1.45	—	—	1.36	10.6
4.0	7.27	26.7	2.3	26.7	1.40	2.06	10.35	1.35	10.3
3.5	6.00	23.9	2.05	23.9	1.35	—	—	1.32	9.95
3.0	4.70	20.9	1.90	20.9	1.30	1.95	9.70	1.30	9.60
2.60	4.0	18.3	1.92	18.25	1.25	—	—	1.28	9.24
2.55	—	—	1.93	17.8	1.20	1.87	9.00	1.28	8.88
2.50	3.81	17.6	1.95	17.5	1.15	—	—	1.31	8.42
2.45	—	—	1.97	17.2	1.10	1.84	8.38	1.35	8.00
2.40	3.65	16.9	1.99	16.8	1.05	—	—	1.41	7.60
2.35	—	—	2.00	16.5	1.00	1.81	7.77	1.49	7.25
2.30	3.48	16.3	2.00	16.2	0.95	1.72	7.44	1.65	7.00
2.25	—	—	1.99	16.0	0.90	1.59	7.18	1.73	6.85
2.20	3.30	15.6	1.98	15.6	0.85	1.45	6.78	1.71	6.72
2.15	—	—	1.96	15.3	0.82	—	—	1.65	6.68
2.10	3.13	15.0	1.92	15.0	0.80	1.32	6.60	1.59	6.65
2.05	—	—	1.88	14.8	0.76	1.19	6.31	1.33	6.50
2.00	2.97	14.5	1.84	14.6	0.75	1.17	6.26	1.26	6.45
1.93	—	—	1.78	14.2	0.74	1.13	6.18	1.20	6.40
1.90	2.80	13.8	1.75	13.85	0.72	1.07	6.00	1.09	6.20
1.85	—	—	1.70	13.5	0.70	1.01	5.83	0.99	6.00
1.80	2.64	13.1	1.65	13.2	0.65	0.90	5.42	0.835	5.50
1.75	—	—	1.61	12.9	0.60	0.795	5.02	0.77	5.08
1.70	2.49	12.5	1.56	12.6	0.55	0.70	4.70	0.695	4.70
1.65	—	—	1.52	12.2					

Section 4). The results of the measurements in the range 1-2.6 μ obtained using the different versions of apparatus 2 coincided to an accuracy of better than 1%.

The optical constants in the range 0.55-2.6 μ were determined in steps of 0.05 μ, except for the region near the maximum of interband conductivity, at which the step was 0.01-0.02 μ. The optical constants in the long-wave region were determined in steps of 0.5-1.0 μ. The spectral width of the slit was less than or equal to the step in the measurements.

The results of our measurements of the optical constants of n and $\varkappa$ appear in Table 7 and Figs. 24 and 25.

The accuracy of the determination of $\varkappa$ in both the visible and infrared regions was 1%, the error increasing to 1.5-2% at the very edges of the range studied. The accuracy of the n determination was likewise 1% and 2-3% at the edges.

In order to discover the possible influence of oxidation on the properties of the indium films, we measured n and $\varkappa$ both immediately after deposition and also after holding for some days in a cryostat. No changes occurred in n and $\varkappa$. This indicates that the oxidation of the mirrors was very slow, and the role of the oxide layer nonexistent or negligible. A further argument in support of this assertion is the excellent quality of the mirror surface. It is well known that, the better a mirror surface, the more slowly is it oxidized. The behavior of such easily oxidized metals as lead, copper, etc., convinces us of this. The indium layers used in [79] did not possess this property. The authors of [79] indicated that they measured the optical properties of the indium, not immediately after deposition, but 6-7 days later, when the changes in optical constants had become much slower; they ascribed these changes to oxidation.

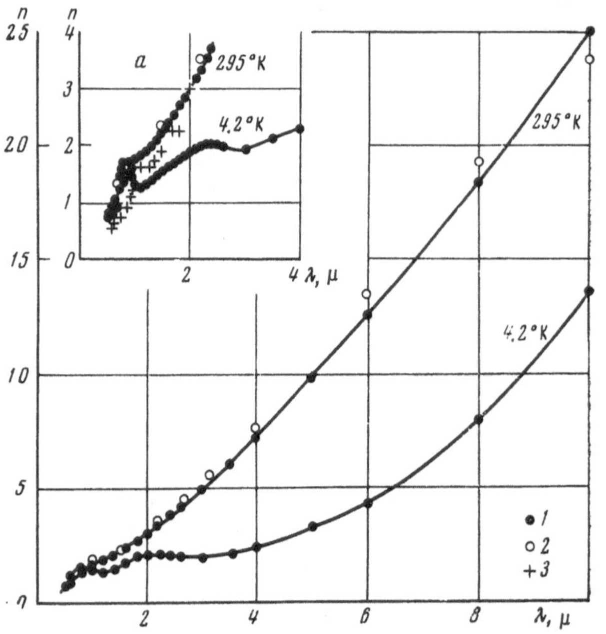

Fig. 24. Dependence of n on the wavelength of the light at two temperatures. 1) Results of [19] relating to layers obtained by method 2; 2) by method 1 [21]; 3) by condensing the indium on a cooled substrate [79].

We consider that the changes were associated with the annealing of the layer deposited on a cooled substrate.

Figures 24 and 25 show the dependence of n and $\varkappa$ on λ and T for the spectral range 0.55-10 μ. In addition to the main figures, the inset illustrates n and $\varkappa$ in the short-wave region.

In addition to the results of [19], which relate to layers prepared by method 2, the same figures indicate the results of [21], relating to layers prepared by method 1, and also to the layers of [79] deposited on a cooled substrate. The latter two cases [21, 79] only applied to room temperature.* In [21], the measurements were made in the range 1-10 μ, and in [79] in the range 0.475-2.0 μ. We see from Figs. 24 and 25 that the results obtained for layers made by methods 1 and 2 were quite close together. The results obtained for the layers deposited on the cooled substrate differed much more from our own data. This confirms that the structure of the metal deposited on the cooled substrate differed from that of the bulk metal.

* The optical constants of indium were also measured by Lenham and Trehern [119], who studied indium single crystals after mechanical and electropolishing. The results were presented in the form of graphs giving the dependence of $2n\varkappa/\lambda$ on $\hbar\omega$ and $\varkappa^2 - n^2$ on λ^2. No precise n and $\varkappa$ values can be determined from these graphs. It would appear that the n and $\varkappa$ values obtained by these authors lay well below our own. On using mechanical polishing, the authors were actually measuring the optical properties of a very work-hardened layer. On using electropolishing, the authors still failed to express the optical constants as functions of the thickness of the layer removed, so that one cannot be entirely convinced that the work-hardened layer had in fact been entirely eliminated.

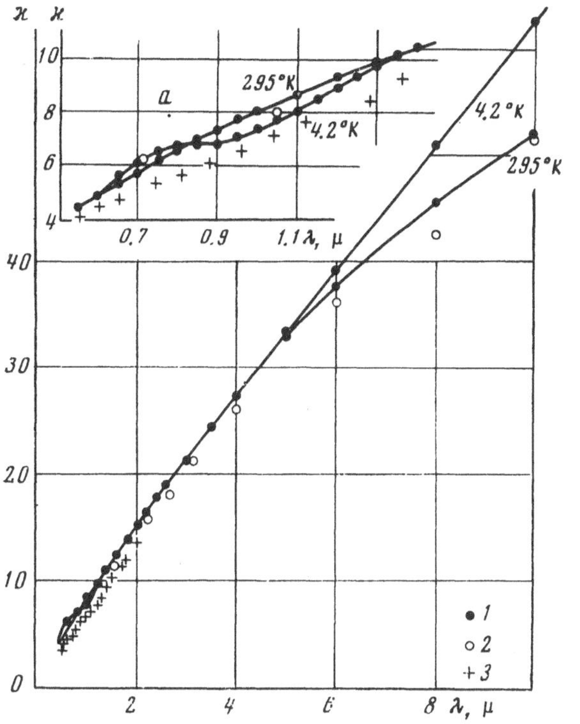

Fig. 25. Dependence of ϰ on the light wavelength for indium at two different temperatures. Notation as in Fig. 24.

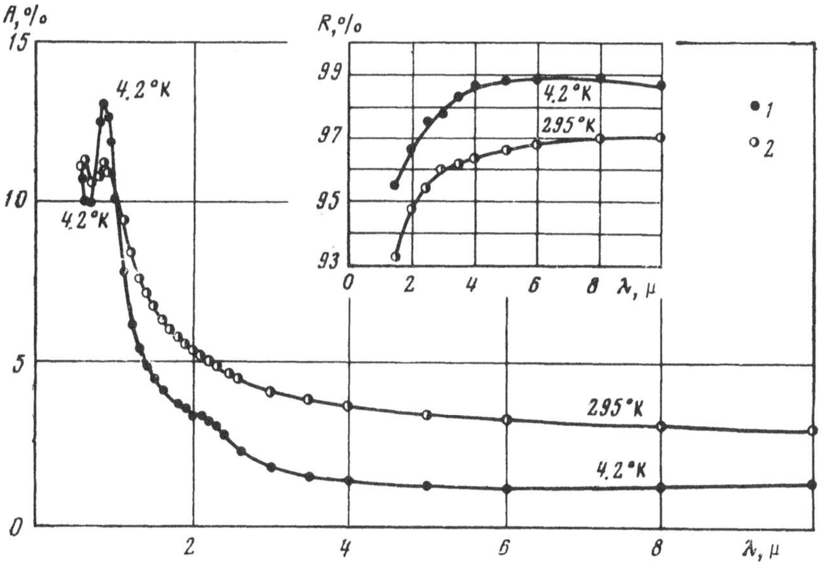

Fig. 26. Dependence of the A and R values of indium on λ at two temperatures: 1) 4.2; 2) 295°K.

We see from the curves of Fig. 24 that, at helium temperatures, there are two clear maxima: one is in the range 0.8-0.9 μ, and the second in the range 2-2.5 μ. At room temperature only a slight change in the form of the curve appears at 1 μ. The qualitative picture of the dependences of n and $\varkappa$ on λ and T agrees with that indicated at the beginning of the present chapter.

Figure 26 shows the dependence of A and R on λ and T. We see from this figure that in the infrared range the light reflection coefficient of indium is large at both temperatures. In the visible region a structure appears in the dispersion of A and R, the values of A rising sharply. This is associated with interband transitions, such as those discussed in Chapters VI and VII.

§ 2. Aluminum

A. Static Characteristics of Aluminum Layers

Aluminum is another Group III metal studied in our experiments. We initially studied unannealed aluminum layers [36]. Much later, after discovering a difference between the conduction-electron concentration N calculated from infrared measurements and that derived from the $|V_g|$ values, respectively, we studied annealed aluminum layers in order to decide whether annealing had increased N by a factor of several times, as in the case of gold [38]. In both cases the conditions of deposition were taken so that the density and conductivity of the layers studied might be as close as possible to those of the bulk metal. The samples were annealed at 300°C in vacuum at a pressure of about $5 \cdot 10^{-6}$ mm Hg for 2 h. We made no special tests relating to the oxide film.

Table 8 shows the characteristics of annealed and unannealed layers. The density of the unannealed layers was 89%, the conductivity at room temperature 71%, and the ratio of the conductivities at nitrogen and room temperatures 30% of the corresponding quantities obtained for the bulk metal.

The whole set of characteristics presented indicates that the unannealed deposited layers differed appreciably from the bulk metal. The main difference lay in the small absolute value and the slight temperature dependence of the static conductivity of the deposited layers. The layers obtained by the spraying technique were more finely dispersed.

The same Table 8 shows the density of the annealed layers. The density agreed with that of the bulk metal within the limits of experimental error. The conductivity of the layers was not measured.

TABLE 8. Static Characteristics of Aluminum Layers

Characteristic	Sprayed layers		Bulk metal	Published data (bulk metal)
	unannealed	annealed		
d, μ	0.3–0.5	0.3	—	—
ρ, g/cm^3	2.4	2.7	2.7	[111]
σ_K, cgse	$2.2 \cdot 10^{17}$	—	$3.1 \cdot 10^{11}$	[111]
σ_N/σ_K	3.5	—	11.5	[110]

TABLE 9. Optical Constants of Aluminum (Unannealed Layers)

λ, μ	295° K		78° K	
	n	ϰ	n	ϰ
0.80	1.12	6.0	0.83	6.0
0.90	1.05	7.0	0.75	7.0
1.20	0.95	9.6	0.63	9.6
1.50	1.14	12.1	0.78	12.1
2.00	1.75	16.1	1.30	16.1
2.50	2.4	19.8	1.7	19.8
3.0	3.2	23.5	2.2	23.5
4.0	4.8	30.0	3.2	30.1
5.0	6.7	37.6	4.4	37.8
6.0	9.5	44.4	6.5	44.9
7.0	12.6	51.0	9.1	52.0
8.0	15.6	58.1	—	—
9.0	21.1	62.1	—	—

B. Optical Constants of Aluminum

The optical constants of aluminum were measured for both unannealed [36] and annealed [38] layers.

The unannealed layers were studied at 295 and 78°K. The spectral range was 0.8-9 μ at 295°K and 0.8-7 μ at 78°K. The measurements at room temperature were carried out in both types of apparatus — apparatus 1 and the low-temperature version of the apparatus described in [16]. At nitrogen temperatures only the low-temperature version was employed. The first version of the low-temperature apparatus differed from the apparatus 2 described above (illustrated in Fig. 16) in that the analyzer rather than the polarizer was rotated. This had the effect that the polarization of the light by the monochromator had to be taken into account and, furthermore, during the measurements at nitrogen temperatures, a slight parasitic signal, associated with the fact that different parts of the apparatus were at different temperatures, had to be compensated in the long-wave part of the spectrum. This arrangement was considerably less convenient than that of apparatus 2. A bolometer acted as radiation receiver in both systems. Subsequently, when studying other metals, we used the apparatus described in Chapter III.

The results of our measurements of the aluminum optical constants are presented in Table 9 and Figs. 27 and 28. It should be noted that the values of ϰ relating to different series of measurements agreed closely with each other. The values of n exhibited a greater spread, although the conditions of deposition, such as the degree of vacuum and the deposition rate, were kept constant. This was evidently associated with the fact that the layers of aluminum deposited were very finely dispersed, and the values of n were particularly sensitive to this characteristic. The error for ϰ in the range 0.9-4.0 μ equalled 1.0-1.5% and in the range 5-9 μ equalled 3-4%. The value of ϰ for λ = 0.8 μ was determined* to an accuracy of 6%. The error for n in the range 3-9 μ was 3%, and in the range 0.9-2 μ was 5%. The value of n for λ = 0.8 μ was determined to an accuracy of 10%.

Comparison between the optical constants of aluminum obtained at room and liquid-nitrogen temperatures shows (Figs. 27 and 28) that ϰ varies very little with temperature, whereas n falls sharply as the temperature is lowered.

*The value of the optical constants at λ = 0.8 μ was only determined in the low-temperature apparatus.

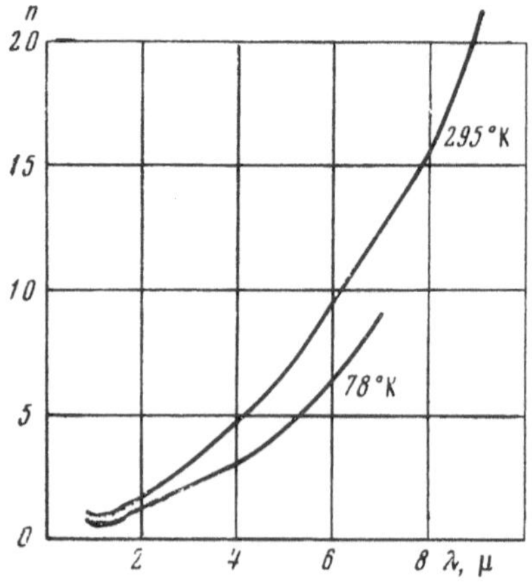

Fig. 27. Dependence of n on the light wavelength for aluminum at two particular temperatures.

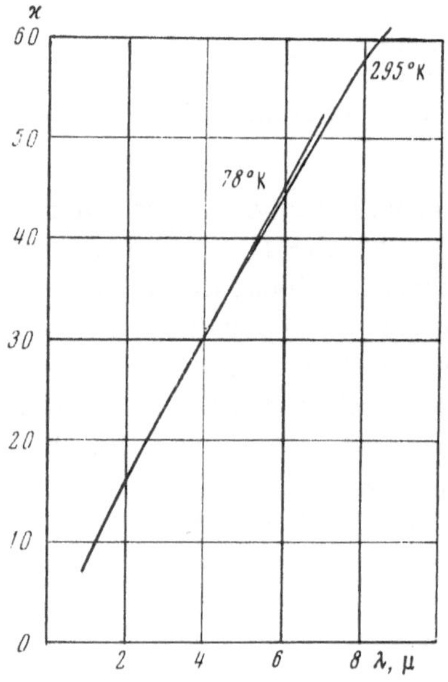

Fig. 28. Dependence of κ on the light wavelength for aluminum at two particular temperatures.

The annealed aluminum layers were only measured at room temperature in apparatus 1 in the spectral range 1.0–10.0 μ. We have already indicated that the aim of these measurements was to discover whether any great change (a factor of several times) took place in the N of aluminum on annealing, such as occurs in the case of gold. Since preliminary experiments showed that annealing had no marked effect on N, we only made one control series of measurements. The results appear in Table 10 and Figs. 29 and 30.

The optical constants of aluminum were only determined at room temperature by other authors. Unfortunately none of them measured the static characteristics of the test layers. Only Beattie determined the conductivity of the samples [89], this being 48% of that of the bulk metal, much smaller than the conductivity of our own samples. The absence of data relating to the conductivity and density of the layers studied prevents us from judging how near the latter were to the bulk metal in their general properties.

TABLE 10. Optical Constants of Aluminum (Annealed Layers at T = 293°K)

λ, μ	n	κ	λ, μ	n	κ
1.00	0.98	7.65	4.0	5.58	29.4
1.50	1.14	11.6	5.0	7.84	35.7
2.00	1.67	15.2	6.0	10.4	41.3
2.50	2.50	18.8	8.0	16.2	52.2
3.00	3.48	22.6	10.0	25.5	60.9

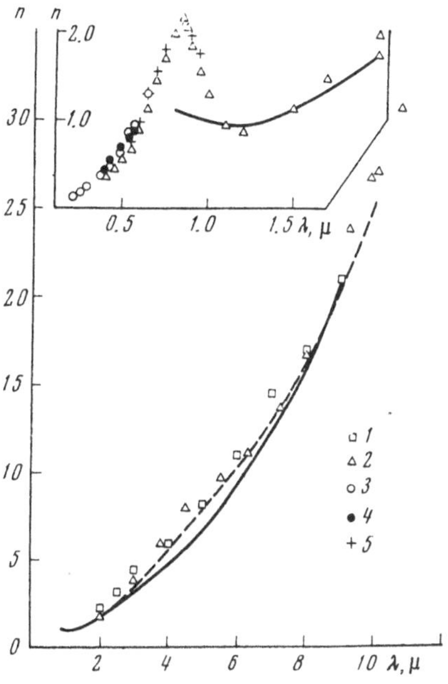

Fig. 29. Dependence of n on the light wavelength for aluminum at room temperature. Solid curve, our own work for unannealed layers [36]; broken curve, for annealed layers [38]. 1) Results of [89]; 2) [79, 118, 120]; 3) [121]; 4) [122]; 5) [103, 123, 124].

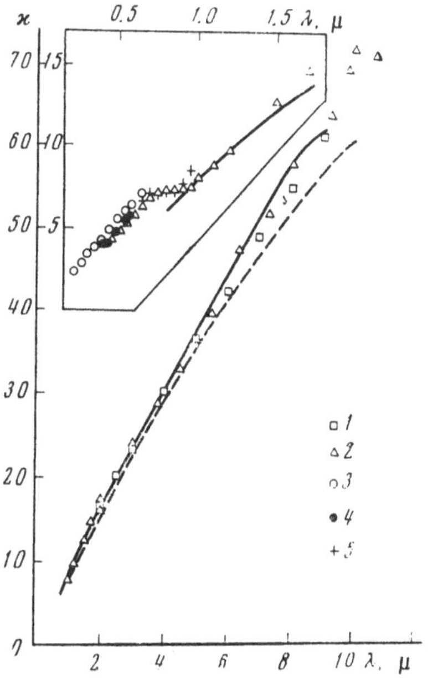

Fig. 30. Dependence of κ on the light wavelength at room temperature (notation as in Fig. 29).

The n and $\varkappa$ values of aluminum measured by various authors at room temperature are illustrated in the same figures.

In the long-wave part of the spectrum, measurements were made earlier by Beattie [89] and Yarovaya [118]. In addition to these, Hodgson [125] presented some results in the form of graphs representing the dependence of $\log (n\varkappa/\lambda)$ and $\log (\varkappa^2 - n^2 + 1)$ on $\log\lambda$. The values of n and $\varkappa$ cannot be determined very accurately from these graphs.

Results relating to the short-wave part of the spectrum are given separately in the insets to Figs. 29 and 30. Shklyarevskii and Yarovaya [79, 120] introduced a special correction for oxidation of the layers in the short-wave region, but not in the long-wave part of the spectrum. The figures omit the Shklyarevskii and Yarovaya data of [87], as the authors themselves considered these less reliable. Hass and Waylonis [121] conducted investigations in the visible and ultraviolet parts of the spectrum, also incorporating a correction for the oxidation of the aluminum, based on [126].

O'Bryan [122] carried out some measurements in vacuum, but only relating to the visible part of the spectrum. Schulz [103, 123] and Schulz and Tangherlini [124] studied the reflection of light from the glass — metal interface, in contrast to the previous authors; the samples were both aged and annealed.

The optical constants of aluminum were similarly measured by reflection at the glass — metal interface in experiments by Shklyarevskii and Miloslavskii [87] as well as by ourselves [38] in the visible [87] and near-infrared regions. In both cases the values of n and $\varkappa$ so obtained were much greater than those obtained by reflection at the outer aluminum — air interface. We consider that this difference was associated with the fact that the inner aluminum layers in contact with the glass had an amorphous structure. This question requires further study, and we shall not develop it any further here. We simply note that the results of Schulz [123, 124] show that, after annealing and aging, the optical properties of the inner layers approach those of the outer layers.

The optical properties of aluminum were also studied by Ehrenreich, Phillip, and Segall [78]. These authors used the results of earlier measurements of the light reflection coefficient R of aluminum for normal incidence obtained in [121, 127, 128]; using the Kramers — Krönig relation [97] they determined* two functions $\varepsilon_1(\omega)$ and $\sigma(\omega)$ in the energy range 0-22 eV. The extrapolation law $R(\omega)$ for high energies was chosen so as to match the calculated and measured values of n and $\varkappa$ [128] for $\hbar\omega = 16.87$ and $\hbar\omega = 21.2$ eV. The optical constants n and $\varkappa$ so obtained were higher than the corresponding values measured in [121, 123, 124] in the visible region. There are also considerable discrepancies for the calculated and directly meaured value of n and $\varkappa$ in the infrared region [36, 79, 89]. These discrepancies are apparently associated with the general failings of the method discussed earlier, and also with the fact that, in plotting the function $R(\omega)$, the results of different authors using aluminum layers with different characteristics were employed. From the results of this investigation for the visible and infrared regions, only a qualitative picture of the $\varepsilon_1(\omega)$ and $\sigma(\omega)$ relationships may be gained. One particular result was a clear maximum of $\sigma(\omega)$ in the region of $\hbar\omega \approx 1.5$ eV, in general agreement with the results of [79, 118, 123, 124].

Lenham and Trehern [119] studied large polycrystalline samples of aluminum, mechanically polished with diamond powder. The authors gave no data for n and $\varkappa$, only indicating the presence of an absorption band at 1.55 eV and a second broad, weak band at 2.85 eV. Unfortunately, the mechanical polishing of aluminum leads to the formation of a thick work-hardened layer. It is therefore not quite clear how far these results correspond to the bulk metal or to layers obtained by vacuum evaporation.

* $\varepsilon_1 \equiv \operatorname{Re} \varepsilon'$.

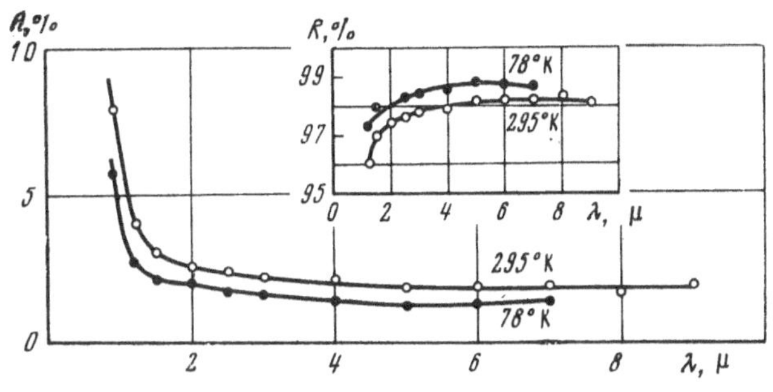

Fig. 31. Dependence of the A and R of unannealed aluminum layers on λ for two different temperatures.

Returning to Figs. 29 and 30, we see that annealing at up to 300°C causes no marked change in the optical constants of aluminum. In the infrared region for $\lambda \geq 1\ \mu$, the results obtained by different authors for $\varkappa$ agree quite satisfactorily with each other. The results obtained for n differ rather more, apparently because of the different conductivities of the layers studied. In the spectral range 0.8-1 μ our own results differ from those of [79, 123, 124]. This difference is particularly substantial for n. According to our own results, we may expect a maximum in the $n(\lambda)$ relation at $\lambda < 0.8\mu$. In [79, 123, 124] a very clear maximum at $\lambda = 0.85\ \mu$ was noted. As our measurements of the optical constants of aluminum in the range 0.8-1.0 μ were carried out in an apparatus having a small dispersion in this range and an insensitive receiver, in subsequent analysis of the experimental data we shall use the position of the maximum obtained for $\sigma(\omega)$ in [79, 123, 124]. For the spectral range 1-9 μ we shall use our own measurements [36]. Averaging the results of different authors seems unreasonable, as we do not know the static characteristics of the layers used in the other investigations.

Figure 31 shows the dependence of R and A on λ and T. We see from the figure that, in the infrared region, the values of R are about 98% at 295°K and about 99% at 78°K. In the visible region R falls and A correspondingly rises, this being due to the interband transitions considered in Chapters VI and VII.

§ 3. Lead

A. Static Characteristics of Lead Layers

Among the metals belonging to the fourth group in the Periodic Table we gave particular attention to tin and lead. Tin has a tetragonal and lead a face-centered cubic lattice.

The lead layers were prepared by the method described in Chapter III; they had very good specular surfaces of a darkish tint without any trace of mat structure. The static characteristics of the layers are given in Table 11 [20, 22, 50, 51].

The density of the layers and also their static conductivity at room and nitrogen temperatures coincided with the corresponding characteristics of the bulk metal. The conductivity at hydrogen temperatures was 93% of that of the bulk metal. The ratio of the conductivity at nitrogen and hydrogen temperatures to the conductivity at room temperature, respectively, equalled 4.15 and 34.

In determining the residual resistance R_{res} of the test layers, we were unable to use the jump in resistance as a measure of R_{res}, as in the region of $T \approx T_c$ the resistance continued varying with temperature. In order to determine R_{res} we studied the temperature dependence of the resistance of the layer over the range 7.3-26°K. This relationship is shown in Fig. 32.

TABLE 11. Static Characteristics of Lead Layers

Characteristic	Test layers	Bulk metal	Published data (bulk metal)
d, μ	0.6–1.6	—	—
ρ, g/cm³	11.3	11.3	[111–113]
σ_K, cgse	$0.41 \cdot 10^{17}$	$0.41 \cdot 10^{17}$	[111–113]
σ_N, cgse	$1.70 \cdot 10^{17}$	$1.70 \cdot 10^{17}$	[111–113]
c_H, cgse	$1.4 \cdot 10^{18}$	$1.5 \cdot 10^{18}$	[111–113]
R_n/R_K	$6 \cdot 10^{-3}$	$1 \cdot 10^{-3}$	[111, 129]
R_{res}/R_K	$\sim 5 \cdot 10^{-3}$	$\sim 10^{-3}$	[111, 129]
T_c, °K	7.3	7.19	[115]
$(\Delta T)_c$, °K	∼0.2	∼0.01	[111, 129]
$\theta_{R,(293,8)}$, °K	85	85	[111–113]
R_X, μΩ-cm/kOe	$+1.5 \cdot 10^{-4}$	$+1.0 \cdot 10^{-4}$	[29]

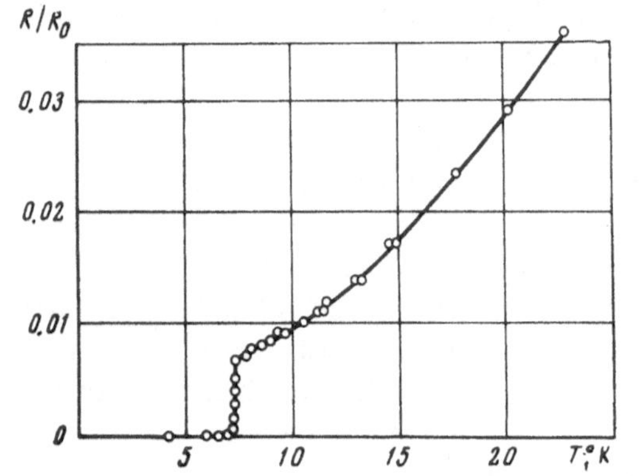

Fig. 32. Dependence of the resistance of the lead test layer on temperature.

The results enable us to determine R_{res} in two ways. The first way lies in extrapolating the temperature dependence to zero, and gives $R_{res} \approx 5 \cdot 10^{-3} R_K$. The second way is as follows. For the temperature range 7.3-20°K, we assume that $R(T) = R_{res} + R'(T)$, where $R'(T)$ is the temperature dependence of an ideal resistance with the same characteristic temperature θ_R as the test samples. By comparing the foregoing formula with the experimental R(T) relationship, we obtain the same value of $R_{res} \approx 5 \cdot 10^{-3} R_K$.

The characteristic temperature was determined by the method indicated earlier for various temperature ranges from 293 to 8°K. We found that θ_R depends only slightly on temperature. The difference in θ_R for various temperature ranges never exceeded 10%. For this reason Table 11 gives the value referred to the whole range (293-8°K). The values of θ_R for our samples agree with the values of θ_R obtained for the bulk metal. We note that the Debye temperature [52] for the bulk metal θ_D = 94°K, i.e., θ_R and θ_D are close together.

The temperature of the superconducting transition practically coincided with that relating to the bulk metal (the difference was smaller than the experimental error). The temperature width of the superconducting transition was $(\Delta T)_c \approx 0.2$°K, i.e., the transition was fairly sharp, although somewhat wider than that of the bulk material.

We also studied the Hall effect of the test layers. The Hall constant was independent of the current density (the measurements were made at $j \leq 7 \cdot 10^3$ A/cm²) and the magnetic field ($H \leq 8$ kOe). The Hall constant of lead was ten times smaller than that of copper. This means that the Hall effects due to the electrons and holes largely compensated each other. The Hall constant was 1.5 times greater than that of the bulk metal. Tables often give the value of

$N_a R_X |e| c$, where N_a is the concentration of the atoms, e is the charge on the electron, and c is the velocity of light. For our own layers this quantity equalled +0.07; for the bulk metal it equalled +0.05. The plus sign indicates the hole character of the Hall effect, as in the bulk metal. Since the Hall effect is very sensitive to slight changes in electron structure, particularly when the resultant effect constitutes the small difference between two large quantities, we may consider that the Hall effect in the test samples is in fact very similar to the Hall effect in the bulk material.

The results presented in Table 11 show that the static characteristics of the test layers of lead coincide with the corresponding characteristics of the bulk metal. We may therefore reasonably consider that our optical constants also represent those of the bulk metal.

B. Optical Constants of Lead

The optical constants of lead were studied in our own earlier investigations [14, 20, 22] and also that of Golovashkin [50]. The original research carried out in [14] related to lead layers with unknown static characteristics. Measurements were made at room temperature over the range 1.3-7 μ. In later investigations [20, 22, 50] the optical constants of lead were studied in far greater detail at room, nitrogen, and helium temperatures. The static characteristics of the layers given in Table 11 were taken* from [20, 22, 50]. We shall use the results of these investigations in what follows. The optical constants obtained in [14] yield good agreement for $\varkappa$, but a slight discrepancy for n. In later investigations, n is slightly lower, this indicating the better conductivity of the samples used in the later work [20, 22, 50].

The optical constants of lead were measured in the spectral range 0.45-12 μ at 293 and 78°K and 0.54-12 μ at T = 4.2°K using apparatuses 1 and 2, and also using a version of apparatus 2 duly modified for measurements in the visible and near-infrared regions (see Chapter III, Sections 3-5). In the overlapping spectral regions the results of the n and $\varkappa$ determinations agreed for the different types of apparatus.

Measurements were made in the spectral range 0.45-1.00 μ with steps of 0.01-0.02 μ, in the range 1.00-2.6 μ with steps of 0.05 μ, and in the range 2.6-12 μ with steps of 0.5-1.0 μ. The spectral widths of the slits was smaller than or equal to the measuring step in these intervals.

We carried out special experiments to discover the influence of any possible oxidation of the samples. For this purpose the optical constants were measured as quickly as possible after sample preparation, and then again after spending one or several days in a cryostat. One day in the cryostat had no effect on the optical constants. Slight changes (2-3% for n and 1-2% for $\varkappa$) only occurred after four days. This shows that good specular lead surfaces only oxidized fairly slowly, so that the effect of oxidation on the results of our measurements could be neglected. We note that mirrors left in air for over a month differed very little in outward appearance from those which had just been prepared.

The results of our measurements of the optical constants of lead are presented in Table 12 and in Figs. 33 and 34. The error in determining n was 1-2% at room temperature; it was 1-2% at nitrogen and helium temperatures in the middle of the spectral range; but it was 2-5% at the ends of the range and at 1.5-2.5 μ. The error in measuring $\varkappa$ at all temperatures was 0.5-1% in the greater part of the spectral range, but a little greater (up to 1.5%) at the ends.

Figures 33-36 relate to both the infrared and the visible parts of the spectrum; they show that qualitatively the dependence of n and $\varkappa$ on λ and T agrees with that indicated at the

*The lead layers used in [20] had $R_{res}/R_K \approx 3 \cdot 10^{-3}$. The remaining characteristics agreed with those of the layers used in [22, 50].

TABLE 12. Optical Constants of Lead

λ, μ	293° K		78° K		4.2° K	
	n	κ	n	κ	n	κ
0.45	1.445	3.18	1.56	3.20	—	—
0.46	1.545	3.20	1.575	3.22	—	—
0.47	1.58	3.23	1.64	3.22	—	—
0.48	1.62	3.25	1.65	3.22	—	—
0.49	1.68	3.28	1.655	3.22	—	—
0.50	1.705	3.30	1.675	3.24	—	—
0.51	1.74	3.31	1.685	3.27	—	—
0.52	1.755	3.34	1.70	3.31	—	—
0.53	1.785	3.36	1.73	3.35	—	—
0.54	1.81	3.37	1.78	3.39	1.82	3.46
0.55	1.835	3.40	1.825	3.43	1.885	3.50
0.56	1.87	3.41	1.875	3.45	1.97	3.51
0.57	1.87	3.43	1.91	3.46	2.04	3.52
0.58	1.90	3.43	1.965	3.47	2.11	3.52
0.59	1.91	3.44	2.01	3.46	2.17	3.52
0.60	1.915	3.45	2.06	3.44	2.26	3.50
0.61	1.945	3.46	2.10	3.42	2.29	3.45
0.62	1.945	3.47	2.11	3.38	2.34	3.38
0.63	1.95	3.48	2.14	3.34	2.39	3.27
0.64	1.945	3.49	2.13	3.28	2.39	3.16
0.65	1.915	3.51	2.09	3.25	2.34	3.09
0.66	1.895	3.51	2.08	3.19	2.28	3.02
0.67	—	—	2.02	3.14	2.18	2.93
0.68	1.855	3.53	1.965	3.10	2.09	2.89
0.69	—	—	1.88	3.08	1.97	2.84
0.70	1.785	3.57	1.80	3.07	1.85	2.82
0.71	—	—	1.685	3.08	1.67	2.83
0.72	1.71	3.64	1.58	3.13	1.545	2.89
0.73	—	—	1.475	3.17	1.40	2.97
0.74	1.635	3.73	1.39	3.24	1.305	3.06
0.75	1.60	3.78	1.34	3.33	1.24	3.15
0.76	1.57	3.84	1.28	3.41	1.175	3.26
0.78	1.525	3.96	1.185	3.58	1.09	3.46
0.79	—	—	1.165	3.65	1.07	3.58
0.80	1.505	4.09	1.13	3.73	1.04	3.61
0.81	—	—	—	—	1.025	3.75
0.82	1.47	4.18	1.09	3.86	1.005	3.78
0.84	—	—	—	—	0.989	3.92
0.85	1.44	4.35	1.055	4.05	0.976	3.97
0.86	—	—	—	—	0.965	4.04
0.88	1.41	4.56	1.01	4.25	0.940	4.16
0.90	1.40	4.68	0.983	4.37	0.904	4.27
0.92	1.385	4.80	0.956	4.49	0.870	4.39
0.95	1.385	4.99	0.917	4.67	0.821	4.57
0.98	1.38	5.16	0.878	4.86	0.771	4.76
1.00	1.38	5.32	0.848	5.00	0.721	4.90
1.05	1.385	5.62	0.801	5.30	0.656	5.22
1.10	1.40	5.98	0.743	5.63	0.574	5.58
1.15	1.415	6.31	0.678	6.00	0.485	5.92
1.20	1.445	6.59	0.651	6.28	0.417	6.29
1.25	1.485	6.93	0.599	6.66	0.346	6.66
1.30	1.50	7.23	0.583	6.96	0.312	6.96
1.35	1.525	7.48	—	—	0.294	7.26
1.40	1.58	7.74	0.566	7.59	0.285	7.59
1.45	1.60	8.01	—	—	0.290	7.90
1.50	1.645	8.30	0.575	8.23	0.275	8.24
1.55	1.695	8.58	—	—	—	—
1.60	1.775	8.90	0.611	8.93	0.300	8.94
1.70	1.895	9.47	0.647	9.59	0.312	9.57

TABLE 12 (Continued)

λ, μ	293° K		78° K		4,2° K	
	n	κ	n	κ	n	κ
1.80	2.05	10.15	0.683	10.2	0.336	10.35
1.90	2.18	10.7	0.728	10.8	0.338	11.0
2.00	2.32	11.2	0.783	11.35	0.387	11.55
2.10	2.47	11.75	0.835	11.95	0.382	12.1
2.20	2.63	12.2	0.901	12.55	0.448	12.75
2.30	2.84	12.8	0.949	13.15	0.409	13.3
2.40	3.03	13.2	0.997	13.75	0.541	13.95
2.50	3.22	13.9	1.05	14.3	0.586	14.55
2.55	3.34	14.15	—	—	0.597	14.75
2.60	3.45	14.45	1.195	14.85	0.614	15.05
3.0	4.27	16.4	1.53	17.3	0.81	17.3
3.5	5.39	18.6	2.01	20.4	1.10	20.1
4.0	6.58	20.8	2.48	22.9	1.49	23.1
5.0	9.04	24.8	3.99	28.7	2.15	28.6
6.0	11.7	28.1	5.41	33.9	2.95	34.4
7.0	14.1	30.9	7.16	38.7	3.75	39.9
8.0	16.4	33.6	8.82	43.9	4.50	45.5
9.0	18.7	35.8	10.5	49.1	5.56	50.6
10.0	21.0	37.4	12.3	54.4	6.70	55.9
11.0	23.2	39.2	14.4	59.1	7.90	61.3
12.0	24.6	40.5	16.3	63.5	9.20	66.5

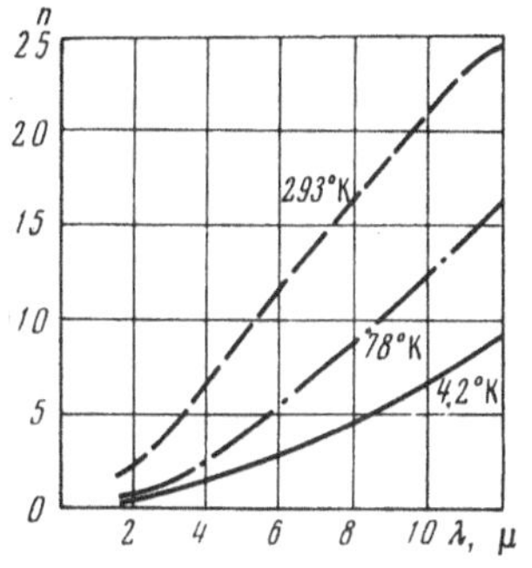

Fig. 33. Dependence of the n of lead on λ for three temperatures (long-wave region).

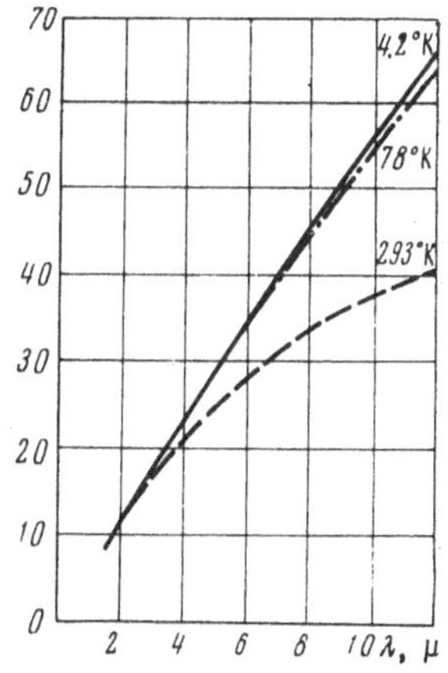

Fig. 34. Dependence of the κ of lead on λ for three temperatures (long-wave region).

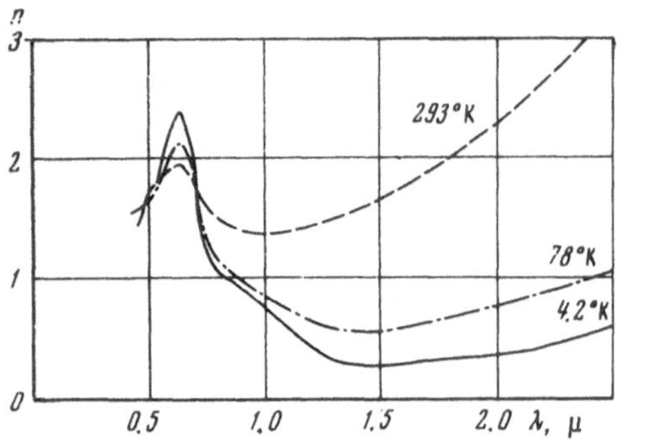

Fig. 35. Dependence of the n of lead on λ for three temperatures (short-wave region).

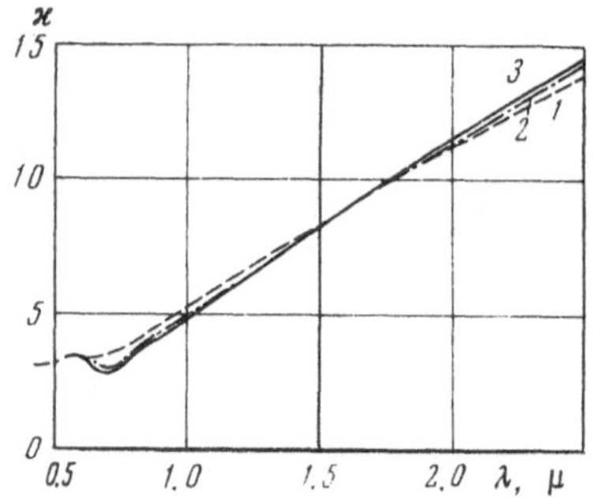

Fig. 36. Dependence of the ϰ of lead on λ for three temperatures (short-wave region): 1) 293°K; 2) 78°K; 3) 4.2°K.

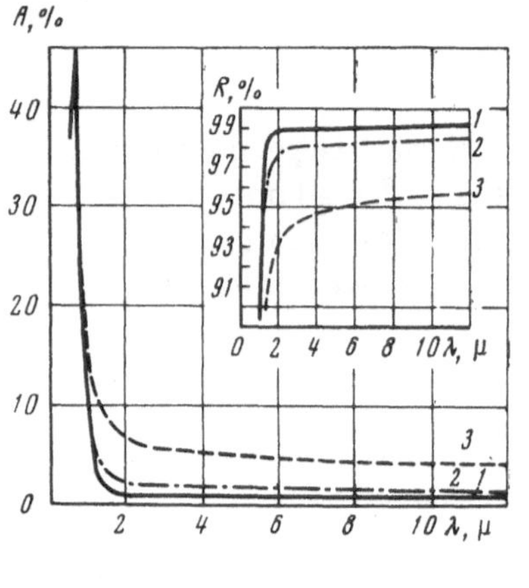

Fig. 37. Dependence of the A and R of lead on λ for three temperatures: 1) 4.2°K; 2) 78°K; 3) 293°K.

beginning of this chapter. The n(λ) relationship exhibits a sharp maximum in the range 0.63-0.64 μ, the height and width of this maximum depending on temperature. Apart from this principal maximum, there is a clearly expressed singularity in the range 0.8-1.5 μ. This singularity appears more sharply at helium temperature, but it is also appreciable at room temperature.

Figure 37 shows the dependence of A and R on λ and T. We see from the figure that for all temperatures the light-reflection coefficient of lead in the infrared (as distinct from the visible) region is large. The value of R is about 99% at 4.2°K, 98% at 78°K, and 95% at 293°K. In the visible region there is a sharp fall in R, and A correspondingly increases. The observed maximum of A is associated with the interband transitions considered in Chapters VI and VII.

The optical properties of lead had hardly ever been studied before. There were no reliable data for either the visible or the infrared regions. In the infrared region we may only mention the work of Ramanathan [130], in which the absorption power of cast and electrically polished lead was measured at helium temperature. The light source was a black body held at room temperature, no monochromatization being incorporated. Ramanathan obtained A = 1.15% for the spectral range 10-12 μ. We ourselves obtained A = 0.83% for the same region. Our lower value of A indicates the better quality of the samples.

For the visible region we may mention some work by Drude [131] relating to one wavelength of $\lambda = 0.589\ \mu$; the results agree fairly closely with our own. In addition to this, Trompette [132] measured the optical constants of thin lead films 0.01-0.03 μ thick for four wavelengths of the mercury spectrum. As might be expected, the optical constants of such thin films differed considerably from those of the bulk metal.

§ 4. Tin

A. Static Characteristics of Tin Layers

White tin is another Group IV metal studied in detail in our experiments. The conditions for preparing the samples were considered in Chapter III. The test layers of tin had excellent mirror surfaces, with no traces of mat structure. The static characteristics [17, 18, 23, 51] of these layers are shown in Table 13.

The density of the layers was 99% of the density of the bulk metal. Allowing for measuring errors, we may consider that the density of the test samples agreed closely with that of the material in bulk form.

The static conductivity at room, nitrogen, and hydrogen temperatures, respectively, equalled 91, 78, and 24% of the corresponding conductivity of the bulk metal. The conductivity of tin layers increased by 4.4 times on cooling to nitrogen temperature and 22.5 times on cooling to hydrogen temperature. The resistance R_{res} was 3.5% of the resistance at room temperature R_K. The resistance of the layer equalled the residual resistance even at T = 4.2°K.

The temperature of the superconducting transition T_c was similar to the corresponding temperature for the bulk metal. The magnitude of T_c was independent of the thickness of the layer, even for samples having d = 0.2 μ. The transition into the superconducting state was very sharp. The temperature width of the transition $(\Delta T)_c$ was of the same order as for the bulk metal. Figure 38 shows the temperature dependence of the resistance of a tin layer close

TABLE 13. Static Characteristics of the Test Layers of Tin

Characteristic	Test layers	Bulk metal	Published data (bulk metal)
d, μ	0.5—1.1	—	—
ρ, g/cm^3	7.2	7.28	[111]
σ_K, cgse	$7.1 \cdot 10^{16}$	$7.8 \cdot 10^{16}$*	[111—113]
σ_N, cgse	$3.1 \cdot 10^{17}$	$4.0 \cdot 10^{17}$*	[111—113]
σ_H, cgse	$1.6 \cdot 10^{18}$	$6.7 \cdot 10^{18}$*	[111—113]
R_{res}/R_K	$3.5 \cdot 10^{-2}$	$1 \cdot 10^{-3}$	[113]
T_c, °K	3.88	3.72	[115]
$(\Delta T)_c$, °K	0.01—0.03	~0.01	—
$\theta_{R,(293,78)}$, °K	187	200	[111—113]
$\theta_{R,(78,20)}$, °K	128	130—135	[111—113]
R_X, $\mu\Omega$-cm/kOe	$+5.0 \cdot 10^{-5}$	$-4.1 \cdot 10^{-5}$	[29]

* The conductivities of the bulk metal given in various reference books differ by 20-30% [111-113, 133, 134]. The table gives the most reliable values.

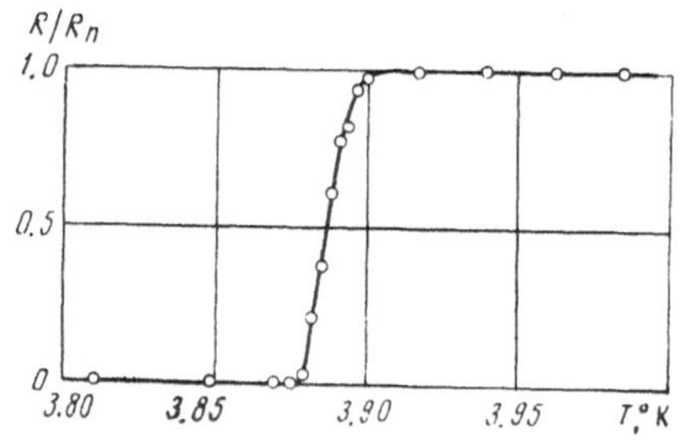

Fig. 38. Temperature dependence of the resistance of a tin layer close to T_c.

to T_c for one of the samples. The small temperature width of the superconducting transition indicates the excellent homogeneity of the samples employed.

Measurements of the temperature dependence of the conductivity of the tin layers enabled us to determine the characteristic temperature θ_R of these layers. The value of θ_R varied considerably with temperature for our layers, as for the bulk metal. We therefore determined the value of this quantity for two temperature ranges: 293-78 and 78-20°K. For the bulk metal θ_R was found in the same way from the temperature dependence of the resistance of the polycrystalline tin samples. The values of θ_R for our layers were slightly lower than the corresponding values for the bulk metal. We note that the Debye temperature [52] for white tin is $\theta_D = 189$°K.

A study of the Hall effect showed that, as in the bulk metal, the Hall effects associated with the electrons and holes in our test layers nearly compensated each other. The Hall constant R_X for tin was thirty times smaller than for copper. The value of R_X was independent of the current density (measurements were made for $j \leq 4.5 \cdot 10^3$ A/cm^2) and also of the magnetic field ($H \leq 7$ kOe). In our layers the hole effect was slightly greater than that associated with the electrons. In bulk polycrystalline samples the electron effect was dominant. In tin single crystals the sign of the effect differed for different directions [135]. In all cases the total effect was almost zero.

The results presented in Table 13 show that the static characteristics of the layers of tin studied are very close to the corresponding characteristics of the bulk metal. We may therefore consider that our own values of the optical constants also characterize the metal in the bulk state.

B. Optical Properties of Tin

We studied the optical constants of tin on a number of occasions [14, 17, 18, 23]. The original investigations [14] concerned tin layers for which the static characteristics had not been determined. The measurements were carried out at room temperature over the spectral range 1.3-6.3 μ. In the later investigations [17, 18, 23] the optical constants of tin were studied in much greater detail. The measurements were carried out over the spectral range 0.73-12 μ at three temperatures: 293, 78, and 4.2°K. The static characteristics given in Table 13 relate to the layers used in the later work. The optical constants obtained in [14] agree closely with those obtained in [17, 18, 23].

The optical-constant measurements for tin at room temperature were carried out in apparatuses 1 and 2. The results of the two sets of measurements were in complete agreement. The measurements at nitrogen and helium temperatures were carried out in apparatus 2.

The results of our measurements of the optical constants of tin are presented in Table 14 and Figs. 39 and 40. The error in the determination of n was 1-2% at room and nitrogen temperatures and 1-3% at helium temperature. The error in determining $\varkappa$ was 0.5-1% at room and nitrogen temperatures and 1-2% at helium temperature. The maximum error occurred at

TABLE 14. Optical Constants of Tin

λ, μ	293° K		78° K		4.2° K	
	n	ϰ	n	ϰ	n	ϰ
0.73	2.18	6.29	2.24	6.19	—	—
0.80	2.40	6.62	2.27	6.42	—	—
0.93	3.15	7.28	3.43	7.17	2.95	7.62
0.99	3.44	7.34	3.92	6.94	3.70	7.15
1.20	3.76	7.63	3.53	6.45	3.05	5.98
1.35	3.57	8.04	2.76	6.99	2.55	6.64
1.50	3.31	8.67	2.09	7.98	1.99	7.80
1.70	3.13	9.88	1.75	9.29	1.51	9.35
2.00	3.10	11.8	1.65	11.4	1.38	11.4
2.50	3.63	14.8	1.69	14.6	1.39	14.6
3.0	4.41	17.8	1.88	18.0	1.58	18.0
3.5	5.27	20.5	2.13	21.1	1.95	21.1
4.0	6.19	23.2	2.46	24.2	2.13	24.2
5.0	8.49	28.5	3.75	29.7	2.75	30.0
6.0	11.0	33.1	4.97	35.5	3.73	35.8
7.0	13.8	37.1	6.51	41.4	4.89	41.6
8.0	16.6	40.6	8.17	47.0	6.05	47.4
9.0	19.3	43.8	10.0	51.7	7.90	53.3
10.0	22.0	46.4	12.4	55.8	10.1	58.7
11.0	24.8	49.0	15.7	59.8	12.6	63.4
12.0	27.8	51.6	18.2	63.8	15.3	67.0

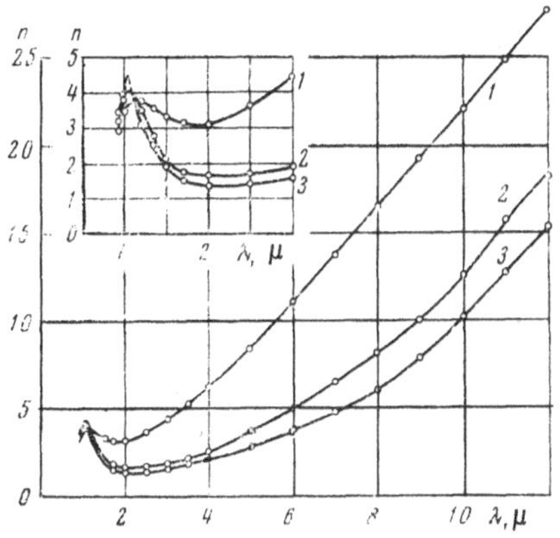

Fig. 39. Dependence of the n of tin on λ for three temperatures: 1) 293°K; 2) 78°K; 3) 4.2°K.

the ends of the range studied and in the region of the absorption band (λ ≈1.2 μ). Figures 39 and 40 illustrate the dependence of n and ϰ on λ for three test temperatures. In the insets to these figures we show the short-wave region on its own. The n(λ, T) and ϰ(λ, T) relationships have the same character as those of lead.

Figure 41 shows the R(λ, T) and A(λ, T) relationships. We see from this figure that the reflecting power of tin is large in the spectral range 2 < λ < 12 μ. At helium temperatures the reflection coefficient reaches 99%. In the short-wave part of the spectrum studied there is a strong absorption band with a maximum at 1-1.2 μ. This band is associated with quantum interband transitions — these will be considered in Chapters VI and VII.

We also studied the possible effects of the surface oxidation of the tin layers, in the same way as in the case of lead. The optical constants of the tin layers only started changing slightly after spending several days in the air. The oxidation of good tin surfaces took place even more slowly than that of good lead surfaces. These experiments indicate that the oxidation of the tin layers has no material effect on the results of the optical-constant measurements.

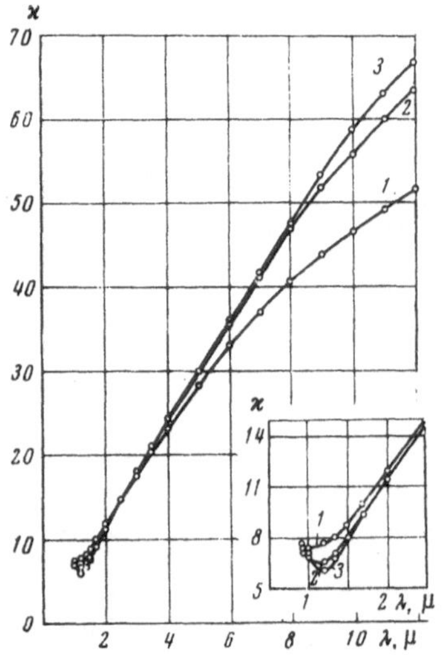

Fig. 40. Dependence of the ϰ of tin on λ for three temperatures. Notation as in Fig. 39.

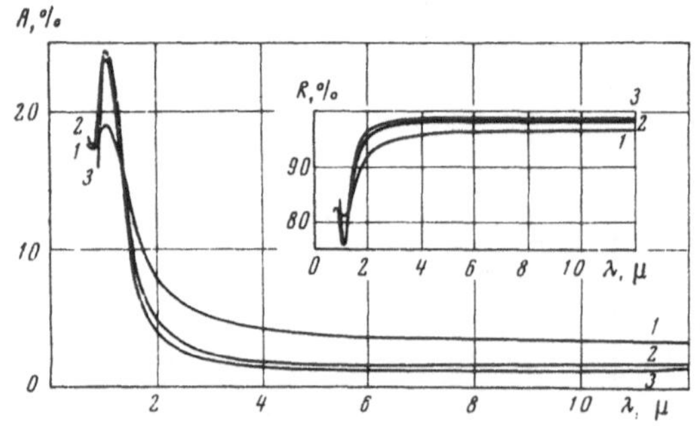

Fig. 41. Dependence of the A and R of tin on λ. Notation as in Fig. 39.

We also checked the effects of annealing on the optical properties of tin layers. Annealing was carried out in vacuum at 150-200°C for several hours. No effect on the optical constants was in fact observed. This agrees with the results of Kirillova, Noskov, and Charikov [109], who also found that for metals with fairly low melting points annealing had hardly any effect on the optical constants.

The optical constants of tin in the infrared region were earlier studied by Hodgson at room temperature [125]. The measurements were only carried out for one semitransparent layer, without determining its static characteristics. As in the case of aluminum, the results were presented in the form of certain graphs, from which it was impossible to determine n and ϰ with any reasonable accuracy.

As regards low-temperature measurements, we may only mention that of Ramanathan [130], who determined the absorption of tin at helium temperature. The light source was a black body at room temperature, no monochromatization being applied. Ramanathan obtained A = 1.24%, considering that this value related to the spectral range 10-12 μ. Our own values for this range gave A = 1.21%.

TABLE 15. Static Characteristics of the Gold Layers

Characteristic	Test layers	Bulk metal		Published data (bulk metal)
d, μ	0.5–1.0	—	—	—
ρ, g/cm^3	19.2	19.3	—	[111–113]
σ_K, cgse	$3.4 \cdot 10^{17}$	$3.78 \cdot 10^{17}$	Original material	[24]
		$4.07 \cdot 10^{17}$	Ideal metal	[110]
σ_N, cgse	$10.0 \cdot 10^{17}$	$16.15 \cdot 10^{17}$	Original material	[24]
		$20.5 \cdot 10^{17}$	Ideal metal	[110]
R_{res}/R_K	$17.2 \cdot 10^{-2}$	not measur.	Original material	—
		0	Ideal metal	—
$\theta_{R,(293,78)}$, °K	184	192	—	[110, 113]
R_X, μΩ-cm/kOe	$0.72 \cdot 10^{-3}$	$0.74 \cdot 10^{-3}$	—	[113]

§ 5. Gold

The effect of the periodic potential of the lattice on the optical properties of monovalent metals is considerably less than on those of polyvalent metals, since the Bragg planes do not intersect the sphere of free electrons; it is nevertheless still appreciable, particularly for such metals as silver, gold, and copper, and it is quite interesting to study. Of the Group I metals we studied silver [14] and gold [24].

In the present treatment we shall simply consider the optical properties of gold, as it was only for this metal that we carried out a complex series of investigations into the optical and other properties.

A. Static Characteristics of Gold Layers

The original material for depositing the gold layers was 99.99% pure gold wire. The deposition was conducted by evaporation from several tungsten boats. The other conditions of deposition corresponded to those described in Chapter III. After deposition, the samples were annealed in vacuum of $1 \cdot 10^{-6}$ mm Hg. The annealing temperature was 400°C. The annealing time was over 3 h. Annealing considerably altered both the static and the optical properties of the test layers.* Our gold layers had excellent mirror surfaces of a yellow color.

The static characteristics of the test layers are shown in Table 15. The density of the gold layers was 99.5% of the density of the bulk metal. Allowing for experimental error, we may consider that this value agreed with that of the metal in bulk form.

The static conductivities at room and nitrogen temperatures were measured both for the test layers and for the original gold wire. The conductivities σ_K and σ_N of the original gold appear in the column labelled "bulk metal." The same column presents data relating to the ideal metal obtained by subtracting the small residual resistance from the measured results of very pure gold samples (see Grüneisen [110]). We see from the table that the conductivity of the test layers at room temperature was 90% of the corresponding conductivity of the original gold wire and 83.5% of that of the ideal metal. The conductivity at nitrogen temperature was 62% of the corresponding conductivity of the original gold wire and 49% of that of the ideal metal. We see that, even after annealing, the physical impurities arising from evaporation and condensation of the metal make its conductivity smaller than the conductivity of the original

* The effect of the annealing of the deposited layers on their properties was noted by Kirillova, Noskov, and Charikov [109] and Schulz and Tangherlini [124].

TABLE 16. Optical Constants of Gold at T = 293°K

λ, μ	n	ϰ	λ, μ	n	ϰ
1.00	0.224	6.71	5.0	3.27	35.2
1.50	0.357	10.4	6.0	4.70	41.7
2.00	0.546	13.9	8.0	7.82	54.6
2.50	0.82	17.3	10.0	11.5	67.5
3.0	1.17	21.0	12.0	15.4	80.5
4.0	2.04	27.9	—	—	—

material. The conductivity ratio σ_N/σ_K equals 2.94 for the samples studied, 4.27 for the original wire, and 5.04 for the ideal metal.

The residual resistance R_{res} was measured at T = 4.2°K. On lowering the temperature to 2°K, the resistance remained constant, which evidently indicated the absence of any anomalies in the temperature dependence of the resistance of the samples studied. For the original gold wire R_{res} was not determined. The gold samples used in [110] had a residual resistance of about 0.03% of the resistance at room temperature.

The characteristic temperature θ_R was determined both for the samples studied and for the bulk metal, using the method described earlier (Chapter III, Section 8). In calculating θ_R, we took no account of the small quadratic terms in the temperature dependence of the resistance of gold, since these only started having an appreciable effect at high temperatures. We see from Table 15 that the characteristic temperatures of the test samples were close to those of the bulk metal. For comparison, we may point out that the Debye temperature [52] for the bulk metal θ_D = 165°K, i.e., θ_R and θ_D are close together.

We also studied the Hall effect of the gold test layers. The measurements were made in magnetic fields of $2 \cdot 10^3$–$6 \cdot 10^3$ Oe at current densities of 170-360 A/cm^2 through the sample. We found that for the fields and currents employed, the Hall constant R_X was independent of field and current. This indicates the validity of the approximation based on the assumption of small H. The value of the Hall constant R_X for our test samples was 97% of the corresponding quantity for the bulk metal. Allowing for experimental accuracy,* we may reasonably consider that the Hall constant for the samples studied coincided with that of the bulk material. Subsequently we shall require the quantity $(1/R_X ec)$, which we shall call N_X. For the layers studied, $N_X = (8.7 \pm 0.4) \cdot 10^{22}$ cm^{-3} and for the bulk metal, $N_X = (8.5 \pm 0.1) \cdot 10^{22}$ cm^{-3}.

The results given in Table 15 show that the static characteristics of the layers studied were close to those of the bulk metal. We may therefore consider that our optical constants also represent those of the metal in the bulk state.

B. Optical Constants of Gold

The optical constants of gold were studied in [24]. The optical constants were measured in the spectral range 1-12 μ, using apparatus 1 at room temperature. The results appear in Table 16. The errors obtained for n and ϰ were, respectively, 2-3 and 1% for most of the spectral range and 4-5 and 2-4% at the ends.

The optical constants of gold in the infrared region were studied considerably less intensively than those of copper and silver. The measurements in this part of the spectrum were carried out on samples obtained by vacuum evaporation. Unfortunately, in no case except [24]

*For the bulk metal the spread in R_X values obtained by different authors was about 1%.

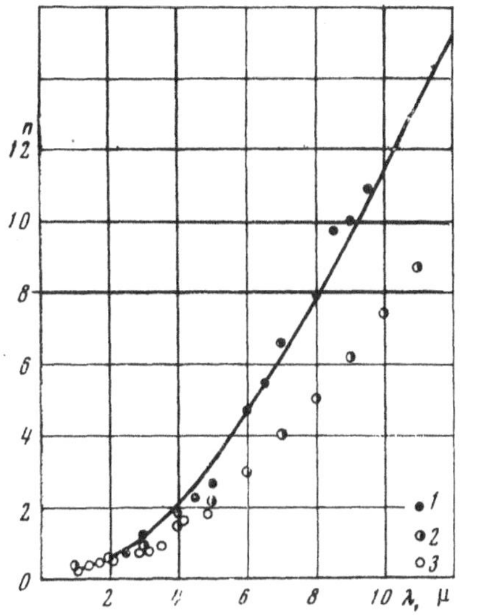

Fig. 42. Dependence of the n of gold on λ at room temperature. The curve represents the results of [24]: 1) [136]; 2) [137]; 3) [138].

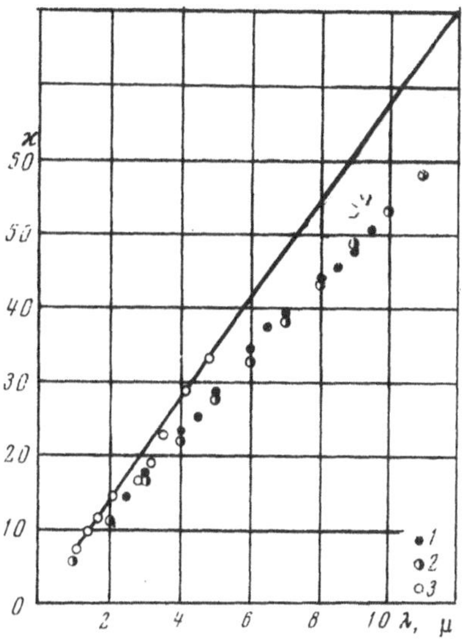

Fig. 43. Dependence of the κ of gold on λ at room temperature. Notation as in Fig. 42.

were the static characteristics of the test layers determined. It is therefore not entirely clear how closely these layers approximated the bulk material.

The measured values of n and κ for gold in the infrared part of the spectrum obtained at room temperature by various authors are shown in Figs. 42 and 43. The continuous line represents our own work [24]. Bolotin et al. [136] used tungsten for the atomizer. The samples were vacuum-annealed at 150°C for 15 h after deposition. Padalka and Shklyarevskii [137] used tantalum for the atomizer, but did not anneal the test layers.

We see from Figs. 42 and 43 that the results of [136, 137] differ from our own data. The difference is particularly sharp for κ. Our own values of κ are much higher, which indicates a greater concentration of conduction electrons. As we shall later see, our measurements give $N = 5.5 \cdot 10^{22}$ cm^{-3}, while those of [136] and [137], respectively, give $3.8 \cdot 10^{22}$ and $3.5 \cdot 10^{22}$ cm^{-3}. Apparently by evaporating the gold from tungsten boats and annealing the samples for a long time in vacuum at high temperatures we obtained samples much closer in their properties to the bulk metal than those of [136, 137]. Particularly unfavorable for producing good gold layers is the use of tantalum boats for evaporation, since molten gold dissolves tantalum.

Apart from the investigations already mentioned, the optical constants of gold in the infrared region were also studied by Hodgson [125]. In this case also, layers obtained by vacuum evaporation were employed, but no annealing was carried out. The results were presented in the form of graphs relating $\log(n\varkappa\omega/2\pi)$ and $\log(\varkappa^2 - n^2 + 1)$ to $\log\lambda$; it is hard to read n and $\varkappa$ from these. The results of this investigation are therefore not shown in Figs. 42 and 43. We may only indicate that the n and $\varkappa$ measured in this investigation were much smaller than the corresponding values obtained in our own measurements.

Cooper, Ehrenreich, and Philipp [100] calculated the real ε_1 and imaginary ε_2 parts of the complex dielectric constant of gold for a wide spectral range. In order to calculate these, data relating to the reflection coefficient of gold at normal incidence, obtained by a number of authors in the spectral range up to $\hbar\omega = 50$ eV, were employed. These data were analyzed by means of the Kramers – Krönig relation (Chapter III, Section 1). The results were presented as two functional relationships, $\varepsilon_1(\hbar\omega)$ and $\varepsilon_2(\hbar\omega)$, giving a qualitative picture of the dispersion of ε_1 and ε_2. The values of n and $\varkappa$ in the range of present interest could not be derived from these curves, and the results of this investigation are therefore also omitted from Figs. 42 and 43. We may simply mention that a value of $N = 0.96N_{val}$ was obtained in [100], quite close to our own value of $N = 0.95N_{val}$.

Figures 42 and 43 show the results of Försterling and Fréedericksz [138] for the optical constants of gold in the near-infrared region up to $\lambda = 4.8\ \mu$. This work was carried out on gold layers obtained by electrodeposition. The static characteristics of the layers were not given. We see from the figures that the values of $\varkappa$ obtained in this investigation were close to the corresponding values obtained in our own work. The values of n differed fairly substantially.

Recently, Shklyarevskii and Yarovaya [139] measured the optical constants of gold in the visible and near-infrared region (up to 1.9 μ). The samples were prepared by evaporation from a tungsten atomizer. After deposition, the samples were annealed in vacuum at 120°C for 10 h. Increasing the annealing temperature to 200°C damaged the samples, thus indicating the inadequacy of the vacuum used in preparation. (The samples were prepared at a pressure of $5 \cdot 10^{-5}$ mm Hg.) The static characteristics of these layers are not known. In the spectral range 1-1.9 μ the values of n and $\varkappa$ for the best samples obtained in the investigation in question approached those of our own work. Unfortunately no tables of n and $\varkappa$ values were given in [139]. The graph presented reproduced the dependence of n and $\varkappa$ on λ on a very small scale and was therefore inconvenient for determining the values of n.

Since the results obtained in [24] relate to layers with properties closely resembling those of the bulk metal, we shall use these results in subsequent analysis of the experimental data.

Figure 44 gives the dependence of R and A on λ and T. We see from this figure that, in the infrared region at T = 293°K, the reflection coefficient $R \approx 99\%$. In the short-wave part of the spectrum studied, R falls and A rises, owing to the effects of interband transitions.

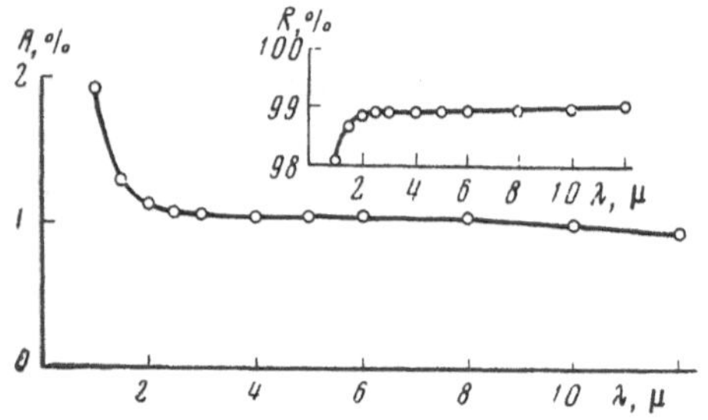

Fig. 44. Dependence of the A and R of gold on λ at room temperature.

CHAPTER V

ANALYSIS OF THE RESULTS OF THE MEASUREMENTS IN THE LONG-WAVE PART OF THE SPECTRAL RANGE STUDIED

§ 1. Analysis of the Experimental Values of n and $\varkappa$ in the Infrared Region for Metals with Different Types of Skin Effect

It has already been mentioned (in Chapter I) that, in the optical range incorporating the visible and infrared parts of the spectrum, we are mainly concerned with the normal or weakly anomalous skin effects. We must therefore first give some detailed consideration to the best methods of analyzing the experimental data corresponding to these cases. At the end of the section we shall also consider the case of the anomalous skin effect.

We shall later show that, for any type of skin effect, measurements of the optical constants n and $\varkappa$ in the infrared part of the spectrum, together with measurements of the density ρ, static conductivity σ_{st}, and ratio of the residual resistance to the resistance at a specified temperature R_{res}/R, enable us to determine the following electron characteristics of the metal: N, $\langle v_F \rangle$, S_F, ν, ν_{ep}, ν_{ep}^{cl}, ν_{ed}, l, and δ.

A. Weakly Anomalous Skin Effect

In the infrared region the optical constants are determined by the conduction electrons, and usually the inequality $|\varepsilon'| \ll 1$ is satisfied. Hence the terms $1/|\varepsilon'|$ may be neglected. Using Eqs. (I.36), (I.51), and (I.54), and separating the real and imaginary parts, we obtain [19]

$$N = \frac{0.1115 \cdot 10^{22}}{\lambda^2} \varkappa^2 \frac{(1 + n^2/\varkappa^2)^2}{1 - n^2/\varkappa^2} \frac{1}{1 - \beta_1}, \tag{V.1}$$

$$\nu = \frac{3.767 \cdot 10^{15}}{\lambda} \frac{n/\varkappa}{1 - n^2/\varkappa^2} \frac{1 - \beta_2}{1 - \beta_1}, \tag{V.2}$$

$$\beta_1 = \frac{3}{8} \frac{\langle v_F \rangle}{c} \varkappa \frac{(1 + n^2/\varkappa^2)(\nu/\omega - n/\varkappa)}{(1 + \nu^2/\omega^2)(1 - n^2/\varkappa^2)} = 0.1251 \cdot 10^{-10} \langle v_F \rangle \varkappa \frac{(1 + n^2/\varkappa^2)(\nu/\omega - n/\varkappa)}{(1 - n^2/\varkappa^2)(1 + \nu^2/\omega^2)}, \tag{V.3}$$

$$\beta_2 = \frac{3}{16} \frac{\langle v_F \rangle}{c} \varkappa \frac{(1 + n^2/\varkappa^2)}{n/\varkappa} \frac{[1 + (n/\varkappa)(\nu/\omega)]}{(1 + \nu^2/\omega^2)} = 0.06254 \cdot 10^{-10} \langle v_F \rangle \varkappa \frac{(1 + n^2/\varkappa^2)}{n/\varkappa} \frac{[1 + (n/\varkappa)(\nu/\omega)]}{(1 + \nu^2/\omega^2)}, \tag{V.4}$$

$$\omega = 1.884 \cdot 10^{15}/\lambda. \tag{V.5}$$

Here N is the concentration of conduction electrons in cm^{-3}; ν is the effective electron collision frequency in sec^{-1}; λ is the light wavelength in μ; ω is the cyclical frequency of the light in sec^{-1}; $\langle v_F \rangle$ is the mean velocity of the electrons on the Fermi surface in cm/sec; c is the velocity of light in cm/sec. The corrections β_1 and β_2 allow for the character of the skin effect. Equations (V.3) and (V.4) are actually calculated for a spherical Fermi surface. The slight inaccuracy associated with this fact is, however, unimportant, since usually β_2 and β_2 are small and may be calculated to a lower accuracy. We note that usually β_2 is much greater than β_1, i.e., the corrections are more important for ν than for N.

In calculating the corrections β_1 and β_2, we must know $\langle v_F \rangle$. This quantity has to be obtained from the same optical measurements. For this purpose we may use Eq. (II.33a)

$$\langle v_F \rangle / v_F^0 \approx \sqrt{N/N_{val}}, \tag{V.6}$$

where v_F^0 is the velocity of the free electrons for the concentration N_{val}; N_{val} is the concentration of the valence electrons.

We shall later show that the value of $\langle v_F \rangle$ calculated by this approximate formula differs from the more accurate calculation of (II.31)-(II.31c) by 3-10%; this has hardly any effect on the determination of N and ν from (V.1) and (V.2).

Equations (V.1)-(V.6) enable us to determine the microcharacteristics of the metal under consideration. For this purpose we must know N_{val} and v_F^0 as well as the optical constants n and $\varkappa$. In order to determine N_{val} we must know the density of the test layers. The value of v_F^0 is to be taken for the bulk metal. We note that, since the values of N_{val} and v_F^0 are only used to determine the $\langle v_F \rangle$ in the equations for β_1 and β_2, the values of N and ν are not very sensitive toward N_{val} and v_F^0. Hence these latter two quantities may be determined to a lower accuracy.

The calculations based on Eqs. (V.1)-(V.6) may be carried out by successive approximations. As the zero approximation we put $\beta_1 = \beta_2 = 0$. Using Eqs. (V.1) and (V.2), we find the N and ν to the zero approximation.

First we determine $\langle v_F \rangle$, β_1, and β_2 to the first approximation from (V.6), (V.3), and (V.4) by using the values of N and ν derived in the zero approximation. Then we use these values of β_1 and β_2, and find N and ν to the first approximation from Eqs. (V.1) and (V.2), and so on. This process converges rapidly. Thus, for example, in order to obtain an accuracy of 0.1% in solving Eqs. (V.1)-(V.6), two or three approximations are usually sufficient. This method of solving the system of equations enables us very simply to set up a program for analyzing experimental data on a computer.

The determination of N enables us to calculate not only $\langle v_F \rangle$ but also the total area of the Fermi surface from the equation

$$S_F/S_F^0 \approx \sqrt{N/N_{val}}. \tag{V.6a}$$

It should be noted that the value of N calculated from (V.1) only has the sense of the concentration of the conduction electrons when it is determined for a spectral region in which the effect of interband transitions may be neglected. In this region N does not depend on the wavelength. Usually this applies to the infrared region. The quantities ν, β_1/λ^2, and β_2 become constant in approximately the same region as N.*

In order to determine the electron–phonon collision frequency ν_{ep}, the classical electron — phonon collision frequency ν_{ep}^{cl}, the frequency of the collisions between electrons and impurities or lattice defects ν_{ed}, the mean free path l, the depth of the skin layer δ, and the mean diameter of the metal crystals L, it is convenient to use the following equations [19]:

$$\nu = \nu_{ep} + \nu_{ed}, \tag{V.7}$$

$$\nu_{ep}^{cl} + \nu_{ed} = \frac{e^2}{m}\frac{N}{\sigma_{st}} = 2.533\cdot 10^8 \frac{N}{\sigma_{st}}, \tag{V.8}$$

$$\nu_{ed}/(\nu_{ep}^{cl} + \nu_{ed}) = R_{res}/R, \tag{V.9}$$

$$l = \langle v_F \rangle / \nu, \tag{V.10}$$

*The dependence of β_1 on λ may be obtained from Eq. (V.3). Usually, $n^2/\varkappa^2 \ll 1$, so that $(1 + n^2/\varkappa^2)/(1 - n^2/\varkappa^2) \approx 1$. We may then approximately consider that $\varkappa \sim \lambda$, $n \sim \lambda^2$, whence $\lambda \sim (\nu/\omega - n/\varkappa)$. For metals at low temperatures and metals of the first group, $\nu^2/\omega^2 \ll 1$, and we may approximately put $1 + \nu^2/\omega^2 \approx 1$. It thus follows that $\beta_1 \sim \lambda^2$. For polyvalent metals at room temperature, in the spectral range in which $\nu^2 \geq \omega^2$, a more complicated relationship is obtained.

Carrying out an analogous estimation of the dependence of β_2 on λ by Eq. (V.4), we find that $\beta_2 \approx$ const for $\nu^2/\omega^2 \ll 1$.

$$\delta = \lambda/(2\pi\varkappa), \tag{V.11}$$

$$L = \langle v_F \rangle / \nu_{ed}. \tag{V.12}$$

Here e and m are the charge and mass of a free electron, σ_{st} is the static conductivity at temperature T in cgse units, R_{res}/R is the ratio of the residual resistance to the resistance at T.

In Eqs. (V.7)-(V.9) we take no account of the interelectron collision frequency ν_{ee}, since this is much smaller than ν for the metals studied [18-20, 22, 23, 50, 51].

B. Normal Skin Effect

For the normal skin effect we must put $\beta_1 = \beta_2 = 0$, and we may then use the foregoing equations. However, for the normal skin effect we may find cases in which $1/|\varepsilon'|$, while remaining smaller than unity, nevertheless becomes large enough to warrant consideration. Under these conditions we must also allow for terms of the order of $\sin^2\varphi/|\varepsilon'|$ (φ is the angle of incidence of the light on the surface of the metal), which leads to a more complicated relationship between n and $\varkappa$, on the one hand, and the phase shift Δ and azimuth ρ on the other [(III.2)-(III.5)]. Subsequently we shall use n and $\varkappa$ to denote the optical constants obtained with due allowance for the terms of order $\sin^2\varphi/|\varepsilon'|$.

Analysis of the experimental data relating to the case of the normal skin effect may reasonably be based on Eqs. (I.36)-(I.38). Here these equations are presented in a form convenient for calculating the microcharacteristics of the metal:

$$N = \frac{0.1115}{\lambda^2} \cdot 10^{22}(1 + \varkappa^2 - n^2)\left[1 + \frac{4n^2\varkappa^2}{(1 + \varkappa^2 - n^2)^2}\right], \tag{V.13}$$

$$\nu = \frac{18.84}{\lambda} \cdot 10^{14} \frac{2n\varkappa}{1 + \varkappa^2 - n^2}. \tag{V.14}$$

The determination of N and ν enables us to calculate the microcharacteristics: $\langle v_F \rangle$, S_F, ν_{ep}, ν_{ep}^{cl}, ν_{ed}, l, δ, and L from (V.6)-(V.12).

C. Anomalous Skin Effect

As already indicated in Chapter I, for the optical range the anomalous skin effect only occurs in monovalent metals at low (helium) temperatures. In these cases, first, the Fermi surface is almost spherical and, second, the inequalities ν/ω, $\langle v_F \rangle/(\omega\delta) \ll 1$ are satisfied. This enables us to carry out all the calculations for a spherical Fermi surface and to use an expansion in two parameters [ν/ω and $\langle v_F \rangle/(\omega\delta)$]. Thus, in the long-wave region in which the frequency of the light is smaller than the threshold frequency of the interband transitions, we may use Eqs. (I.15)-(I.17) and (V.6), which we re-write in a form convenient for calculation of the microcharacteristics:

$$N = \frac{1.79}{\lambda^2} \cdot 10^{22} \frac{(1 + P^2G)^2}{\left(\frac{4\varkappa}{n^2 + \varkappa^2}\right)^2}, \tag{V.15}$$

$$v_F = v_F^0 \sqrt{N/N_{val}},$$

$$\nu = 2.82 \cdot 10^2 \sqrt{N}\left(\frac{4n}{n^2 + \varkappa^2} 10^2 - 0.250 \cdot 10^{-8} v_F + 0.334 \cdot 10^{-8} v_F P^2 D\right), \tag{V.16}$$

$$P = 0.861 \cdot 10^{-21} \sqrt{N} v_F \lambda, \tag{V.17}$$

$$q = 6.13 \cdot 10^5 \frac{\nu}{v_F \sqrt{N}}, \tag{V.18}$$

$$G = 0.0865 + 0.216q + 0.125q^2, \tag{V.19}$$

$$D = 0.215 + 0.748q + 0.750q^2 + 0.216q^3. \tag{V.20}$$

The microcharacteristics are calculated from the measured optical constants by successive approximations. In the zero approximation we put P = 0 in (V.15) and find N. Then we determine v_F from (V.6). In the same approximation (with P = 0), we obtain ν from (V.16), after which we calculate P, q, G, and D from (V.17)-(V.20). Using the values of P and G, we determine N in the first approximation. Then we find v_F and ν in the first approximation by using (V.6) and (V.16), and so on. In this way we obtain N, v_F, and ν. Then we use the earlier scheme to calculate the remaining microcharacteristics from (V.6a), (V.7)-(V.12).

§ 2. Application of the Foregoing Scheme of Analyzing the Experimental Values of n and $\varkappa$ for the Weakly Anomalous Skin Effect to the Optical Constants of the Metals Studied in the Present Investigation

We shall now proceed to analyze our own experimental data for a number of metals. As we shall shortly see, the weakly anomalous skin effect holds for all these metals which we studied. The calculation of the microcharacteristics is accordingly based on Eqs. (V.1)-(V.6).*

In the case of gold, for which $\nu/\omega \ll 1$ over the whole spectral range studied, the experimental results were also analyzed by the formulas of the anomalous skin effect, i.e., (V.15)-(V.20). The results obtained by the two methods agreed closely.

In all cases the calculations were carried out by computer, using the method of successive approximations just described. The number of approximations was chosen so as to give an accuracy of 0.1% in the calculations of N and ν (usually 2-4).

In order to determine N, ν, $\langle v_F \rangle$, β_1, and β_2, we have to know the values of N_{val} and v_F^0 as well as the optical constants. In calculating N_{val} we allowed for the density of our samples. As v_F^0 we took the value for the bulk metal. It was pointed out earlier that the values of N and ν were not very sensitive to N_{val} and v_F^0. Thus, for example, a change of 10% in N_{val} or v_F^0 has hardly any effect on N, and changes ν by only 1-2%, which is smaller than the error committed in determining this quantity. Naturally, in the determination of $\langle v_F \rangle$, β_1, and β_2 a change in N_{val} and v_F^0 has a much greater effect. Thus, on changing N_{val} by 10%, the values of $\langle v_F \rangle$, β_1, and β_2 alter by about 5%. On changing v_F^0 by 10% they alter by about 10%.

The results of our calculations of the microcharacteristics of the test metals are presented in the corresponding tables. After carrying out the calculations, we verified the character of the skin effect on every occasion for the metal under consideration. For this purpose we considered the corrections β_1 and β_2 associated with the anomalous skin effect and also calculated the complex expansion parameter 2α used in obtaining Eqs. (I.51), (I.52), and (I.54). According to the latter equation,

$$\begin{aligned} 2\alpha &= \frac{3}{8}\,\frac{\langle v_F \rangle}{c}\,\varkappa\,\frac{n/\varkappa - i}{1 - i\nu/\omega} = a - ib, \\ a &= \frac{3}{8}\,\frac{\langle v_F \rangle}{c}\,\varkappa\,\frac{n/\varkappa + \nu/\omega}{1 + (\nu/\omega)^2}, \\ b &= \frac{3}{8}\,\frac{\langle v_F \rangle}{c}\,\varkappa\,\frac{1 - (n/\varkappa)(\nu/\omega)}{1 + (\nu/\omega)^2} \end{aligned} \tag{V.21}$$

*In [17, 18, 23, 24, 37, 38] we used the formulas for the weakly anomalous skin effect. The value of $\langle v_F \rangle$ was determined, however, by a method less accurate than that used in the present investigation.

Table 17 shows the coefficients β_1, β_2, a, and b. The same table gives the expansion parameters ν/ω and $\langle v_F\rangle/(\omega\delta)$, used in the well-known formulas for the anomalous skin effect.

It is quite clear from the table that for all the polyvalent metals β_1 and $\beta_2 \ll 1$, which indicates the applicability of Eqs. (V.1)-(V.6). This is confirmed by the smallness of the complex expansion parameter 2α. The neglect of second-order terms gives an error of $\leqslant 1\%$. Hence the accuracy of the microcharacteristics given below is determined by the accuracy of the measurements of n and $\varkappa$, and not by that of the equations employed. An exception is the quantity $\langle v_F\rangle$. For this we use the approximate formulas, the accuracy of which is 5-10%, as indicated in Chapter VII.

The coefficients β_1 and β_2 determine the influence of the degree of anomaly of the skin effect on N and ν, respectively. In the majority of cases $\beta_1 \ll \beta_2$, i.e., the anomalous nature of the skin effect has a much greater influence on ν than on N.

We see from Table 17 that the use of the equations of the normal skin effect in the infrared region leads to serious errors, particularly when analyzing the low-temperature measurements. At helium temperature, in fact, the coefficient β_2 reached 0.27 for indium, 0.29 for lead, and 0.23 for tin. For aluminum even at nitrogen temperature $\beta_2 \approx 0.26$. At room temperature β_2 is much smaller than at helium temperature, but even in this case it approximately equals 0.08 for indium, 0.17 for aluminum, 0.03 for lead, and 0.06 for tin. The coefficient β_1 reached 0.04 at room temperature for indium and tin and 0.03 for aluminum and lead. On reducing the temperature this coefficient diminishes.

Let us consider the expansion parameters ν/ω and $\langle v_F\rangle/(\omega\delta)$. We see from Table 17 that these are larger than the analogous expansion parameters for the equations of the weakly anomalous skin effect. Then the ratio $\langle v_F\rangle/(\omega\delta)$ remains much smaller than unity at all temperatures. The parameter ν/ω becomes of the order of unity for indium, tin, and lead at room temperature in the spectral range 6-12 μ, so that none of the methods of calculating the microcharacteristics using ν/ω or ω/ν as expansion parameters may be used. The well-known equations for the anomalous skin effect (V.15)-(V.20), in particular, are unsuitable.

Thus the most suitable equations for analyzing the experimental results in the case of indium, aluminum, lead, and tin are those relating to the weakly anomalous skin effect.

For gold at room temperature, β_2 reached 0.37. The equations of the weakly anomalous skin effect may still be used for this. Reducing the temperature usually leads to a considerable rise in β_2. Hence, in analyzing experimental results at helium temperatures it may be better to use the equations relating to the anomalous skin effect. In the present investigation we only measured the optical constants of gold at room temperature. Since $\nu/\omega \ll 1$, we may therefore analyze the experimental data equally well by using Eqs. (V.1)-(V.6) or (V.15)-(V.20). We have already indicated that the two methods gave almost identical values for the microcharacteristics of gold.

The smallness of the complex expansion parameter 2α is one of the necessary conditions for determining the microcharacteristics relating to the conduction electrons from Eqs. (V.1)-(V.6). Another necessary condition is that interband transitions should not have any effect on the optical constants. On satisfying this condition, the values of N and ν should be independent of λ. For each metal we chose a spectral range satisfying this condition. For this range we calculated the average values of N, ν, $\langle v_F\rangle$, β_1/λ^2, and β_2. The last two quantities were used in estimating the contribution of the conduction electrons to ε_1 and σ in the visible and near-infrared spectral ranges.

The errors in determining the principal microcharacteristics of the metals are given in Table 18. The accuracy of N_a is related to the accuracy of the density measurement.

TABLE 17. Coefficients β_1 and β_2 Allowing for the Character of the Skin Effect and Expansion Parameters $2\alpha = a - ib$, ν/ω, and $\langle v_F\rangle/(\omega\delta)$ for the Metals Studied

λ, μ	$\beta_1\cdot10^2$	$\beta_2\cdot10^2$	$a\cdot10^2$	$b\cdot10^2$	$\frac{\nu}{\omega}\cdot10^2$	$\frac{\langle v_F\rangle}{\omega\delta}\cdot10^2$
			In, T=295° K			
10	4.1	7.2	6.0	1.5	120	23
8	2.9	8.0	5.4	2.6	92	20
6	1.9	8.6	4.3	3.2	70	17
4	1.1	7.2	2.9	3.0	55	12
2	0.47	6.0	1.3	2.0	40	7
			In, T=4.2° K			
10	1.3	21.9	4.9	7.8	36	25
8	0.59	25.4	2.7	6.9	23	20
6	0.30	26.6	1.5	5.5	16	15
4	0.19	23.7	0.9	4.0	13	11
2	0.26	9.7	0.8	2.3	23	7
			Al, T=295° K, unannealed layers			
9	3.1	15.1	7.6	5.8	68	29
7	1.6	17.2	5.0	6.4	44	23
5	0.78	17.5	2.8	5.4	30	17
3	0.40	14.6	1.4	3.6	24	11
1.2	0.14	8.2	0.5	1.6	18	4
			Al, T=78° K, unannealed layers			
7	0.84	24.5	3.6	7.5	28	23
5	0.35	26.5	1.7	5.8	17	16
3	0.22	21.0	0.9	3.8	15	11
1,2	0.08	12.4	0.3	1.6	12	4
			Pb, T=293° K			
12	3.0	2.6	2.7	−0.2	194	13
10	2.6	2.9	2.8	0.1	164	13
8	2.1	3.3	2.9	0.6	126	11
6	1.5	3.6	2.6	1.1	98	10
4	0.80	3.8	1.8	1.4	68	7
2	0.28	3.3	0.7	1.1	42	4
1.5	0.19	2.5	0.5	0.8	41	3
			Pb, T=78° K			
12	1.6	14.9	4.6	5.6	48	21
10	1.2	14.6	3.7	5.2	41	18
8	0.85	13.2	2.7	4.4	37	14
6	0.52	13.0	1.7	3.7	29	11
4	0.23	12.8	0.8	2.6	19	7
2	0.08	9.8	0.3	1.3	12	4
1.5	0.06	6.8	0.2	1.0	13	3
			Pb, T=4.2° K			
12	0.50	28.9	2.6	7.5	20	21
10	0.37	28,0	1.9	6.4	18	18
8	0.25	27.9	1.3	5.3	14	15
6	0.19	24.4	0.9	4.0	13	11
4	0.10	21.8	0.5	2.7	10	7
2	0.03	20.9	0.1	1.4	5	4
1.5	0.02	14.2	0.1	0.9	6	2

Table 17 (Continued)

λ, μ	$\beta_1 \cdot 10^2$	$\beta_2 \cdot 10^2$	$a \cdot 10^2$	$b \cdot 10^2$	$\frac{\nu}{\omega} \cdot 10^2$	$\frac{\langle v_F \rangle}{\omega\delta} \cdot 10^2$
			Sn, $T = 293°$ K			
12	3.8	4.7	4.5	0.4	150	19
10	2.9	5.2	4.3	1.1	120	17
8	2.2	5.7	3.9	1.8	94	15
6	1.4	6.1	3.1	2.2	71	12
4	0.77	5.5	1.9	2.0	55	8
2	0.40	2.8	1.0	1.0	55	4
1.5	0.48	1.4	0.9	0.5	88	3
			Sn, $T = 78°$ K			
12	1.9	13.4	5.1	5.1	55	21
10	1.2	15.3	3.7	5.4	40	18
8	0.73	17.0	2.6	5.2	30	16
6	0.44	15.8	1.6	4.0	24	12
4	0.22	15.0	0.8	2.9	18	8
2	0.18	4.7	0.5	1.2	28	4
1.5	0.25	1.7	0.6	0.6	55	3
			Sn, $T = 4.2°$ K			
12	1.4	18.1	4.5	6.6	40	22
10	0.79	21.6	3.1	6.5	28	20
8	0.43	23.2	1.9	5.5	20	16
6	0.27	21.4	1.2	4.3	17	12
4	0.18	17.3	0.7	2.9	15	8
2	0.15	5.6	0.3	1.3	23	4
1.5	0.23	1.7	0.6	0.6	54	2
			Au, $T = 293°$ K			
12	0.87	36.6	5.7	12.3	25	37
10	0.72	34.0	4.4	10.4	23	30
8	0.51	32.5	3.0	8.6	20	25
6	0.31	31.6	1.9	6.8	16	19
4	0.13	32.2	0.8	4.6	10	12
2	0.04	29.5	0.2	2.3	6	6

TABLE 18. Errors in Determining the Microcharacteristics of the Metals Studied (in %)

Characteristic	In		Al, unannealed layers		Pb			Sn			Au
	295° K	4.2° K	295° K	78° K	293° K	78° K	4.2° K	293° K	78° K	4.2° K	293° K
N_a	4	4	3	3	3	3	3	5	5	5	4
N	1	2	2	2	1	1	1	1	1	1	1
$\langle v_F \rangle$	5–10	5–10	5–10	5–10	5–10	5–10	5–10	5 10	5–10	5–10	5–10
ν	3	5–6	4	6	1	3	4	1	3	3	6
β_1/λ^2	14	14	10	10	30	10	16	10	10	10	8
β_2	5–6	5–6	5	5	10	6	10	6–7	6–7	6–7	4

The mean velocity of the electrons on the Fermi surface $\langle v_F \rangle$ was calculated by Eq. (V.6). As already indicated, the accuracy of the determination of this quantity was mainly determined by the accuracy of the equation in question.

The error inherent in N and ν is related to the errors committed in measuring n and $\varkappa$, and may readily be calculated from the formulas

$$\Delta N = \frac{\partial N}{\partial n}\Delta n + \frac{\partial N}{\partial \varkappa}\Delta\varkappa, \tag{V.22}$$

$$\Delta\nu = \frac{\partial\nu}{\partial n}\Delta n + \frac{\partial\nu}{\partial\varkappa}\Delta\varkappa. \tag{V.23}$$

Equations (V.1) and (V.2) clearly indicate that in the infrared region N is mainly determined by $\varkappa$ and ν by $n/\varkappa$. Hence in calculating ΔN and $\Delta\nu$ we may assume that $N \sim \varkappa^2$ and $\nu \sim n/\varkappa$. Hence we obtain

$$\Delta N/N \approx 2\Delta\varkappa/\varkappa, \tag{V.24}$$

$$\Delta\nu/\nu \approx \Delta n/n + \Delta\varkappa/\varkappa. \tag{V.25}$$

Since usually $\Delta n/n > \Delta\varkappa/\varkappa$, we may say that $\Delta\nu/\nu > \Delta N/N$. The error in question affects the calculation of N and ν in respect of one pair of values of n and $\varkappa$, relating to one value of λ. In the present investigation we took the average values over the spectral range (determined in the manner indicated above) for N and ν. In this case, ΔN and $\Delta\nu$ were found from the scatter in the values relating to different wavelengths. In the same way we determined the error for β_1/λ^2 and β_2. It should be noted that in the ranges indicated β_2 was almost independent of λ, while β_1/λ^2 was only approximately constant. This follows from Eq. (V.3), and is in fact observed experimentally. The greatest value of β_1/λ^2 with varying λ occurs at room temperature, for which ν reaches a maximum. For these reasons the average value of β_1/λ^2 can only be determined to a relatively low accuracy, the error rising as high as 30%. However, even this error has no marked effect on our calculations of the contribution made by the conduction electrons to ε_1 and σ in the visible and near-infrared parts of the spectrum, owing to the smallness of β_1 in these spectral regions.

After determining N and ν, we calculated the frequencies ν_{ep}, ν_{ep}^{cl}, and ν_{ed} from Eqs. (V.7)-(V.9). These calculations, which required measurements of the static conductivity σ_{st} and the ratio R_{res}/R, were carried out for indium, lead, tin, and gold. For aluminum the ratio R_{res}/R of the test layers was not measured. However, the value of this ratio was approximately estimated by using the measured values of the conductivities $\sigma_{st}(T_K)$ and $\sigma_{st}(T_N)$ and the tabulated conductivities of the bulk (mass) metal $\sigma_m(T_K)$ and $\sigma_m(T_N)$, after which the frequencies in question were also calculated for aluminum, although to a lower accuracy.

In should be noted that the change in N with temperature leads to a corresponding change in ν_{ed}. However, since $\nu_{ed} \ll \nu$, the slight inaccuracy associated with the fact that Eq. (V.9) contains $\nu_{ed}(T_{He})$ rather than $\nu_{ed}(T)$ is none too serious. In cases in which the temperature dependence of N was substantial (indium and tin) the value of ν_{ed} was so much smaller than ν that the optical properties of the samples were hardly affected at all.

The determination of $\langle v_F \rangle$ and ν enabled us to calculate the range of the electrons l from Eq. (V.10). For the long-wave part of the spectral range studied, the quantity so calculated was almost independent of the wavelength, as might be expected.

For all the metals studied we used Eq. (V.11) to find the depth of the skin layer δ, which varied slightly with the wavelength of the incident light. Comparison of l with δ showed that for polyvalent metals $l < \delta$; however, at low temperatures the ratio l/δ tended to approach unity. For gold, l was slightly greater than δ.

TABLE 19. Results of an Analysis of the Optical Constants of Indium, Using the Equations of the Weakly Anomalous Skin Effect ($N_{val} = 10.3 \cdot 10^{22}$ cm^{-3}; $v_F^0 = 1.74 \cdot 10^8$ cm/sec)

λ, μ	295° K			4.2° K		
	$N \cdot 10^{-22}$, cm^{-3}	$\nu \cdot 10^{-14}$, sec^{-1}	$\langle v_F \rangle \cdot 10^{-8}$, cm/sec	$N \cdot 10^{-22}$, cm^{-3}	$\nu \cdot 10^{-14}$, sec^{-1}	$\langle v_F \rangle \cdot 10^{-8}$, cm/sec
10.0	6.12	2.26	1.34	5.04	0.678	1.22
8.0	5.99	2.17	1.33	4.79	0.535	1.19
6.0	6.07	2.19	1.34	4.82	0.507	1.19
5.0	6.17	2.34	1.35	4.91	0.535	1.20
4.0	6.26	2.58	1.36	5.09	0.624	1.22
3.5	6.33	2.68	1.36	5.32	0.728	1.25
3.00	6.34	2.77	1.36	5.56	0.941	1.28
2.60	6.41	3.11	1.37	5.69	1.33	1.29
2.50	6.39	3.21	1.37	5.68	1.49	1.29
2.40	6.39	3.34	1.37	5.71	1.67	1.30
2.30	6.45	3.45	1.38	5.80	1.84	1.31
2.20	6.44	3.58	1.38	5.90	1.99	1.32
2.10	6.51	3.70	1.38	5.99	2.14	1.33
2.00	6.67	3.80	1.40	6.25	2.18	1.36
1.90	6.68	3.97	1.40	6.23	2.32	1.35
1.80	6.69	4.17	1.40	6.30	2.43	1.36
1.70	6.81	4.37	1.42	6.43	2.55	1.37
1.60	6.84	4.61	1.42	6.47	2.74	1.38
1.50	6.77	4.98	1.41	6.19	3.03	1.35

Since we used pure metals for preparing the test layers (99.99% or better), the measured electron–impurity collision frequency ν_{ed} was mainly determined by the crystal boundaries. This enabled us to calculate the mean crystal size L of the metal layers from (V.12). In every case we found that $L \gg l, \delta$, i.e., the test layers consisted of large crystals. For lead the crystal size L coincided with the layer thickness d, i.e., the crystal "grew through" the whole thickness of the deposited layer.

In deriving Eqs. (V.1) and (V.2) we neglected terms of the order of $1/|\varepsilon'|$ compared with unity. For all the metals studied the value of $1/|\varepsilon'|$ was of the order of 10^{-3}-10^{-4} over the spectral range in which N and ν were measured. Neglect of this quantity introduced hardly any error.

The values found for l and δ enabled us to compare these quantities with the thickness of the test layers d. For all the metals examined, $l, \delta < 4 \cdot 10^{-6}$ cm whereas $d > 30 \cdot 10^{-6}$ cm. Thus, $d \gg l, \delta$, confirming the "bulk" nature of our samples in relation to optical measurements.

§ 3. Results of the Analysis of the Experimental Data in the Infrared Region

The results of our analysis of the experimental data in the infrared region are presented in Tables 19-25. The captions of the tables indicate the values of N_{val} and v_F^0 used in calculating the corrections associated with the degree of anomaly of the skin effect. In calculating N_{val} we remembered that for lead, tin, gold, and annealed aluminum the density of the layers practically coincided with that of the bulk metal. For indium* and unannealed aluminum the value

*In our earlier paper [19] we used the value of N_{val} corresponding to the bulk metal for indium. This had hardly any effect on N and ν, but caused an error of roughly 5% in $\langle v_F \rangle$, β_1, and β_2.

TABLE 20. Results of an Analysis of the Optical Constants of Unannealed Aluminum Layers, Using the Equations of the Weakly Anomalous Skin Effect ($N_{val} = 16.1 \cdot 10^{22}$ cm^{-3}; $v_F^0 = 2.02 \cdot 10^8$ cm/sec)

λ, μ	295°K			78°K		
	$N \cdot 10^{-22}$, cm^{-3}	$\nu \cdot 10^{-14}$, sec^{-1}	$\langle v_F \rangle \cdot 10^{-8}$, cm/sec	$N \cdot 10^{-22}$, cm^{-3}	$\nu \cdot 10^{-14}$, sec^{-1}	$\langle v_F \rangle \cdot 10^{-8}$, cm/sec
9	7.71	1.41	1.40	—	—	—
8	7.43	1.14	1.37	—	—	—
7	7.21	1.19	1.35	6.80	0.740	1.31
6	7.08	1.18	1.34	6.69	0.697	1.30
5	6.99	1.15	1.33	6.66	0.656	1.30
4	6.81	1.32	1.31	6.55	0.784	1.29
3	7.26	1.49	1.36	7.04	0.939	1.34
2.5	7.33	1.60	1.36	7.16	1.05	1.35
2	7.50	1.81	1.38	7.38	1.27	1.37
1.5	7.46	2.13	1.38	7.35	1.37	1.37
1.2	7.36	2.88	1.37	7.24	1.81	1.35

TABLE 21. Results of an Analysis of the Optical Constants of Annealed Aluminum Layers, Using the Equations of the Weakly Anomalous Skin Effect ($N_{val} = 18.1 \cdot 10^{22}$ cm^{-3}; $v_F^0 = 2.02 \cdot 10^8$ cm/sec, T = 293°K)

λ, μ	$N \cdot 10^{-22}$, cm^{-3}	$\nu \cdot 10^{-14}$, sec^{-1}	$\langle v_F \rangle \cdot 10^{-8}$, cm/sec	λ, μ	$N \cdot 10^{-22}$, cm^{-3}	$\nu \cdot 10^{-14}$, cm^{-1}	$\langle v_F \rangle \cdot 10^{-8}$, cm/sec
10	7.21	1.79	1.27	3	6.82	1.77	1.24
8	6.45	1.45	1.21	2.5	6.69	1.83	1.23
6	6.46	1.51	1.21	2	6.69	1.88	1.23
5	6.63	1.54	1.22	1.5	6.87	2.27	1.24
4	6.75	1.65	1.23	1	6.86	4.89	1.24

was 90 and 89% of the density of the bulk metal, respectively. For v_F^0 we in all cases used the value corresponding to the bulk metal.

Let us determine the spectral ranges for each metal in which the role of the interband transitions is slight, so that we may calculate the characteristics of the conduction electrons from (V.1)-(V.6). For this purpose we use Figs. 45-56.

We see from Figs. 45-56 that for indium, lead, tin, and gold these intervals are easily determined as the regions in which N and ν are independent of λ. For aluminum the interval in question is less reliably determined than for the other metals. We see from Figs. 47-50 that the N and ν calculated from (V.1)-(V.5) vary slightly throughout the whole spectral range employed. This applies to both annealed and unannealed layers. It will be shown in Chapter VI that, in the near-infrared region, aluminum has a band of interband transitions associated with the pseudopotential Fourier component V_{111}. This is a weak band, and we may reasonably consider that its influence only extends to about 4 μ. Hence we may take the range 4-7 μ as the required interval for both temperatures in the case of unannealed layers. We specially chose the same interval for both temperatures, so as to be able to establish the temperature dependence of N and ν for aluminum more accurately. For the annealed layers the corresponding interval is 4-8 μ.

TABLE 22. Results of an Analysis of the Optical Constants of Lead, Using the Equations of the Weakly Anomalous Skin Effect (N_{val} = 13.2 · 10^{22} cm^{-3}; v_F^0 = 1.82 · 10^8 cm/sec)

λ, μ	293°K $N \cdot 10^{-22}$, cm^{-3}	293°K $\nu \cdot 10^{-14}$, sec^{-1}	293°K $\langle v_F \rangle \times 10^{-8}$, cm/sec	78°K $N \cdot 10^{-22}$, cm^{-3}	78°K $\nu \cdot 10^{-14}$, sec^{-1}	78°K $\langle v_F \rangle \times 10^{-8}$, cm/sec	4.2°K $N \cdot 10^{-22}$, cm^{-3}	4.2°K $\nu \cdot 10^{-14}$, sec^{-1}	4.2°K $\langle v_F \rangle \times 10^{-8}$, cm/sec
12.0	3.89	3.04	0.99	3.86	0.746	0.98	3.64	0.316	0.96
11.0	4.09	3.13	1.01	3.90	0.768	0.99	3.66	0.322	0.96
10.0	4.05	3.08	1.01	3.89	0.776	0.99	3.65	0.331	0.96
9.0	4.02	2.98	1.00	3.84	0.816	0.98	3.67	0.338	0.96
8.0	4.04	2.98	1.01	3.82	0.863	0.98	3.72	0.340	0.97
7.0	4.08	3.05	1.01	3.80	0.907	0.98	3.73	0.380	0.97
6.0	4.13	3.09	1.02	3.86	0.900	0.98	3.75	0.411	0.97
5.0	4.11	3.08	1.02	3.91	0.936	0.99	3.72	0.439	0.97
4.0	4.09	3.21	1.01	3.79	0.901	0.98	3.77	0.478	0.97
3.5	4.07	3.30	1.01	3.91	0.936	0.99	3.71	0.460	0.97
3.0	4.10	3.39	1.01	3.80	0.987	0.98	3.73	0.458	0.97
2.6	4.10	3.55	1.01	3.71	1.04	0.97	3.76	0.460	0.97
2.55	4.07	3.57	1.01	—	—	—	3.75	0.467	0.97
2.5	4.06	3.57	1.01	3.71	0.983	0.97	3.80	0.474	0.98
2.4	3.96	3.69	1.00	3.72	1.01	0.97	3.79	0.476	0.97
2.3	4.01	3.70	1.00	3.71	1.06	0.96	3.74	0.472	0.97
2.2	3.95	3.75	1.00	3.69	1.11	0.96	3.76	0.470	0.97
2.1	3.99	3.82	1.00	3.67	1.13	0.96	3.71	0.436	0.97
2.0	3.98	3.95	1.00	3.65	1.18	0.96	3.73	0.500	0.97
1.9	4.01	4.09	1.00	3.65	1.21	0.96	3.75	0.477	0.97
1.8	4.01	4.28	1.00	3.63	1.28	0.96	3.70	0.550	0.96
1.7	3.91	4.50	0.99	3.60	1.38	0.96	3.55	0.598	0.94
1.6	3.89	4.77	0.99	3.52	1.50	0.94	3.49	0.668	0.94
1.55	3.85	4.88	0.98	—	—	—	—	—	—
1.5	3.84	5.06	0.98	3.41	1.64	0.92	3.38	0.720	0.92

TABLE 23. Results of an Analysis of the Optical Constants of Tin, Using the Equations of the Weakly Anomalous Skin Effect (N_{val} = 14.8 · 10^{22} cm^{-3}; v_F^0 = 1.89 · 10^8 cm/sec)

λ, μ	293°K $N \cdot 10^{-22}$, cm^{-3}	293°K $\nu \cdot 10^{-14}$, sec^{-1}	293°K $\langle v_F \rangle \times 10^{-8}$, cm/sec	78°K $N \cdot 10^{-22}$, cm^{-3}	78°K $\nu \cdot 10^{-14}$, sec^{-1}	78°K $\langle v_F \rangle \times 10^{-8}$, cm/sec	4.2°K $N \cdot 10^{-22}$, cm^{-3}	4.2°K $\nu \cdot 10^{-14}$, sec^{-1}	4.2°K $\langle v_F \rangle \times 10^{-8}$, cm/sec
12.0	5.03	2.36	1.10	4.09	0.861	0.99	4.12	0.628	1.00
11.0	4.86	2.29	1.08	4.11	0.846	1.00	4.21	0.572	1.01
10.0	4.79	2.25	1.08	4.07	0.754	0.99	4.23	0.528	1.01
9.0	4.80	2.22	1.08	4.15	0.708	1.00	4.20	0.493	1.01
8.0	4.80	2.23	1.08	4.24	0.706	1.01	4.13	0.471	1.00
7.0	4.80	2.22	1.08	4.22	0.728	1.01	4.12	0.502	1.00
6.0	4.77	2.23	1.07	4.16	0.758	1.00	4.11	0.521	1.00
5.0	4.76	2.34	1.07	4.14	0.828	1.00	4.12	0.556	1.00
4.0	4.67	2.58	1.06	4.22	0.824	1.01	4.18	0.692	1.00
3.5	4.69	2.63	1.06	4.19	0.956	1.01	4.16	0.861	1.00
3.0	4.74	3.18	1.07	4.16	1.19	1.00	4.11	0.971	1.00
2.5	4.70	3.80	1.06	3.97	1.64	0.98	3.91	1.32	0.97
2.0	4.78	5.18	1.07	3.86	2.66	0.97	3.79	2.19	0.96
1.7	5.09	7.68	1.11	3.71	4.21	0.95	3.65	3.56	0.94
1.5	5.75	11.1	1.18	3.88	6.96	0.97	3.67	6.75	0.94
1.35	7.10	15.3	1.31	4.75	13.0	1.07	4.18	12.5	1.00
1.2	9.26	20.4	1.50	7.81	24.5	1.37	5.97	21.6	1.20
0.99	11.8	22.7	1.68	14.1	31.5	1.85	12.9	26.8	1.76

TABLE 24. Results of an Analysis of the Optical Constants of Gold, Using the Equations of the Weakly Anomalous Skin Effect ($N_{val} = 5.89 \cdot 10^{22}$ cm^{-3}; $v_F^0 = 1.39 \cdot 10^8$ cm/sec; T = 293°K)

λ, μ	$N \cdot 10^{-22}$, cm^{-3}	$\nu \cdot 10^{-14}$, sec^{-1}	$\langle v_F \rangle \cdot 10^{-8}$, cm/sec	λ, μ	$N \cdot 10^{-22}$, cm^{-3}	$\nu \cdot 10^{-14}$, sec^{-1}	$\langle v_F \rangle \cdot 10^{-8}$, cm/sec
12.0	5.64	0.399	1.36	3.0	5.52	0.479	1.34
10.0	5.58	0.439	1.35	2.50	5.38	0.499	1.33
8.0	5.55	0.467	1.35	2.00	5.41	0.523	1.33
6.0	5.61	0.492	1.36	1.50	5.38	0.646	1.33
5.0	5.68	0.478	1.36	1.00	5.04	1.06	1.29
4.0	5.52	0.470	1.35				

TABLE 25. Results of an Analysis of the Optical Constants of Gold, Using the Equations of the Anomalous Skin Effect ($N_{val} = 5.89 \cdot 10^{22}$ cm^{-3}; $v_F^0 = 1.39 \cdot 10^8$ cm/sec; T = 293°K)

λ, μ	$N \cdot 10^{-22}$, cm^{-3}	$\nu \cdot 10^{-14}$, sec^{-1}	$\langle v_F \rangle \cdot 10^{-8}$, cm/sec	λ, μ	$N \cdot 10^{-22}$, cm^{-3}	$\nu \cdot 10^{-14}$, sec^{-1}	$\langle v_F \rangle \cdot 10^{-8}$, cm/sec
12.0	5.87	0.445	1.39	3.0	5.55	0.483	1.35
10.0	5.73	0.471	1.37	2.50	5.40	0.502	1.33
8.0	5.64	0.488	1.36	2.00	5.43	0.525	1.34
6.0	5.67	0.504	1.36	1.50	5.40	0.649	1.33
5.0	5.72	0.486	1.37	1.00	5.07	1.06	1.29
4.0	5.55	0.476	1.35				

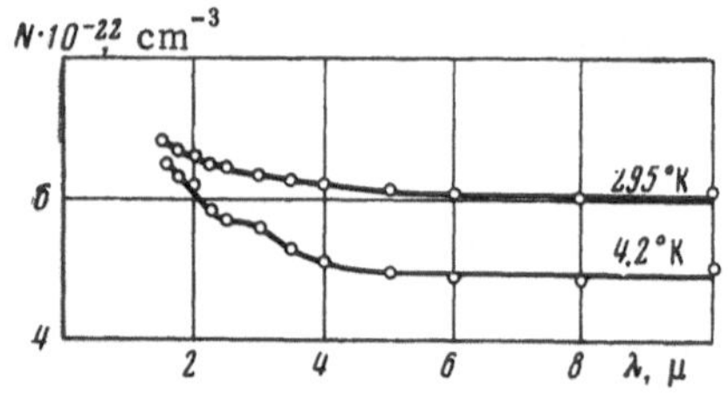

Fig. 45. Dependence of N on λ for indium at 295 and 4.2°K – Eq. (V.1).

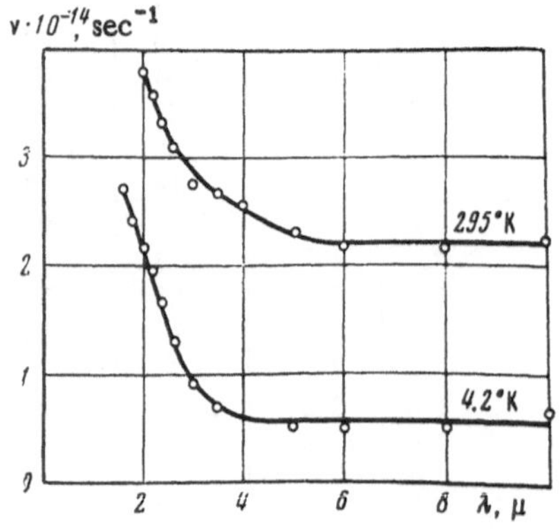

Fig. 46. Dependence of ν on λ for indium at 295 and 4.2°K – Eq. (V.2).

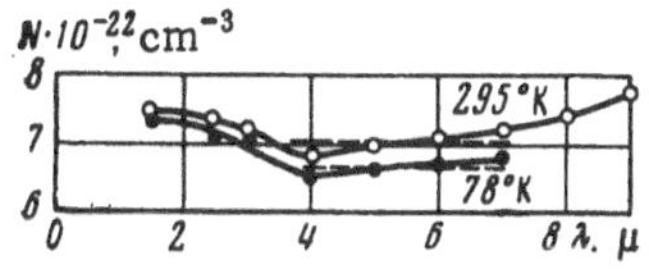

Fig. 47. Dependence of N on λ for unannealed aluminum layers at two temperatures. Broken line gives the mean value of N in the range 4-7 μ.

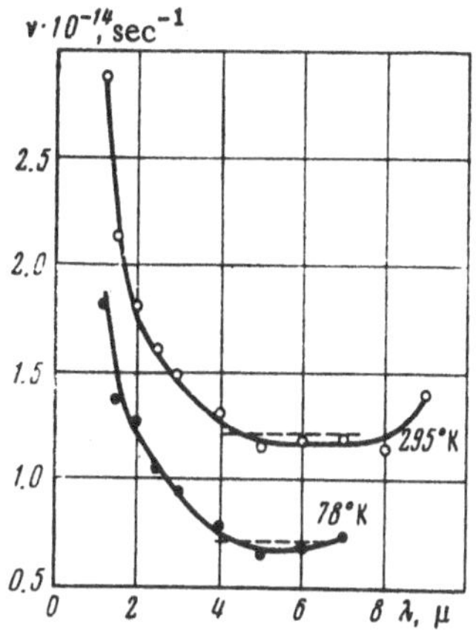

Fig. 48. Dependence of ν on λ for unannealed aluminum layers at two temperatures. Broken line gives the mean value of ν for the range 4-7 μ.

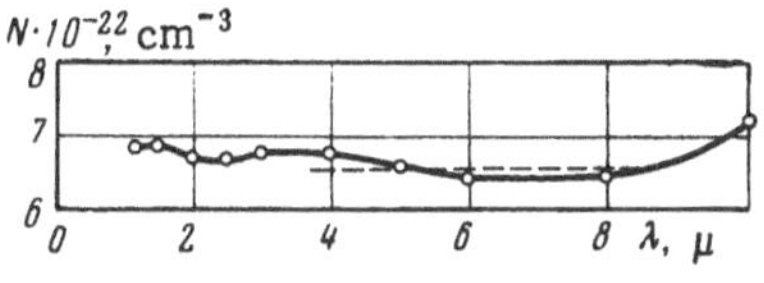

Fig. 49. Dependence of N on λ for annealed aluminum layers at 293°K. Broken line gives the mean value of N for the range 4-8 μ.

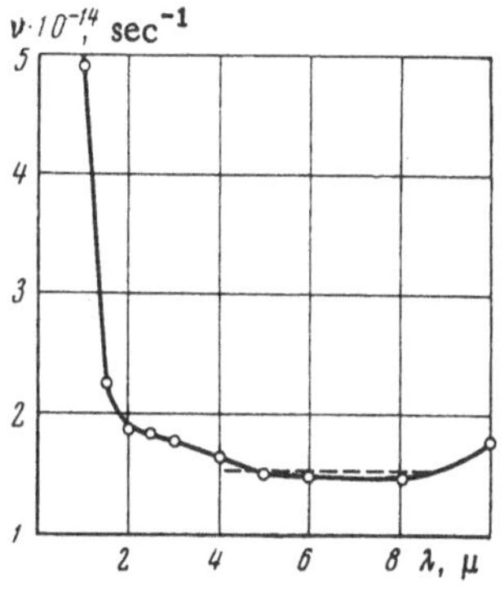

Fig. 50. Dependence of ν on λ for annealed aluminum layers at 293°K. Broken line gives the mean value of ν for the range 4-8 μ.

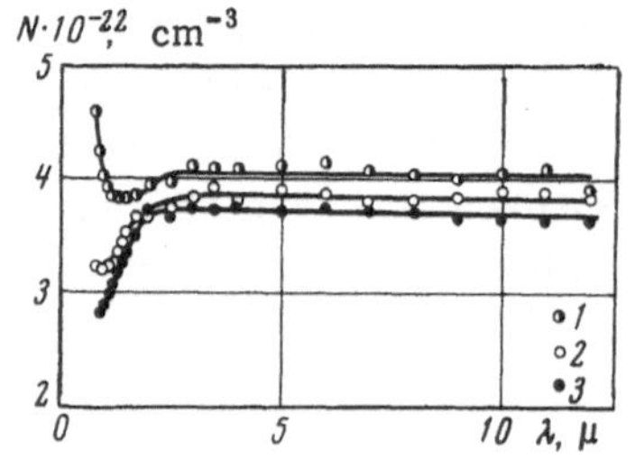

Fig. 51. Dependence of N on λ for lead at three temperatures: 1) 293°K; 2) 78°K; 3) 4.2°K.

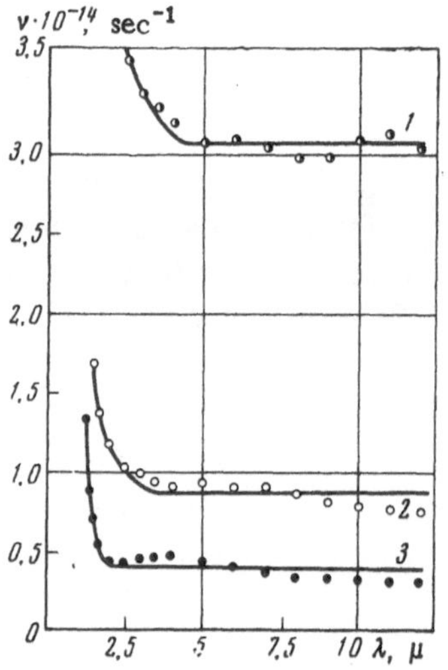

Fig. 52. Dependence of ν on λ for lead at three temperatures: notation as in Fig. 51.

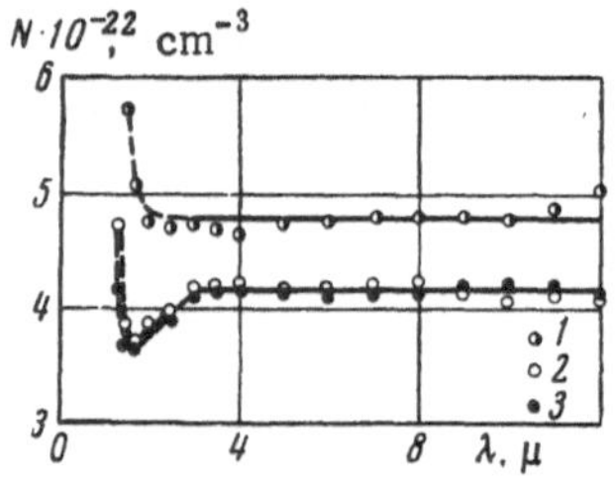

Fig. 53. Dependence of N on λ for tin at three temperatures: notation as in Fig. 51.

The results of our analysis of Figs. 45-56 appear in Table 26.

The spectral ranges for $\langle v_F \rangle$ coincide with those for N; the spectral ranges for β_1/λ_2 and β_2 coincide for those for ν.

The effective electron collision frequency is more sensitive to the presence of interband transitions than the concentration of the conduction electrons. Tin and lead have strong transitions in the near-infrared region; the spectral ranges in which the mean value of ν was determined were accordingly narrower than the corresponding ranges for N (Figs. 51-54).

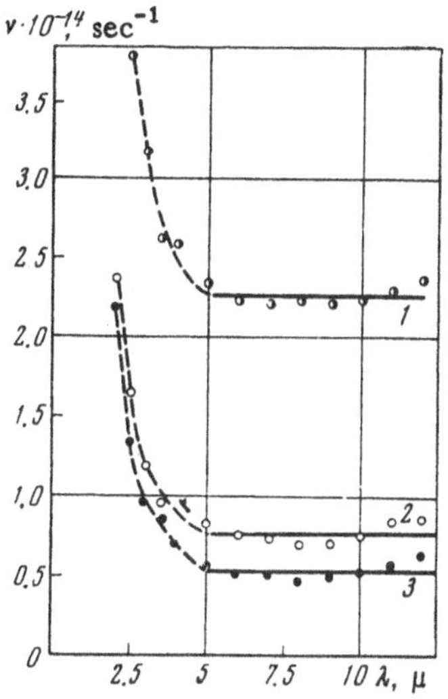

Fig. 54. Dependence of ν on λ for tin at three temperatures: notation as in Fig. 51.

Fig. 55. Dependence of N on λ for gold at 293°K.

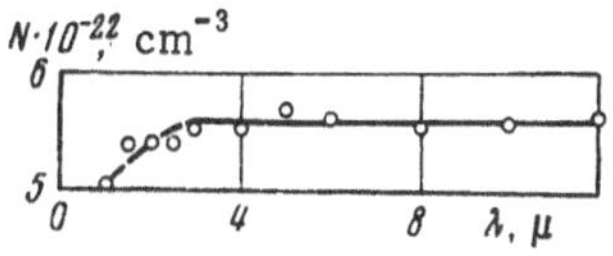

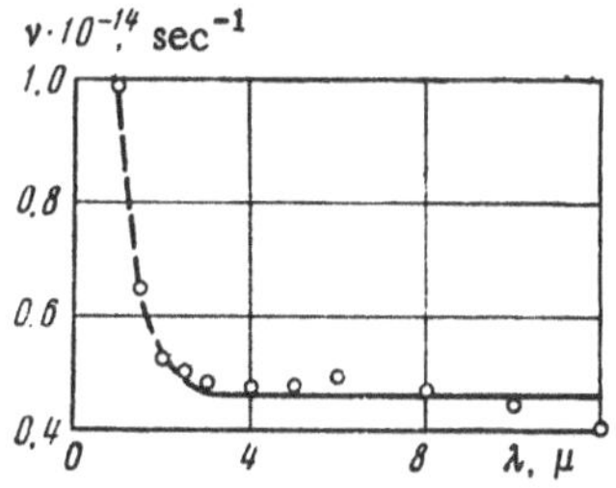

Fig. 56. Dependence of ν on λ for gold at 293°K.

TABLE 26. Spectral Ranges Used for Determining the Characteristics of the Conduction Electrons

Metal	T, °K	Spectral range, μ	
		for N	for ν
In	295	5–10	5–10
	4.2	4–10	4–10
Al Unannealed layers	295	4–7	4–7
	78	4–7	4–7
Al Annealed layers	293	4–8	4–8
Pb	293	2.5–12	4–12
	78	3–12	3–12
	4.2	1.8–12	2–12
Sn	293	3–12	5–12
	78	3–12	5–12
	4.2	3–12	5–12
Au	293	3–12	3–12

TABLE 27. Microcharacteristics of the Metals

Characteristic	In		Al			Pb			Sn			Au
			unannealed layers		annealed layers							
	295° K	4.2° K	295° K	78° K	293° K	293° K	78° K	4.2° K	293° K	78° K	4.2° K	293° K
$N_a \cdot 10^{-22}$, cm^{-3}	3.46	3.46	5.36	5.36	6.03	3.29	3.34	3.36	3.70	3.70	3.70	5.89
$N \cdot 10^{-22}$, cm^{-3}	6.09	4.93	7.02	6.68	6.57	4.06	3.85	3.73	4.79	4.16	4.15	5.59
$\langle v_F \rangle \cdot 10^{-8}$, cm/sec	1.34	1.20	1.33	1.30	1.22	1.01	0.98	0.97	1.08	1.00	1.00	1.35
$\nu \cdot 10^{-14}$, sec^{-1}	2.24	0.58	1.21	0.72	1.54	3.07	0.87	0.41	2.27	0.77	0.53	0.46
$\nu_{ep} \cdot 10^{-14}$, sec^{-1}	2.19	0.54	1.05	—	—	3.06	0.86	0.40	2.21	0.72	0.48	0.39
$\nu_{ep}^{cl} \cdot 10^{-14}$, sec^{-1}	1.58	—	0.65	—	—	2.49	0.56	—	1.65	0.29	—	0.34
$\nu_{ed} \cdot 10^{-14}$, sec^{-1}	0.05	0.04	0.16	—	—	0.01	0.01	0.01	0.06	0.05	0.05	0.07
$(\beta_1/\lambda^2) \cdot 10^4$, μ^{-2}	5.0	1.0	3.3	1.6	3.8	4.0	1.4	0.5	3.3	1.2	0.8	0.8
$\beta_2 \cdot 10^2$	8.0	24.7	16.8	24.8	12.0	3.4	13	24	5.5	15.4	21.2	33.0
$l \cdot 10^6$, cm	0.60	2.1	1.1	1.8	0.8	0.3	1.1	2.5	0.5	1.3	1.9	2.9
$\delta \cdot 10^6$, cm	2.8	2.5	2.1	2.1	2.3	3.5	2.9	2.8	3.1	2.7	2.7	2.3
$L \cdot 10^6$, cm	30	30	8	—	—	100	100	100	20	20	20	20

The average values of N, ν, $\langle v_F \rangle$, β_1/λ^2, and β_2 for the ranges* indicated in Table 26 are presented in Table 27. The same table gives the other microcharacteristics† obtained from Eqs. (V.7)-(V.12).

In the case of lead, we allowed for the change taking place in the concentration of the atoms with temperature, using the temperature dependence of the density determined by White [140]. Since this change was very slight and had hardly any effect on the electron characteristics, we neglected it in the case of the other metals, for which the temperature dependence of the density was still less marked than for lead.

A discussion of the principal electron characteristics of the metals will be presented in Chapter VII. Here we simply note that for indium, lead, and tin, $\nu_{ed} \ll \nu$, ν_{ep} even at helium temperatures, i.e., ν_{ed} has practically no influence on the optical constants of these metals. For aluminum at room and nitrogen temperatures, ν_{ed}/ν, respectively, equals 0.13 and 0.22. For gold at room temperature, even after annealing, $\nu_{ed}/\nu = 0.15$. This means that ν_{ed} has a marked influence on the optical properties of aluminum and gold layers, particularly at low temperatures.

Furthermore, for indium, aluminum, lead, and tin, $l/\delta < 1$ at all temperatures. However, at low temperatures, this ratio is almost equal to unity. Gold even has $l/\delta > 1$ at room temperature, further indicating the inapplicability of the equations of the normal skin effect.

The ratios L/δ and L/l for indium, lead, tin, and gold are of the order of 10 or over. This means that the dimensions of the crystals are fairly large and that the scattering of the electrons by the boundaries of the crystals is only slight. For aluminum, $L/\delta \approx 4$, $L/l > 4$, so that the crystals may still be regarded as large.

*It should be noted that for aluminum the values of N calculated as the mean values in the spectral ranges 1.2-9, 1.2-7, and 1-10 μ were very close to those calculated in the ranges indicated in Table 26. Thus the effect of the interband transitions of aluminum on this quantity was only slight.

† For aluminum the ratio $R_{res}/R(T_K)$ required in order to calculate ν_{ep}, ν_{ep}^{cl}, and ν_{ed}, was estimated by using the ratio of the static conductivities $\sigma_{st}^m(T_N)/\sigma_{st}^m(T_K)$, $\sigma_{st}(T_N)/\sigma_{st}(T_K)$, and $\sigma_{st}^m(T_K)/\sigma_{st}(T_K)$. The index m means that the corresponding conductivity relates to the bulk ("massive") metal. This estimate gave $R_{res}/R(T_K) \approx 0.20$.

CHAPTER VI

ANALYSIS OF EXPERIMENTAL RESULTS IN THE SHORT-WAVE PART OF THE SPECTRAL RANGE STUDIED

Let us now consider the results of our measurements in the short-wave part of the spectrum. Together with the long-wave data, these measurements enable us to study the interband transitions in metals. We indicated in Chapter II that, in metals, transitions associated with the Bragg energy splitting played a major role in the visible and near-infrared parts of the spectrum. A study of these transitions enables us to determine the Fourier components of the pseudopotential V_g and the broadening of the energy levels for the electrons close to the Bragg planes.

§ 1. Determination of the Fourier Components of Pseudopotential from Optical Measurements

In order to determine the Fourier components of the pseudopotential, we must first separate the influence of the conduction electrons on the optical properties of the metal from that of the electrons taking part in the interband transitions. For this purpose we make use of the additivity of the complex dielectric constant

$$\varepsilon'(\omega) \equiv \varepsilon_1(\omega) - i\varepsilon_2(\omega) \equiv \varepsilon_1(\omega) - i\,\frac{4\pi\sigma(\omega)}{\omega},$$

$$\varepsilon_1(\omega) = 1 + \varepsilon_{1e}(\omega) + \tilde{\varepsilon}_1(\omega), \tag{VI.1}$$

$$\sigma(\omega) = \sigma_e(\omega) + \tilde{\sigma}(\omega).$$

Here ε_1 and ε_2 are the real and imaginary parts of the complex dielectric constant; σ is the optical conductivity; ε_{1e} and σ_e relate to the conduction electrons, $\tilde{\varepsilon}_1$ and $\tilde{\sigma}$ to the interband transitions.

For ε_{1e} and σ_e using Eqs. (I.50), (I.52), and (I.54), we have

$$\varepsilon_{1e} = -\left(\frac{4\pi e^2}{m}N\right)\frac{1}{\omega^2}\,\frac{(1-\beta_1)}{1+(\nu^2/\omega^2)(1-\beta_1)^2/(1-\beta_2)^2}, \tag{VI.2}$$

$$\frac{4\pi\sigma_e}{\omega} = \left(\frac{4\pi e^2}{m}N\right)\frac{1}{\omega\nu}\cdot\frac{(1-\beta_2)}{1+(\omega^2/\nu^2)(1-\beta_2)/(1-\beta_1)}. \tag{VI.3}$$

Confining attention to terms of the first order in β_1 and β_2, we shall have

$$\varepsilon_{1e} = -\left(\frac{4\pi e^2}{m}N\right)\frac{1}{(\omega^2+\nu^2)}(1-B_1), \tag{VI.4}$$

$$B_1 = \beta_1 + \frac{2\nu^2}{\omega^2+\nu^2}(\beta_2-\beta_1). \tag{VI.4a}$$

$$\sigma_e = \left(\frac{e^2}{m}N\right)\frac{\nu}{\omega^2+\nu^2}(1-B_2). \tag{VI.5}$$

$$B_2 = \beta_2 + \frac{2\omega^2}{\omega^2+\nu^2}(\beta_1-\beta_2). \tag{VI.5a}$$

In Eqs. (VI.4)-(VI.5) the quantities N, ν, β_1/λ^2, and β_2 are average values obtained in the long-wave region, in which they are independent of λ. It is these values which determine the contribution of the free electrons to both ε_1 and σ in the short-wave range also.

The quantities $\tilde{\varepsilon}_1 + 1$ and $\tilde{\sigma}$ are equal to the difference between the experimental values of ε_1 and σ and those calculated from Eqs. (VI.4) and (VI.5). We remember that

$$\varepsilon_1 = n^2 - \varkappa^2,$$

$$\sigma = \frac{n\varkappa\omega}{2\pi} = 2.998 \cdot 10^{14} \frac{n\varkappa}{\lambda}.$$

Here λ is the wavelength of the light in microns, and σ is the conductivity in cgse units.

Of greatest interest is the function $\tilde{\sigma}$. It should be noted that measurements relating to helium and nitrogen temperatures enable us to separate out the value of $\tilde{\sigma}$ with a greater accuracy than measurements at room temperature since, at low temperatures, the contribution of the conduction electrons to σ is only slight.

When subsequently analyzing experimental data, we shall use the characteristics of the conduction electrons obtained in the preceding chapter (Table 27).

It was shown in Sections 8 and 9 of Chapter II that the maxima of the function $\tilde{\sigma}(\omega)$ enabled us to determine the Fourier components of the pseudopotential.* If the band $\tilde{\sigma}_g$ is fairly well isolated from other bands, we may determine the corresponding Fourier component of the pseudopotential from (II.101)

$$2\,|\,V_g\,| = \hbar\omega_{\max}/t_g.$$

To an accuracy of about 10% we may put $t_g = 1$. If we require a greater accuracy, we may find the value of t_g from Fig. 9 by using the relative expansion of the energy levels γ_g'. The latter quantity may be determined from the shape of the band $\tilde{\sigma}_g(\omega)$ using the equation

$$\gamma_g' = (\omega_{g\,\max} - \omega_\gamma)\, t_g/\omega_{g\,\max}, \tag{VI.6}$$

where ω_γ is the abscissa of the point having the ordinate indicated in Fig. 12 for the desired value of γ_g'. It is convenient to use the method of successive approximations. In the zero approximation we consider $t_g = 1$ and determine ω_g as the abscissa of the point with an ordinate of $0.7\sigma_{\max}$. This enables us to find γ_g' in the zero approximation. Then, using the resultant value of γ_g', we consider Fig. 9 and determine the displacement of the maximum of t_g in the first approximation; we then refine the value of the ordinate from which ω_γ is to be found in Fig. 12. After this, we obtain ω_γ to a first approximation as the abscissa of the point with an ordinate of $\tilde{\sigma}_{\max}(I_\gamma/I_{\max})$. This enables us to determine γ_g' in the next approximation, and so on. Having determined γ_g', we obtain the final value of t_g.

If the bands corresponding to different Fourier components of the pseudopotential partly overlap, then in order to separate them it is convenient to make use of a method of successive approximations, based on a linear extrapolation of the sharpest section of the function $\tilde{\sigma}(\omega)$ in the direction of the threshold† (i.e., for $\omega \ll \omega_{\max}$) and Eq. (II.99) for $\omega > \omega_{\max}$. For $\omega/\omega_{\max} > (1 + 2\gamma')$ we may use an extrapolation based on the $1/\omega^2$ law.

It is interesting to compare the absolute experimental and theoretical values of $\tilde{\sigma}_g(\omega)$, particularly the maximum values of this function. The theoretical value is given by (II.99). Let us present this in a form convenient for calculation:

$$(\tilde{\sigma}_g)_{\max} = 1.753 \cdot 10^{33} n_g p_g l_{\max}. \tag{VI.7}$$

*Optical measurements only enable us to determine $|V_g|$. The sign remains indeterminate.
†A justification for this extrapolation is presented in Section 3 of this chapter.

All the quantities are given in cgse units. The value of I_{max} depends on $\gamma_g^!$; it may be found from Fig. 10.

We may compare the asymptotic value of σ_g for $\omega \gg \omega_g$ with theory. The theoretical value is determined by Eq. (II.106). We present this in a form suitable for calculation:

$$\tilde{\sigma}_g(\omega) \approx 5.52 \cdot 10^{33} n_g p_g (\omega_g/\omega)^2, \quad \omega \gg \omega_g. \tag{VI.8}$$

If $n_g p_g$ is unknown, but experiment enables us to determine $\tilde{\gamma}_g^!$, then it is convenient to use yet another form of this equation

$$\tilde{\sigma}_g(\omega) \approx \pi \frac{(\tilde{\sigma}_g)_{max}}{I_{max}} \left(\frac{\omega_g}{\omega}\right)^2, \tag{VI.8a}$$

where the value of $(\tilde{\sigma}_g)_{max}$ is taken from experiment and I_{max} from Fig. 10.

In addition to the Fourier components of the pseudopotential V_g, optical measurements also enable us to determine the relative broadening of the energy levels $\gamma_g^!$ by the method indicated earlier. This quantity is also of considerable interest, since it characterizes the interaction of the electrons close to the Bragg plane with other electrons and phonons.

Our experimental work [17, 19, 20] showed that γ_g' had a minimum value at helium temperature, increasing sharply on rising to room temperature. At low temperatures the bands corresponding to different V_g were therefore much better resolved than at high temperatures. Hence the Fourier-components of the pseudopotential could be determined far more accurately at low temperatures.

Thus optical measurements in the region of interband transitions enable us to determine the Fourier components of the pseudopotential V_g and from these to calculate N, S_F, and $\langle v_F \rangle$. It will be shown in Chapter VII that the values of these parameters determined from measurements in the long- and short-wave parts of the spectrum agree closely with one another.

§ 2. Application of the Scheme for Analyzing the Optical Constants in the Short-Wave Region to the Metals of Present Interest

After analyzing the results of the measurements carried out in the long-wave region for indium, aluminum, lead, and tin, we also analyzed the results of optical measurements relating to the short-wave part of the spectrum. In this region the contribution of the conduction electrons to the dielectric constant $\varepsilon_1(\omega)$ and conductivity $\sigma(\omega)$ is separated from that of the interband transitions. First we determined the contribution of the conduction electrons $\varepsilon_{1e}(\omega)$ and $\sigma_e(\omega)$ from Eqs. (VI.4) and (VI.5), then the contribution of the interband transitions $\tilde{\varepsilon}_1(\omega)$ and $\tilde{\sigma}(\omega)$ from Eq. (VI.1). All the calculations were carried out on an electronic computer. For the existing accuracy of the optical-constant measurements, no more than three significant figures can be retained in the calculations of ε_1, σ, ε_{1e}, and σ_e, since even the third figure involves an error. In comparing the differences $\varepsilon_1 - \varepsilon_{1e}$ and $\sigma - \sigma_e$, the error frequently passes into the second significant figure. Sometimes even the first becomes uncertain. In this case we indicate the corresponding results with an asterisk in the tables.

The maxima of the function $\tilde{\sigma}_g(\omega)$ enabled us to obtain the Fourier components of the pseudopotential $|V_g|$ from Eq. (II.101). The error in establishing the position of the maxima of $\tilde{\sigma}_g(\omega)$ is mainly due to the width and superposition of the bands. A slight change in the parameters of the conduction electrons (within the limits of the errors involved in calculating these parameters) has hardly any effect on the positions of the maxima. As we shall see later, for the metals under consideration the value of ω_{max} was determined to an accuracy of 1% for strong bands and 10% for weak bands.

For the majority of bands $\tilde{\sigma}_g(\omega)$ the shape of the curves in the $\omega < \omega_{max}$ region yielded the relative width of the energy levels γ_g' on using Eq. (VI.6). The error in obtaining γ_g' was usually about 10%, being mainly due to inaccuracy in separating overlapping bands.

Then we obtained the maximum values of $\tilde{\sigma}_{max}$ relating to different bands. The error was 1-5%, also being mainly due to the inaccuracy involved in separating the bands.

For indium we studied the threshold of the interband transitions. The error for the threshold frequency ω_t was determined as follows.* We varied the parameters of the conduction electrons within the range of accuracy of their determination and then calculated $\tilde{\varepsilon}_1(\omega)$ and $\tilde{\sigma}(\omega)$ once again. The threshold moved by 4-5%. This quantity represents the error in question. Since the accuracy of ν is lower than that of N, the overall error is mainly associated with the error in ν. Although at helium temperature the contribution of the conduction electrons to ε_1 and σ is much smaller than at room temperature, the accuracy of the determination of ν at room temperature is higher than the accuracy of the determination of ν at helium temperature. The error in ω_t was therefore roughly the same in both cases.

§ 3. Indium

The results of our analysis of the experimental data for indium are presented in Table 28 and Figs. 57-59. We see from Table 28 and Fig. 57 that at helium temperature there are two sharply separated bands $\tilde{\sigma}_g$ not too severely overlapping. The band corresponding to the higher frequencies may be identified with the $\tilde{\sigma}_{111}$ band, and the second one with the $\tilde{\sigma}_{200}$ band. The basis for this identification is presented in the next chapter.†

The separation of the bands for the case of T = 4.2°K was carried out in the following manner (Fig. 57). Since the longer-wave band $\tilde{\sigma}_{200}$ lay a reasonable way from the $\tilde{\sigma}_{111}$ band, in the direction of the threshold side, we may consider that the effect of the $\tilde{\sigma}_{111}$ band in the region of the maximum of the $\tilde{\sigma}_{200}$ band was negligibly small. The contribution of the $\tilde{\sigma}_{200}$ band in the region of the maximum of the $\tilde{\sigma}_{111}$ band was taken into account by means of Eq. (VI.8a). The difference $\tilde{\sigma}(\omega) - \tilde{\sigma}_{200}(\omega)$ determines $\tilde{\sigma}_{111}(\omega)$ in the region of the maximum of this band and for $\omega > \omega_{max}$ also. The steepest part of the threshold side of the band $\sigma_{111}(\omega)$ we extend linearly until it intersects the horizontal axis. The difference $\tilde{\sigma}(\omega) - \tilde{\sigma}_{111}(\omega)$ gives $\tilde{\sigma}_{200}(\omega)$. The bands $\tilde{\sigma}_{111}(\omega)$ and $\tilde{\sigma}_{200}(\omega)$ separated in this way are shown as broken lines in Fig. 57.

It follows from Table 28 that the maximum of the $\tilde{\sigma}_{200}(\omega)$ band has a slight structure, comprising two almost merging maxima. This structure repeats itself in all the measurements, and does not constitute an experimental error. It may well be that the structure is due to the overlapping of two bands corresponding to different Bragg planes. The parameters of the bands after separation are given in Table 29.

At room temperature both bands became wider and their maxima diminished. This made separation more difficult (Fig. 58). Remembering that the asymptotic value of $\tilde{\sigma}_g$ at $\omega \gg \omega_{max}$ is independent of γ' [see (II.94)], we may to a first approximation consider the contribution of the $\tilde{\sigma}_{200}$ band in the region of the maximum of the $\tilde{\sigma}_{111}$ band as being the same as at 4.2°K. This enables us to determine the $\tilde{\sigma}_{111}$ band for $\omega \geq 0.88$ eV. Then we extrapolate the $\tilde{\sigma}_{111}$ band to zero in accordance with a linear law in the region $\omega < 0.88$ eV. We then determine the band $\tilde{\sigma}_{200}(\omega) = \tilde{\sigma}(\omega) - \tilde{\sigma}_{111}(\omega)$. The bands $\tilde{\sigma}_{111}$ and $\tilde{\sigma}_{200}$ so obtained are shown in Fig. 58 as broken lines. The parameters of the two bands after separation are also given in Table 29. Comparison between the theoretical and experimental values of $\tilde{\sigma}_{max}$ is carried out in the next chapter.

*In [19] we determined the error in ω_t for indium less precisely.

†In [19] we took the opposite identification of the bands, on the basis of an analogy with aluminum; this, however, was not really justified.

TABLE 28. Contribution of Interband Transitions to the Dielectric Constant and Conductivity of Indium

$\hbar\omega$, eV	λ, μ	295° K				4.2° K			
		$-\varepsilon_1$	$\sigma \cdot 10^{-14}$, cgse	$\tilde{\varepsilon}_1+1$	$\tilde{\sigma} \cdot 10^{-14}$, cgse	$-\varepsilon_1$	$\sigma \cdot 10^{-14}$, cgse	$\tilde{\varepsilon}_1+1$	$\tilde{\sigma} \cdot 10^{-14}$, cgse
2.26	0.55	21.6	17.9	−5.2	14.7	21.6	17.8	−8.2	17.0
2.07	0.60	24.6	19.9	−5.1	16.1	25.2	19.5	−9.3	18.6
1.91	0.65	28.6	22.5	−5.7	18.1	29.6	21.2	−10.9	20.1
1.77	0.70	33.0	25.2	−6.5	20.1	35.0	25.4	−13.4	24.2
1.72	0.72	34.9	26.7	−6.8	21.3	37.3	28.1	−14.4	26.8
1.68	0.74	36.9	28.3	−7.3	22.6	39.5	31.1	−15.3	29.7
1.65	0.75	37.8	29.3	−7.4	23.4	40.0	32.5	−15.1	31.1
1.63	0.76	38.4	29.6	−7.2	23.6	40.5	34.1	−15.0	32.6
1.55	0.80	41.8	32.6	−7.2	25.9	41.7	39.6	−13.4	38.0
1.51	0.82	—	—	—	—	41.9	40.3	−12.2	38.6
1.46	0.85	43.9	34.7	−4.9	27.2	42.2	40.5	−10.3	38.7
1.38	0.90	49.0	38.0	−5.3	29.6	43.9	39.5	−8.1	37.4
1.31	0.95	52.4	40.4	−3.8	31.1	46.3	36.4	−6.4	34.1
1.24	1.00	57.1	42.2	−3.4	31.9	50.3	32.4	−6.1	29.9
1.18	1.05	—	—	—	—	55.8	30.6	−7.1	27.8
1.12	1.10	66.8	42.0	−2.0	29.5	62.2	29.4	−8.8	26.3
1.08	1.15	—	—	—	—	69.2	28.8	−10.8	25.4
1.03	1.20	77.5	42.0	−0.7*	27.2	77.2	28.4	−13.7	24.7
0.993	1.25	—	—	—	—	83.7	28.4	−14.8	24.4
0.955	1.30	90.3	43.6	−0.5*	26.3	90.5	28.8	−16.0	24.5
0.919	1.35	—	—	—	—	97.3	29.2	−16.9	24.6
0.886	1.40	103	45.7	+1*	25.8	104	29.8	−18	24.8
0.856	1.45	—	—	—	—	111	29.8	−18	24.5
0.827	1.50	116	48.1	+2	25.3	117	30.3	−18	24.6
0.800	1.55	—	—	—	—	129	32.1	−23	26.0
0.775	1.60	134	51.5	0	25.7	139	33.0	−26	26.5
0.752	1.65	—	—	—	—	147	33.7	−27	26.8
0.730	1.70	150	54.9	0	25.9	156	34.7	−29	27.4
0.709	1.75	—	—	—	—	164	35.6	−29	27.8
0.689	1.80	165	57.6	+3	25	172	36.3	−29	28.1
0.671	1.85	—	—	—	—	179	37.2	−28	28.5
0.653	1.90	183	61.0	+3	25.3	189	38.2	−30	29.1
0.643	1.93	—	—	—	—	198	39.3	−34	29.9
0.620	2.00	201	64.6	+4	25.2	210	40.3	−34	30.2
0.605	2.05	—	—	—	—	216	40.7	−31	30.1
0.591	2.10	215	67.0	+9	23.9	221	41.1	−27	30.0
0.577	2.15	—	—	—	—	230	41.8	−27	30.1
0.564	2.20	232	70.1	+12	23.1	239	42.1	−27	29.9
0.552	2.25	—	—	—	—	252	42.4	−30	29.6
0.540	2.30	254	73.9	+11	22.9	258	42.2	−26	28.8
0.528	2.35	—	—	—	—	268	42.1	−26	28.2
0.517	2.40	272	77.0	+15	21.9	278	41.8	−25	27.2
0.506	2.45	—	—	—	—	292	41.5	−29	26.3
0.496	2.50	295	80.4	+14	21.1	302	40.9	−28	25.1
0.487	2.55	—	—	—	—	313	40.4	−28	24.0
0.477	2.60	319	84.4	+12	20.7	329	40.4	−33	23.4
0.412	3.0	415	98.2	+12	16.3	433	39.7	−40	17.1
0.355	3.5	535	123	+19	17	567	42.0	−35	11.4
0.310	4.0	660	145	+28	13	708	46.0	−17	6.2
0.248	5.0	941	189	+20*	6*	1060	60.8	+10*	−0.5*
0.207	6.0	1230	230	+0*	−3*	1480	81.2	+30*	−6.0*

* See text.

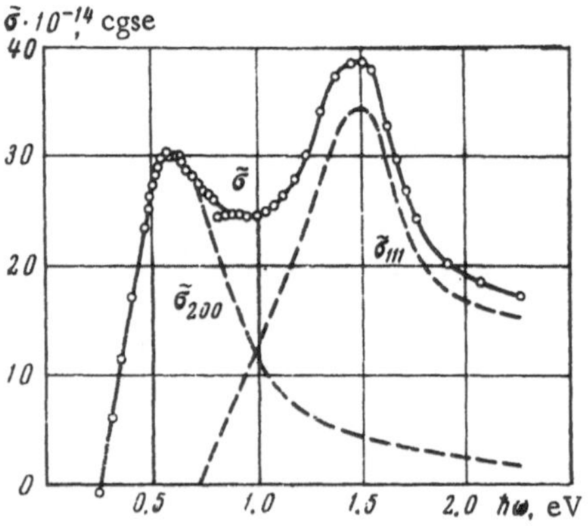

Fig. 57. Dependence of the interband conductivity $\tilde{\sigma}$ on $\hbar\omega$ for indium at 4.2°K. Solid line: $\tilde{\sigma}(\omega)$ (experimental values); broken lines: $\tilde{\sigma}_{111}$ and $\tilde{\sigma}_{200}$.

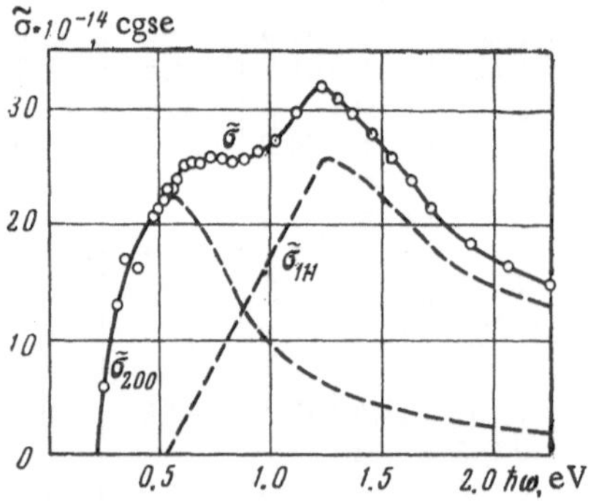

Fig. 58. Dependence of the interband conductivity $\tilde{\sigma}$ on $\hbar\omega$ for indium at 295°K.

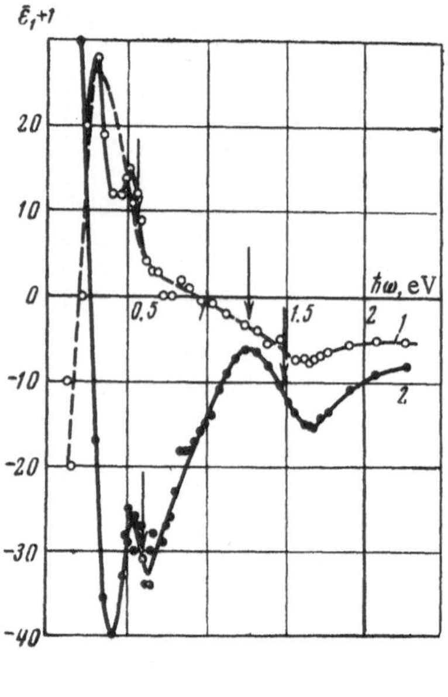

Fig. 59. Dependence of the interband dielectric constant $\tilde{\varepsilon}$ on $\hbar\omega$ for indium at two temperatures: 1) 295; 2) 4.2°K. Arrows indicate the positions of the $\tilde{\sigma}_g$ maxima.

TABLE 29. Parameters of the Indium Interband-Conduction Bands

T°, K	$\tilde{\sigma}_{111}$			$\tilde{\sigma}_{200}$		
	$\hbar\omega_{max}$, eV	$\tilde{\sigma}_{max}\cdot 10^{-14}$, cgse	γ'	$\hbar\omega_{max}$, eV	$\tilde{\sigma}_{max}\cdot 10^{-14}$, cgse	γ'
4.2	1.48±0.01	34.3±0.4	0.16±0.02	0.60±0.01	30.1±0.7	0.25±0.02
295	1.26±0.02	25.7±0.3	0.22±0.02	0.56±0.04	22.4±1.0	0.36±0.04

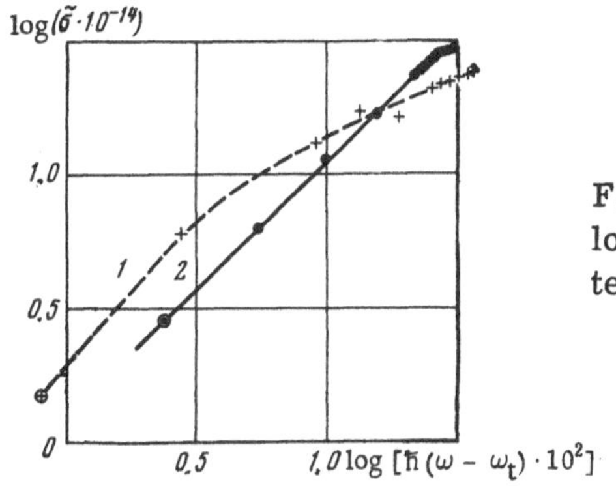

Fig. 60. Dependence of log $\tilde{\sigma}$ on log [$\hbar(\omega-\omega_t)$] for indium at two temperatures: 1) 295°K; 2) 4.2°K.

Table 28 and Fig. 59 give the dependence of $\tilde{\varepsilon}_1(\omega)+1$ on $\hbar\omega$ for both temperatures. In the figure the arrows indicate the positions of the $\tilde{\sigma}_g(\omega)$ maxima. We see that the $\tilde{\sigma}_g(\omega)$ maxima correspond to the region of anomalous dispersion of $\tilde{\varepsilon}_1(\omega)$, as might be expected for interband transitions.

The resultant data enable us to determine the threshold of the interband transitions. The values of the threshold frequencies at helium and room temperatures, respectively, equal $\hbar\omega_t$ = 0.255 ± 0.01 and 0.22 ± 0.01 eV. Figure 60 shows the dependence of log $\tilde{\sigma}$ on log [$\hbar(\omega-\omega_t)$] for both temperatures. Since the measurements in the neighborhood of the threshold were not carried out in very great detail, we also used the values of $\tilde{\sigma}$ for λ = 4.5 μ at 4.2°K and λ = 5.5 μ at 295°K obtained by linear interpolation of the values corresponding to λ = 4 and 5 μ at helium and λ = 5 and 6 μ at room temperature. These points shown are surrounded by circles. We shall consider, in the usual way, that close to the threshold of interband transitions $\tilde{\sigma}\sim(\omega-\omega_t)^x$. It then follows from the figure that at helium temperature x = 0.96 ± 0.05, this relation being satisfied over a wide spectral range 0.25-0.53 eV. The result confirms the correctness of the principle of linearly extrapolating the values of $\tilde{\sigma}_g(\omega)$ close to the threshold. At room temperature, x does not remain constant. Close to the threshold, x ≈ 1; it transforms smoothly to 0.4 for the range 0.3-0.5 eV.

§ 4. Aluminum

As we shall later see, aluminum has two $\tilde{\sigma}_g$ bands which may be identified with $\tilde{\sigma}_{200}$ and $\tilde{\sigma}_{111}$. The question of identifying the bands is discussed in the next chapter. The $\tilde{\sigma}_{200}$ band is much greater than the $\tilde{\sigma}_{111}$. The bands lie in the region of 1.5 and 0.4 eV, respectively.*

We did not ourselves carry out any very detailed measurements in the region of 1.5 eV, so in studying this band we shall make use of published data. Unfortunately, all the measurements of which we are aware in this spectral range relate to room temperature only. This band was studied most fully by Shklyarevskii and Yarovaya [79], whose results agree qualitatively with those of Schulz and Tangherlini [123, 124]. The same band was also studied by Ehrenreich, Philipp, and Segall, but their results [78] were presented in the form of graphs from which neither the intensity nor the shape of the band could be determined with adequate accuracy. In what follows we shall use the results of [79].

*A certain anomaly in the optical properties of aluminum in the near infrared was first noted by Shklyarevskii and Yarovaya [79], who correctly attributed it to interband transitions.

TABLE 30. Contribution of Interband Transitions to the Dielectric Constant ε_1 and Conductivity of Aluminum σ in the Range 0.6-3 eV

λ, μ	$\hbar\omega$, eV	$-\varepsilon_1$	$\sigma \cdot 10^{-14}$, cgse	$\tilde{\varepsilon}_1+1$	$\tilde{\sigma} \cdot 10^{-14}$, cgse
0.40	3.10	15.6	10.7	−5.5	9.6
0.45	2.76	18.6	13.3	−5.9	11.9
0.475	2.61	20.7	14.4	−6.5	12.9
0.500	2.48	23.5	17.0	−7.8	15.2
0.525	2.36	25.5	18.9	−8.2	16.9
0.550	2.26	28.0	19.8	−8.9	17.6
0.575	2.16	30.7	23.4	−9.9	21.0
0.600	2.07	33.1	26.2	−10.4	23.6
0.625	1.99	36.7	29.2	−12.2	26.4
0.650	1.91	39.4	33.3	−12.9	30.3
0.70	1.77	45.6	43.2	−14.8	39.7
0.75	1.65	48.2	48.9	−12.9	44.9
0.80	1.55	50.2	54.6	−10.1	50.1
0.85	1.46	48.9	55.8	−3.6	50.7
0.875	1.42	48.8	51.3	−0.8	45.7
0.90	1.38	51.4	45.3	−0.6	39.6
0.95	1.31	55.5	37.5	+1.0	31.1
1.00	1.24	62.6	31.5	0.0	24.5
1.10	1.12	77.2	22.9	−1.6	14.4
1.20	1.03	93.9	21.4	−4.0	11.3
1.50	0.827	161	28.8	−21	13.1
1.70	0.730	215	38.8	−36	18.6
2.00	0.620	295	51.1	−49	23.4

Note: ε_1 and σ are experimental values [79]; $\tilde{\varepsilon}_1$ and $\tilde{\sigma}$ are the contributions of the interband transitions.

Table 30 gives the $\tilde{\varepsilon}_1 + 1$ and $\tilde{\sigma}$ of aluminum for the range 0.6-3 eV calculated from Eqs. (VI.1)-(VI.5). In allowing for the contributions of the conduction electrons, we used the microcharacteristics of aluminum which we obtained in the infrared region (Table 27).

Although the aluminum layers of [79] may have differed slightly in their properties from our own, this is not particularly important in the present case, since the contribution of the conduction electrons to the spectral range in question is only about 10%.

The second $\tilde{\sigma}_{111}$ band of aluminum has a maximum at around 0.4 eV. The contribution of this to the region under the maximum of the $\tilde{\sigma}_{200}$ band is slight, of the order of 10%. The shortwave edge of the $\tilde{\sigma}_{111}$ band was obtained by extrapolating the part of the experimental curve passing through the points 0.62, 0.73, and 0.83 eV on a $1/\omega^2$ law. On subtracting the edge of the $\tilde{\sigma}_{111}$ bands from $\tilde{\sigma}$ we obtain $\tilde{\sigma}_{200}$. Figure 61 illustrates $\tilde{\sigma}(\omega)$ and $\tilde{\sigma}_{200}(\omega)$.

The $\tilde{\sigma}_{111}$ band was found from our measurements of the unannealed aluminum layers in the infrared part of the spectrum. The results obtained for the interband conductivity $\tilde{\sigma}$ in this region are given in Table 31 and Fig. 62. The $\tilde{\sigma}_{111}$ band is much weaker than the $\tilde{\sigma}_{200}$. The separation of the former cannot be achieved with any great accuracy in view of the large corrections associated both with the conduction electrons and the effects of the stronger band. The contribution of the conduction electrons to $\tilde{\sigma}$ for the region of present interest is several times greater than that of the interband transitions. The correction associated with the $\tilde{\sigma}_{200}$ band is also large. The long-wave edge of the $\tilde{\sigma}_{200}$ band was obtained by linearly extrapolating the straight line passing through the points with abscissas 1.38 and 1.03 eV (Fig. 62). By subtracting $\tilde{\sigma}_{200}$ from $\tilde{\sigma}$ we obtain $\tilde{\sigma}_{111}$. The $\tilde{\sigma}_{111}$ band at room and nitrogen temperatures is shown

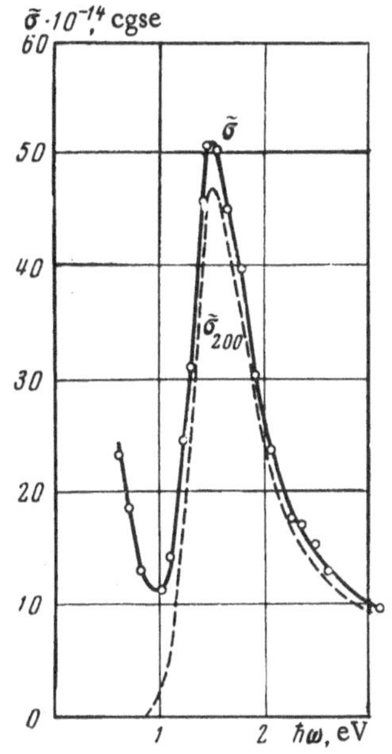

Fig. 61. Dependence of the interband conductivity $\tilde{\sigma}$ on ω for aluminum at room temperature [79]. Solid curve: $\tilde{\sigma}(\omega)$ (experimental values); broken curve: $\tilde{\sigma}_{200}(\omega)$.

TABLE 31. Contribution of the Interband Transitions to the Conductivity of Aluminum

λ, μ	ħω, eV	295° K		78° K	
		$\sigma \cdot 10^{-14}$, cgse	$\tilde{\sigma} \cdot 10^{-14}$, cgse	$\sigma \cdot 10^{-14}$, cgse	$\tilde{\sigma} \cdot 10^{-14}$, cgse
0.90	1.38	24.5	18.8	17.5	14.0
1.20	1.03	22.8	12.7	15.1	9.0
1.50	0.827	27.6	11.8	18.9	9.3
2.00	0.620	42.2	14.6	31.4	14.4
2.50	0.496	57.0	14.3	40.4	14.0
3.00	0.412	75.1	14.6	51.7	13.9
4.0	0.310	108	4*	72.2	6.1
5.0	0.248	151	3*	99.7	—1.5*
6.0	0.207	211	1*	146	+3*

Note: σ are the experimental values obtained from our measurements of unannealed aluminum layers; $\tilde{\sigma}$ is the contribution of the interband transitions.

* See text.

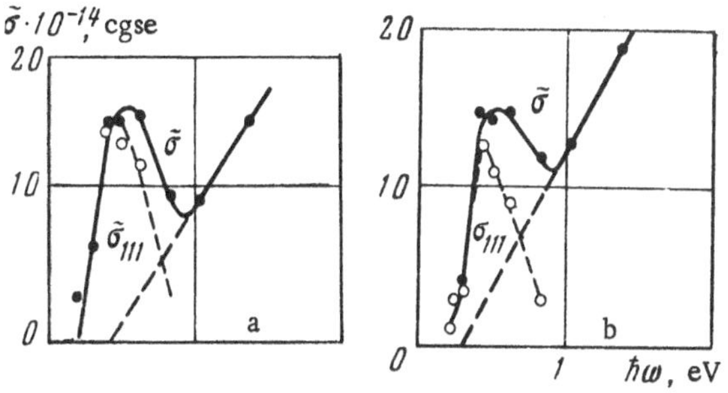

Fig. 62. Dependence of the interband conductivity $\tilde{\sigma}$ on $\hbar\omega$ for aluminum at two temperatures (our own results for unannealed aluminum layer): a) 78; b) 295°K. Solid curves: $\tilde{\sigma}$; curves with light circles: $\tilde{\sigma}_{111}$; broken lines: short-wave edge of $\tilde{\sigma}_{200}$.

TABLE 32. Parameters of the Interband-Conductivity Bands of Aluminum

T, °K	$\tilde{\sigma}_{111}$	$\tilde{\sigma}_{200}$		
	$\hbar\omega_{max}$, eV	$\hbar\omega_{max}$, eV	$\tilde{\sigma}_{max}\cdot 10^{-14}$, cgse	γ'
78	0.44 ± 0.06	—	—	—
295	0.43 ± 0.06	1.50 ± 0.01	46.8 ± 0.6	0.09 ± 0.004

in Fig. 62 by curves with white circles. In what follows we shall only use the position of the maximum of this band, since the other parameters can only be obtained to a very poor accuracy.

The parameters of the two $\tilde{\sigma}_g$ bands of aluminum are given in Table 32. A discussion of the results and a comparison with theory will be presented in the next chapter.

§ 5. Lead

The results of our analysis of the experimental data relating to lead at the three temperatures are presented in Table 33 and Figs. 63-66. Separation of the $\tilde{\sigma}$ is achieved very accurately at all temperatures, since the contribution of the conduction electrons at helium and room temperatures equals 1 and 10%, respectively, for the region near $\hbar\omega = 2$ eV and 6 and 20% for the region near $\hbar\omega = 1.5$ eV.

We see from the table and the figures that lead has two bands $\tilde{\sigma}_g$ which may be identified with $\tilde{\sigma}_{111}$ and $\tilde{\sigma}_{200}$, respectively. The identification of the bands is discussed in the next chapter. The $\tilde{\sigma}_{111}$ band is much greater than the $\tilde{\sigma}_{200}$. The bands lie in the 2-3 and 1-1.5 eV regions, respectively. These bands overlap at all temperatures. The bands were separated in the following way. In the first approximation we neglected the effect of the $\tilde{\sigma}_{200}$ band on the $\tilde{\sigma}_{111}$. By linearly extrapolating the steepest part of the long-wave edge of the $\tilde{\sigma}_{111}$ to zero we determine $\tilde{\sigma}_{200} = \tilde{\sigma} - \tilde{\sigma}_{111}$ to this approximation. We found the position of the maximum, its magnitude, and the parameter γ'_{200}. Then in the next approximation we plotted the dependence of $\tilde{\sigma}_{200}$ on ω in the overlap region of the bands from the formula

$$\tilde{\sigma}_g = \text{const}\ \mathcal{I}(\omega), \tag{VI.9}$$

where $\mathcal{I}(\omega)$ is determined, for the γ' value so found, from Eq. (II.100). In calculating $\tilde{\sigma}_{200}$ we matched both the position and the height of the theoretical and experimental maxima. This enabled us to determine $\tilde{\sigma}_{111} = \tilde{\sigma} - \tilde{\sigma}_{200}$ to this approximation. The results of the separation procedure are shown in Figs. 63-65 by the broken lines.

It follows from Table 33 and Fig. 64 that the measurements at 78°K, carried out up to $\hbar\omega = 2.8$ eV, indicate the existence of structure in the $\tilde{\sigma}_{111}$ maximum. This maximum would appear to be double. At room temperature the structure only appears very weakly. However, it repeats itself systematically for all series of measurements and cannot therefore be attributed to experimental error. The existence of structure even at room temperature is indicated by the great width of the $\tilde{\sigma}_{111}$ band. At helium temperature the structure fails to appear as the measurements only extend to $\hbar\omega \approx 2.15$ eV. Analysis of the structure of the $\tilde{\sigma}_{111}$ maximum requires detailed measurements in the near-ultraviolet part of the spectrum.

We consider that this structure is associated with spin-orbital interaction which, in the case of lead, is moderately strong.

If our assumption as to the spin-orbital splitting of the $\tilde{\sigma}_{111}$ maximum is correct, then, in determining V_{111} from (II.101) we ought to replace ω_{max} by the value of ω_0 corresponding to

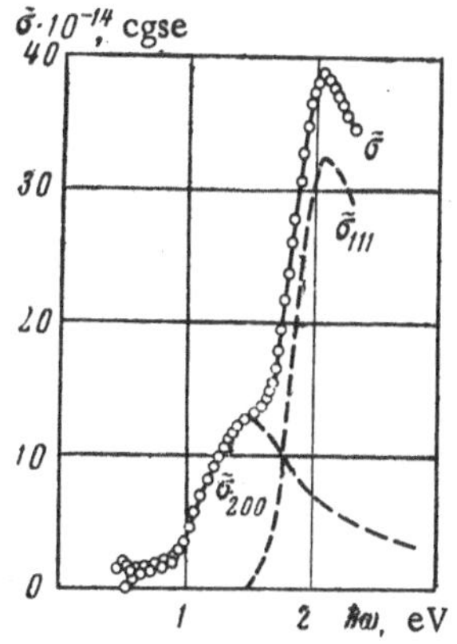

Fig. 63. Dependence of the interband conductivity $\tilde{\sigma}$ on $\hbar\omega$ for lead at 4.2°K. Solid curve: $\tilde{\sigma}(\omega)$ (experimental values); broken curve: $\tilde{\sigma}_{111}(\omega)$ and $\tilde{\sigma}_{200}(\omega)$.

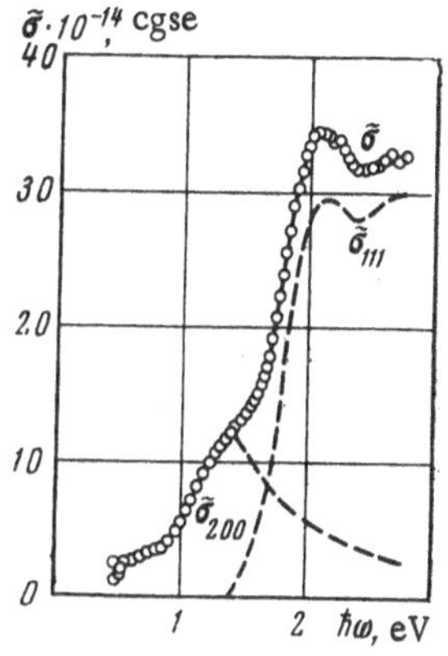

Fig. 64. Dependence of the interband conductivity $\tilde{\sigma}$ on $\hbar\omega$ for lead at 78°K.

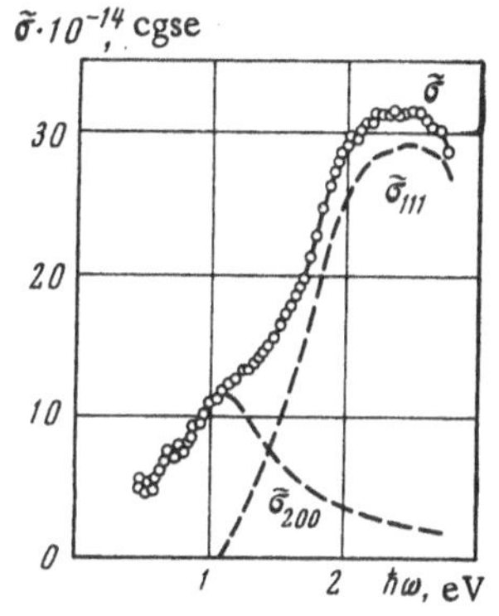

Fig. 65. Dependence of the interband conductivity $\tilde{\sigma}$ on $\hbar\omega$ for lead at 293°K.

the center of the band. At room temperature, $\omega_0 \approx \omega_{max}$. At nitrogen temperature we may expect that ω_0 coincides with the position of the local minimum between the two maxima. At helium temperature, unfortunately, we cannot indicate the position of ω_0. Later, in Table 34, we shall give both the values of ω_0 and the values of ω_{1max} corresponding to the position of the left-hand maximum. In order to determine ω_{1max} at room temperature we used the aforementioned slight trace of structure in the $\tilde{\sigma}_{111}$ maximum.

The parameters of the $\tilde{\sigma}_{111}$ and $\tilde{\sigma}_{200}$ bands for the three temperatures are presented in Table 34. For the band $\tilde{\sigma}_{200}$, $\omega_0 \approx \omega_{1max} \approx \omega_{max}$. The value of γ'_{111} was derived from the left-hand maximum. In determining $\tilde{\gamma}'_{200}$, we took no account of the slight additional absorption ob-

TABLE 33. Contribution of the Interband Transitions to the Dielectric Constant and Conductivity of Lead

λ, μ	$\hbar\omega$, eV	293° K				78° K				4.2° K			
		$-\varepsilon_1$	$\sigma\cdot10^{-14}$, cgse	$\tilde{\varepsilon}_1+1$	$\tilde{\sigma}\cdot10^{-14}$, cgse	$-\varepsilon_1$	$\sigma\cdot10^{-14}$, cgse	$\tilde{\varepsilon}_1+1$	$\tilde{\sigma}\cdot10^{-14}$, cgse	$-\varepsilon_1$	$\sigma\cdot10^{-14}$, cgse	$\tilde{\varepsilon}_1+1$	$\tilde{\sigma}\cdot10^{-14}$, cgse
0.45	2.76	8.02	30.6	−0.69	28.8	7.81	33.3	−0.82	32.7	—	—	—	—
0.46	2.70	7.85	32.2	−0.19	30.3	7.89	33.1	−0.58	32.5	—	—	—	—
0.47	2.64	7.94	32.6	0.06	30.5	7.68	33.7	−0.05	33.1	—	—	—	—
0.48	2.58	7.94	32.9	0.40	30.8	7.65	33.2	0.31	32.6	—	—	—	—
0.49	2.53	7.94	33.7	0.75	31.5	7,63	32.6	0.66	32.0	—	—	—	—
0.50	2.48	7.98	33.7	1.06	31.5	7.69	32.5	0.94	31.9	—	—	—	—
0.51	2.43	7.93	33.9	1.47	31.5	7.85	32.4	1.12	31.7	—	—	—	—
0.52	2.39	8.08	33.8	1.70	31.3	8.07	32.4	1.27	31.7	—	—	—	—
0.53	2.34	8.10	33.9	2.04	31.4	8.23	32.8	1.46	32.0	—	—	—	—
0.54	2.30	8.08	33.9	2.45	31.2	8.32	33.5	1.74	32.7	8.66	35.0	1.10	34.6
0.55	2.26	8.19	34.0	2.73	31.2	8.43	34.1	2.00	33.3	8.70	36.0	1.42	35.6
0.56	2.22	8.13	34.1	3.19	31.3	8.39	34.6	2.43	33.8	8.44	37.0	2.05	36.6
0.57	2.18	8.27	33.7	3,46	30.8	8.32	34.8	2.89	33.9	8.23	37.8	2,64	37.3
0.58	2.14	8.15	33.7	3.98	30.6	8.18	35.2	3.43	34.3	7.94	38.4	3.31	37.9
0.59	2.10	8.19	33.4	4.37	30.2	7.93	35.3	4.08	34.4	7.68	38.8	3.96	38.3
0.60	2.07	8.24	33.0	4.74	29.7	7.59	35.4	4,83	34.4	7.14	39.5	4.90	39.0
0.61	2.03	8.19	33.1	5.22	29.7	7.29	35.3	5.55	34.3	6.66	38.8	5,79	38.3
0.62	2.00	8.26	32.6	5.59	29.1	6.97	34.5	6.29	33.4	5.95	38.2	6.91	37.7
0.63	1.97	8.31	32.3	5.98	28.7	6.58	34.0	7.12	32.9	4.98	37.2	8.30	36.7
0.64	1.94	8.40	31.8	6.35	28,1	6.22	32.7	7.91	31.6	4.27	35.4	9.43	34.8
0.65	1.91	8.65	31.0	6.55	27.2	6.19	31.3	8.38	30.2	4.07	33.3	10.1	32.8
0.66	1.88	8.73	30.2	6.94	26.3	5.85	30.1	9.18	29.0	3.92	31.3	10.6	30.7
0.67	1.85	—	—	—	—	5.78	28.4	9.70	27.2	3.83	28.6	11.2	28.0
0.68	1.82	9.02	28.9	7.60	24.7	5.75	26.9	10.2	25.6	3.98	26.6	11.5	26.0
0.69	1.80	—	—	—	—	5.95	25.2	10.5	23.9	4.18	24.3	11.7	23.7
0.70	1.77	9.56	27.3	8.04	22.8	6.18	23.7	10.7	22.3	4.53	22.3	11.9	21.7

0.71	1.75	—	—	—	—	6.65	21.9	10.7	20.6	5.20	20.0	11.7	19.3
0.72	1.72	10.3	25.9	8.3	21.2	7.30	20.6	10.6	19.2	5.97	18.6	11.4	17.9
0.73	1.70	—	—	—	—	7.87	19.2	10.5	17.8	6.86	17.1	11.0	16.4
0.74	1.68	11.2	24.7	8.4	19.7	8.57	18.2	10.3	16.8	7.66	16.2	10.7	15.4
0.75	1.65	11.7	24.2	8.4	19.1	9.29	17.8	10.1	16.3	8.38	15.6	10.4	14.9
0.76	1.63	12.3	23.8	8.4	18.6	9.99	17.2	9.93	15.7	9.25	15.1	10.1	14.3
0.78	1.59	13.4	23.2	8.4	17.7	11.4	16.3	9.6	14.7	10.8	14.5	9.6	13.7
0.79	1.57	—	—	—	—	12.0	16.1	9.6	14.5	11.7	14.5	9.2	13.7
0.80	1.55	14.5	23.1	8.4	17.3	12.6	15.8	9.4	14.1	12.0	14.1	9.5	13.2
0.81	1.53	—	—	—	—	—	—	—	—	13.0	14.2	8.9	13.3
0.82	1.51	15.3	22.5	8.7	16.4	13.7	15.4	9.5	13.6	13.3	13.9	9.2	13.0
0.84	1.48	—	—	—	—	—	—	—	—	14.4	13.8	9.2	12.9
0.85	1.46	16.8	22.1	8.9	15.6	15.3	15.1	9.6	13.1	14.8	13.7	9.4	12.7
0.86	1.44	—	—	—	—	—	—	—	—	15.4	13.6	9.3	12.6
0.88	1.41	18.8	21.9	8.8	14.9	17.0	14.6	9.6	12.5	16.4	13.3	9.5	12.3
0.90	1.38	19.9	21.8	8.9	14.5	18.1	14.3	9.8	12.1	17.4	12.9	9.7	11.8
0.92	1.35	21.1	21.7	9.0	14.1	19.2	14.0	9.9	11.7	18.5	12.4	9.8	11.3
0.95	1.31	23.0	21.8	9.1	13.7	21.0	13.5	10.1	11.1	20.2	11.8	10.0	10.6
0.98	1.27	24.7	21.8	9.3	13.2	22.8	13.1	10.2	10.5	22.1	11.2	10.0	9.9
1.00	1.24	26.4	22.0	9.0	13.1	24.3	12.7	10.2	10.0	23.5	10.6	9.9	9.2
1.05	1.18	29.7	22.2	9.3	12.4	27.4	12.1	10.5	9.2	26.8	9.78	10.0	8.29
1.10	1.13	33.8	22.8	8.8	12.1	31.1	11.4	10.5	8.1	30.8	8.73	9.6	7.09
1.15	1.08	37.8	23.3	8.6	11.6	35.5	10.6	10.0	7.0	34.8	7.48	9.4	5.70
1.20	1.03	41.3	23.8	9.0	11.1	39.0	10.2	10.5	6.3	39.4	6.55	8.7	4.61
1.25	0.993	45.8	24.7	8.6	10.9	44.0	9.57	9.7	5.36	44.2	5.53	8.0	3.41
1.30	0.955	50.0	25.0	8.7	10.2	48.1	9.36	10.0	4.81	48.3	5.01	8.1	2.72
1.35	0.919	53.6	25.3	9.4	9.4	—	—	—	—	52.6	4.74	8.3	2.28
1.40	0.886	57.4	26.2	10.2	9.1	57.3	9.20	10.0	3.93	57.5	4.63	8.0	1.98
1.45	0.856	61.6	26.5	10.6	8.3	—	—	—	—	62.3	4.74	7.9	1.89
1.50	0.827	66.2	27.3	10.8	7.9	67.4	9.46	9.8	3.42	67.8	4.53	7.3	1.49
1.55	0.800	70.7	28.1	11.1	7.5	—	—	—	—	—	—	—	—
1.60	0.775	76.1	29.6	10.8	7.7	79.4	10.2	8.4	3.4	79.8	5.03	5.7	1.57
1.70	0.730	86.1	31.6	11.1	7.1	91.5	10.9	7.4	3.2	91.5	5.27	5.0	1.36

Conclusion of TABLE 33

λ, μ	$\hbar\omega$, eV	293° K				78° K				4.2° K			
		$-\varepsilon_1$	$\sigma\cdot 10^{-14}$, cgse	$\tilde{\varepsilon}_1+1$	$\tilde{\sigma}\cdot 10^{-14}$, cgse	$-\varepsilon_1$	$\sigma\cdot 10^{-14}$, cgse	$\tilde{\varepsilon}_1+1$	$\tilde{\sigma}\cdot 10^{-14}$, cgse	$-\varepsilon_1$	$\sigma\cdot 10^{-14}$, cgse	$\tilde{\varepsilon}_1+1$	$\tilde{\varepsilon}\cdot 10^{-14}$, cgse
1.80	0.689	98.8	34.7	9.1	7.4	104	11.6	8	2.9	107	5.79	1*	1.42
1.90	0.653	110	36.8	9	6.7	116	12.4	7	2.8	121	5.87	0*	0.99
2.00	0.620	120	38.9	11	6.0	128	13.3	8	2.6	133	6.70	0*	1.30
2.10	0.591	132	41.4	11	5.5	142	14.2	8	2.5	146	6.60	1*	0.65
2.20	0.564	142	43.7	13	4.7	157	15.4	8	2.5	162	7.78	—1*	1.25
2.30	0.540	156	47.4	11	5.2	172	16.3	8	2.2	177	7.09	0*	—0.04*
2.40	0.517	165	50.0	15	4.6	188	17.1	8	1.8	194	9.43	—2*	1.66
2.50	0.496	183	53.7	10	5.0	203	18.0	9	1.4	211	10.2	—3*	1.8
2.55	0.487	189	55.6	10	5.3	—	—	—	—	217	10.4	—1*	1.6
2.60	0.477	197	57.5	9	5.5	219	20.5	10	2.5	226	10.7	—1*	1.5
3.0	0.412	251	70.0	10	4.4	297	26.4	6*	2.7	299	14.0	0*	1.9
3.5	0.355	317	85.9	13	2.8	412	35.1	—3*	3.2	403	18.9	3*	2.5
4.0	0.310	389	103	10	2*	518	42.6	10*	1.2	531	25.8	—2*	4.4
5.0	0.248	533	134	—2*	1*	808	68.7	—1*	5.7	813	36.9	8*	3.6
6.0	0.207	653	164	—5*	3*	1120	91.6	8*	3.5	1170	50.7	—2*	3.2
7.0	0.177	756	187	—8*	1*	1450	119	40	3	1580	64.1	3*	0.1*

* See text.

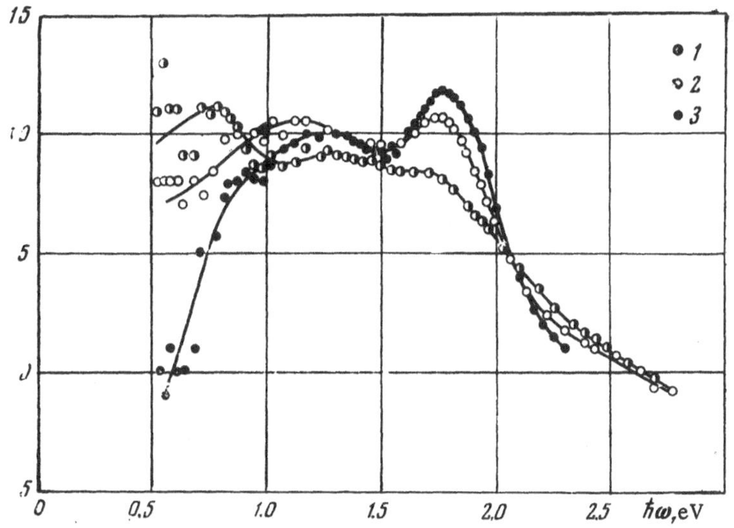

Fig. 66. Interband dielectric constant $\tilde{\varepsilon}_1$ as a function of $\hbar\omega$ at three temperatures: 1) 293°K; 2) 78°K; 3) 4.2°K.

TABLE 34. Parameters of the Interband-Conductivity Bands of Lead

T, °K	$\tilde{\sigma}_{111}$				$\tilde{\sigma}_{200}$		
	$\hbar\omega_{1max}$, eV	$\hbar\omega_0$, eV	$\tilde{\sigma}_{max} \times 10^{-14}$, cgse	γ'	$\hbar\omega_{max}$, eV	$\tilde{\sigma}_{max} \times 10^{-14}$, cgse	γ'
4.2	2.07±0.01	—	32.3±0.7	0.10±0.01	1.48±0.02	12.9±0.3	0.20±0.02
78	2.14±0.01	2.38±0.03	29.4±0.5	0.15±0.01	1.38±0.02	12.1±0.3	0.20±0.02
293	2.24±0.03	2.48±0.01	28.5±0.6	0.21±0.02	1.08±0.03	11.6±0.5	0.27±0.03

served in lead in the region of 0.2-0.9 eV. Allowance for the effect of this absorption on the $\tilde{\sigma}_{200}$ band may slightly reduce $(\tilde{\sigma}_{200})_{max}$ and increase $\tilde{\gamma}'_{200}$, at 293°K.

Apart from the foregoing $\tilde{\sigma}_{111}$ and $\tilde{\sigma}_{200}$ conductivity maxima, lead also exhibits a slight additional absorption in the region of $\hbar\omega < 0.9$ eV, as indicated in Figs. 63-65. This absorption cannot be eliminated by any reasonable change in the parameters relating to the conduction electrons. The effect occurs at all temperatures. The ratio of the $\tilde{\sigma}$ in this region (for $\hbar\omega = 0.6$ eV) to the maximum value of $\tilde{\sigma}$ equals ~3% at 4.2°K, ~8% at 78°K, and ~20% at 293°K. The additional absorption ends in the region of 0.2-0.3 eV. An analysis of the $\tilde{\sigma}(\omega)$ relation close to the threshold in question requires more accurate and detailed measurements of the optical constants in this region. The additional absorption is further discussed in Section 10 of Chapter VII.

Figure 66 shows the dependence of $\tilde{\varepsilon}_1 + 1$ on $\hbar\omega$. At all temperatures there is a clear manifestation of a region of anomalous dispersion associated with the $\tilde{\sigma}_{111}$ band. A weaker but perfectly perceptible region of anomalous dispersion also appears in response to the $\tilde{\sigma}_{200}$ band.

These results will be discussed and compared with theory in the next chapter.

§ 6. Tin

The results of our analysis of the experimental data relating to tin at three temperatures are presented in Table 35 and Figs. 67-69. The separation of $\tilde{\sigma}$ in the range $\hbar\omega \approx$ 1-2 eV may be carried out with great accuracy at all temperatures, since the contribution of the conduction electrons amounts to only 2% at helium, 3% at nitrogen, and 10% at room temperature. The separation of $\tilde{\sigma}$ in the range $\hbar\omega \approx 0.5$ eV cannot be achieved so accurately, as the contribution of the conduction electrons amounts to some 50% at helium, 60% at nitrogen, and 70% at room temperature.

We see from the figures that in the spectral range under consideration tin has two $\tilde{\sigma}_g$ bands. The possible identification of the bands is presented in the next chapter. The band corresponding to the greater value of $\hbar\omega$ (let us call it $\tilde{\sigma}_1$) is much stronger than the $\tilde{\sigma}_2$ band. Let us place the bands in the regions 1.0-1.5 and 0.4-0.6 eV, respectively. The bands overlap at all temperatures. Unfortunately, measurements are insufficiently detailed in the region of the two maxima. The separation of the bands and the determination of their parameters can therefore only be achieved with a relatively poor accuracy.

The bands were separated as follows. We neglected the effect of band $\tilde{\sigma}_2$ on band $\tilde{\sigma}_1$. By linearly extrapolating the steep part of the long-wave edge of $\tilde{\sigma}_1$ to zero, we determined $\tilde{\sigma}_2 = \tilde{\sigma} - \tilde{\sigma}_1$. The results of the separation of the two bands are shown in Figs. 67-69 by broken lines. The inadequately detailed measurements prevent us from discovering whether the σ_1 has a complex structure. The parameters of the two bands are given in Table 36. The coefficient γ_1' is determined from the ordinate of $0.7 \cdot (\tilde{\sigma}_1)'_{max}$. For the second band only the positions of the maxima are determined.*

Figure 70 illustrates the dependence of $\tilde{\varepsilon}_1 + 1$ on $\hbar\omega$ at the three temperatures. At all temperatures there is a clear region of anomalous dispersion associated with the maximum of $\tilde{\sigma}_1$. In the region of the $\tilde{\sigma}_2$ maximum, the separation of $\tilde{\varepsilon}_1$ can only be achieved with a poor degree of accuracy since, in this region, the contribution of the conduction electrons to ε_1 is much greater than that of the interband transitions.

A discussion on the results obtained will be presented in the next chapter, and so will a comparison with theory.

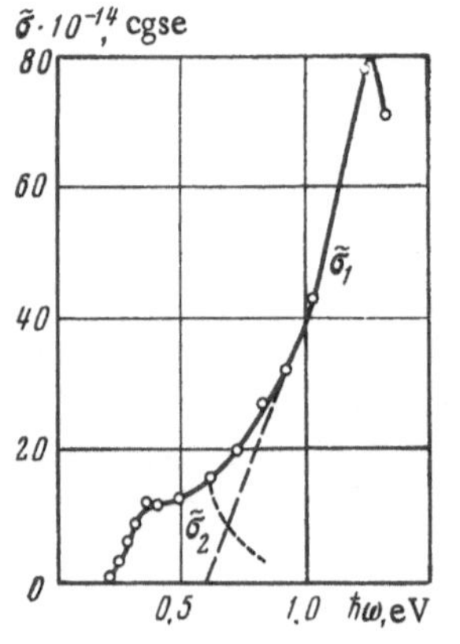

Fig. 67. Dependence of the interband conductivity $\tilde{\sigma}$ on $\hbar\omega$ for tin at 4.2°K. Solid curve: $\tilde{\sigma}(\omega)$ (experimental values); broken curves: $\tilde{\sigma}_1(\omega)$ and $\tilde{\sigma}_2(\omega)$.

* For the second band there is a certain structure at helium temperature. However, the insufficient detail of the measurements in the region of the principal maximum prevents us from discussing this problem.

TABLE 35. Determination of the Contribution of Interband Transitions to the Dielectric Constant and Conductivity of Tin

$\hbar\omega$, eV	λ, μ	293° K				78° K				4.2° K			
		$-\varepsilon_1$	$\sigma\cdot10^{-14}$, cgse	$\tilde{\varepsilon}_1+1$	$\tilde{\sigma}\cdot10^{-14}$, cgse	$-\varepsilon_1$	$\sigma\cdot10^{-14}$, cgse	$\tilde{\varepsilon}_1+1$	$\tilde{\sigma}\cdot10^{-14}$, cgse	$-\varepsilon_1$	$\sigma\cdot10^{-14}$, cgse	$\tilde{\varepsilon}_1+1$	$\tilde{\sigma}\cdot10^{-14}$, cgse
1.70	0.73	34.8	56.3	−12.1	52.0	33.3	56.9	−13.4	55.5	—	—	—	—
1.55	0.80	38.1	59.5	−10.9	54.3	36.1	54.6	−12.3	52.9	—	—	—	—
1.33	0.93	43.1	73.9	−6.5	66.9	39.6	79.3	−7.4	77.0	49.4	72.5	−17.2	70.9
1.25	0.99	42.0	76.5	−0.6	68.6	32.8	82.4	+3.7	79.8	37.4	80.1	−1.0	78.2
1.03	1.20	44.1	71.7	+16.3	60.2	29.1	56.9	+24.5	53.1	26.5	45.6	+27.0	42.9
0.919	1.35	51.9	63.7	+24.1	49.2	41.2	42.8	+26.5	38.0	37.6	37.6	+30.1	34.1
0.827	1.50	64.2	57.4	+29.0	39.6	59.3	33.3	+24.2	27.4	56.9	31.0	+26.6	26.7
0.730	1.70	87.8	54.5	+30.7	31.9	83.2	28.7	+23.9	21.1	85.1	24.9	+22.1	19.4
0.620	2.00	130	54.8	+31	24.1	127	28.2	+21	17.7	128	23.6	+20	16.0
0.496	2.50	206	64.4	+38	18.1	210	29.6	+20	13.3	211	24.3	+20	12.5
0.412	3.0	297	78.4	+40	14.3	320	33.8	+9*	10.5	321	28.4	+10*	11.5
0.355	3.5	392	92.5	+46	9.2	441	38.5	+4*	7.1	441	35.2	+8*	12.2
0.310	4.0	500	108	+44	5*	580	44.6	−4*	3.9	581	38.6	+3*	8.7
0.248	5.0	740	145	+22	1*	868	66.8	+14	4.5	892	49.5	+10*	1.1
0.207	6.0	975	182	−1*	−2*	1240	88.2	0*	0.6*	1270	66.7	+10*	1.1

* See text.

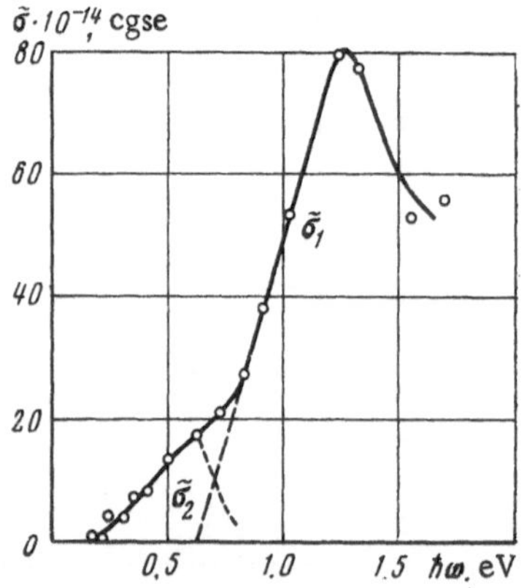

Fig. 68. Dependence of the interband conductivity $\tilde{\sigma}$ on $\hbar\omega$ for tin at 78°K.

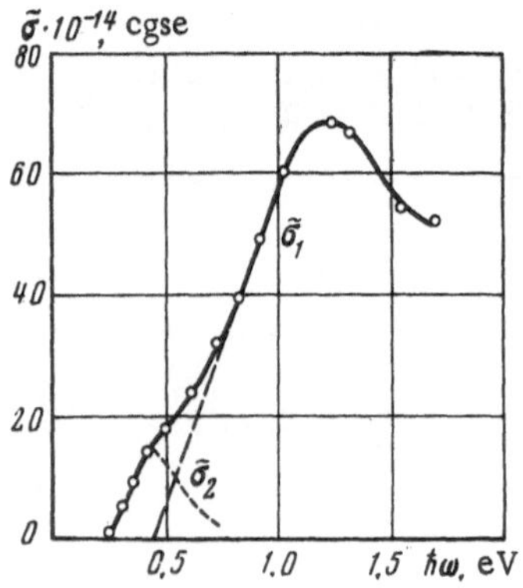

Fig. 69. Dependence of the interband conductivity $\tilde{\sigma}$ on $\hbar\omega$ for tin at 293°K.

TABLE 36. Parameters of the Interband-Conductivity Bands of Tin

T, °K	$\tilde{\sigma}_1$			$\tilde{\sigma}_2$		
	$\hbar\omega_{max}$, eV	$\tilde{\sigma}_{max}\cdot 10^{-14}$, cgse	γ'	$\hbar\omega_{max}$, eV	$\tilde{\sigma}_{max}\cdot 10^{-14}$, cgse	γ'
4.2	1.27 ± 0.10	80 ± 3	0.12 ± 0.03	0.6 ± 0.1	—	—
78	1.27 ± 0.10	80 ± 3	0.17 ± 0.04	0.6 ± 0.1	—	—
293	1.22 ± 0.10	68 ± 3	0.25 ± 0.05	0.45 ± 0.1	—	—

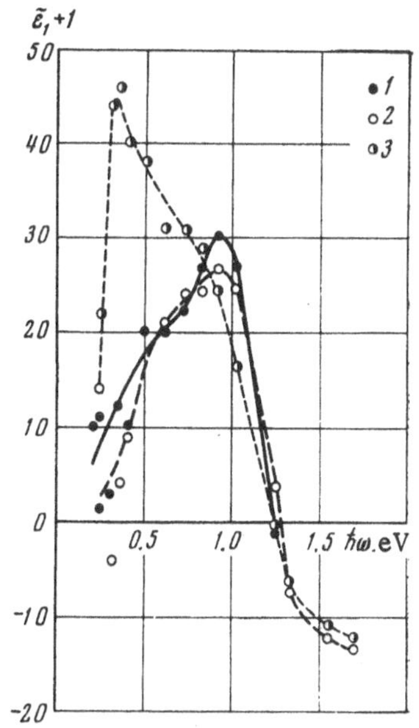

Fig. 70. Dependence of the interband dielectric constant $\tilde{\varepsilon}_1$ on $\hbar\omega$ for tin at three temperatures: 1) 4.2°K; 2) 78°K; 3) 293°K.

CHAPTER VII

DISCUSSION OF THE RESULTS OBTAINED

§ 1. Characteristics of the Conduction Electrons of Metals Obtained from Measurements in the Long-Wave Region

Table 37 presents a summary of the main electron characteristics of the metals studied. In calculating N_a, v_F^0, and S_F^0 for lead we allowed for the temperature dependence of the lattice constant a. For the other metals we used the lattice-constant values relating to room temperatures, since allowing for $a(T)$ made very little difference to the parameters indicated.

We see from Table 37 that, in the case of the monovalent metal gold, for which the Bragg planes fail to intersect the sphere of free electrons, N is close to N_{val}, despite the fact that (as we shall demonstrate in Section 11 of this chapter), the ratio $|V_{111}|/E_F^0 \approx 0.3$.

For all the polyvalent metals studied, the Bragg planes intersect the sphere of free electrons, and for these N is much smaller than N_{val}. This qualitatively confirms the influence of the Bragg planes on N mentioned in Section 3 of Chapter II. A quantitative verification of the conclusions of this section will be presented in Section 5 of the present chapter. Here we simply note one fact. For all the "good" metals the Fourier components of the pseudopotential are such that the ratio N/N_a for T = 4.2°K is close to unity. On increasing the temperature, N rises. A discussion of the temperature dependence of N will be presented in Section 6. It is interesting to note that for molten metals $N \approx N_{val}$ (see Hodgson [141]).

Table 37 also gives the average velocity of the electrons on the Fermi surface $\langle v_F \rangle$ and the total area of the Fermi surface S_F calculated from the approximate equation (II.33a). For comparison we also give the analogous characteristics v_F^0 and S_F^0 relating to free electrons for a concentration of these equal to the valence concentration. It follows from the table that for all our polyvalent metals $\langle v_F \rangle$ and S_F are much smaller than v_F^0 and S_F^0, respectively. This is also simply due to the effect of the Bragg planes. No "phonon overgrowth" of electrons close

TABLE 37. Principal Electron Characteristics of the Metals Studied, Obtained from Measurements in the Long-Wave Region

Metal	$T°$, K	N/N_a	N/N_{val}	$v_F^0 \cdot 10^{-8}$, cm/sec	$\langle v_F \rangle \times 10^{-8}$, cm/sec	$S_F^0 \cdot 10^{37}$, $g^2 \cdot cm^2/sec^2$	$S_F \cdot 10^{37}$, $g^2 \cdot cm^2/sec^2$	$\nu \cdot 10^{-14}$, sec^{-1}
Au	293	0.95	0.95	1.39	1.35	2.01	1.96	0.46
Al	78	1.25	0.417	2.02	1.30	4.26	2.75	0.72
	295	1.31	0.433	2.02	1.33	4.26	2.80	1.21
In	4.2	1.42	0.473	1.74	1.20	3.16	2.17	0.58
	295	1.76	0.587	1.74	1.34	3.16	2.42	2.24
Pb	4.2	1.11	0.278	1.84	0.97	3.52	1.87	0.41
	78	1.15	0.288	1.83	0.98	3.50	1.86	0.87
	293	1.23	0.308	1.82	1.01	3.46	1.94	3.07
Sn	4.2	1.12	0.280	1.89	1.00	3.72	1.97	0.53
	78	1.12	0.280	1.89	1.00	3.72	1.97	0.77
	293	1.30	0.325	1.89	1.08	3.72	2.12	2.27

to the Fermi surface appears in optics. This question was considered in detail by Kogan [142] in particular. In contrast to optical measurements, measurements of the specific heat and effective masses in the de Haas–van Alphen effect should be sensitive to the phonon overgrowth in question. Without allowing for this fact we are unable to compare the results obtained for the average velocity of the electrons on the Fermi surface from optical measurements and from the foregoing effects. *

Finally, Table 37 gives the effective collision frequencies ν. This quantity exhibits a sharp temperature dependence, which we shall consider in Section 9 of this chapter.

§ 2. Determination of the Fourier Components of Pseudopotential from Optical Measurements

It was shown in Chapter II that the presence of Bragg reflections led to the appearance of maxima in the interband conductivity $\tilde{\sigma}$. Our investigations into interband conductivity in the visible and near-infrared part of the spectrum for a number of polyvalent metals confirmed the existence of $\tilde{\sigma}$ maxima. The position of the maxima enabled us to determine the Fourier components of the pseudopotential. For this purpose we used Eq. (II.101).

It clearly follows from the results of the previous chapter that the spectral range which we have been considering usually contains two wide overlapping $\tilde{\sigma}_g$ bands corresponding to the two Fourier components of the potential which play the principal part in optics. The widths of the bands, and hence their degree of overlapping, diminish with falling temperature. Hence the absolute values of the Fourier components V_g may be determined more accurately at helium temperatures. However, the optical method, in contrast to the de Haas–van Alphen effect and cyclotron resonance, enables us to derive the Fourier components of the pseudopotential at high temperatures as well.

So far the most reliable experimental values of V_g have been obtained from the de Haas–van Alphen effect for aluminum and lead. Let us compare the results of the optical method with the foregoing data. First of all, we consider aluminum and lead and then indium and tin, for which no such detailed analyses of V_g by the other (nonoptical) methods have been carried out. In the tables of this section, V_g will be given in electron volts.

*We failed to allow for this in our previous publications.

TABLE 38. Determination of the Fourier Components of Pseudopotential (in eV) for Aluminum (T = 295°K)

From optical measurements		From the de Haas–van Alphen effect		Calculation [28]	
$\|V_{111}\|$	$\|V_{200}\|$	V_{111}	V_{200}	V_{111}	V_{200}
0.22	0.72	+0.24	+0.76	+0.27	+0.78

A. Aluminum

Aluminum has a face-centered cubic lattice. The sphere of free electrons is intersected by the {111} and {200} planes (Fig. 4). This should lead to the appearance of two bands of interband conductivity. Experiment confirms the existence of two $\tilde{\sigma}_g$ bands. In identifying these we used the inequality $|V_{200}|>|V_{111}|$, which follows from the de Haas–van Alphen effect for this particular metal. The observed maxima of $\tilde{\sigma}_{200}$ and $\tilde{\sigma}_{111}$ enable us to find $|V_{200}|$ and $|V_{111}|$.

The values of these Fourier components of pseudopotential obtained by different methods are shown in Table 38. In determining these by the optical method we used the results of Table 32. For V_{200} the coefficient $t_{200} = 1.04$, which corresponds to $\gamma'_{200} = 0.09$. For V_{111} we neglected the deviation of t_{111} from unity. (The difference $t_{111} - 1$ is much smaller than the error obtained for ω_{max}.) The error obtained for $|V_{200}|$ was about 1% and for $|V_{111}|$ about 15%.

Using the de Haas–van Alphen effect, Ashcroft [61] obtained V_{200} and V_{111} to an error of 2-5%.

The results of the calculation were taken from Heine [28]. The error was about 17% for V_{200} and 50% for V_{111}.

The optical measurements of V_{111} and V_{200} relate to room temperature. The de Haas–van Alphen measurements were carried out at helium temperatures. Since the temperature dependence of the Fourier components of the pseudopotential of aluminum is only slight, as indicated by the slight temperature dependence of N, we may compare the quantities presented in the table without making any special allowance* for the temperature dependence of V_{111} and V_{200}.

It follows from Table 38 that the results of the determinations of V_{200} and V_{111} by the two experimental methods agree very closely with each other and with the calculated data.

B. Lead

Like aluminum, lead has a face-centered cubic lattice. The sphere of free electrons is intersected by the Bragg planes {111} and {200}. As in the case of aluminum, this should give two bands of interband conductivity. Experiment in fact confirms two $\tilde{\sigma}_g$ bands; in order to identify these we use the relation $|V_{111}| > |V_{200}|$ obtained from the de Haas–van Alphen effect for this metal [62]. The observed $\tilde{\sigma}_{111}$ and $\tilde{\sigma}_{200}$ bands enable us to find $|V_{111}|$ and $|V_{200}|$.

The Fourier components of the pseudopotential V_{111} and V_{200} obtained by the different methods are shown in Table 39. The components were determined by the optical method at three temperatures using the results of Table 34. The value of $|V_{111}|$ was found both from the frequency corresponding to the center of the band (ω_0) and from the frequency corresponding to the left-hand maximum ($\omega_{1\,max}$). We used the t_g determined by the values of γ'_g given in the

*Allowance for the temperature dependence of V_{111} and V_{200} makes the agreement between the results of the determinations based on the two experimental methods even better.

same Table 34. The error for $|V_{200}|$ was 1.5-3%. The error for V_{111} related to the error in $\hbar\omega_{1\,max}$ or $\hbar\omega_0$ was about 1%. The difference in the values of $|V_{111}|$ derived from $\hbar\omega_{1max}$ and $\hbar\omega_0$ was about 10%.

The V_{111} and V_{200} values were determined from the de Haas–van Alphen effect by Anderson and Gold [62]. The corresponding measurements were carried out at 1-4.2°K. The accuracy in the determination of V_{111} and V_{200} was 2.5 and 5%, respectively.

The results of the calculation were taken from Heine [28]. The error for V_{111} and V_{200} was 12 and 30%, respectively.

It follows from Table 39 that the values of the Fourier component of the pseudopotential V_{111} obtained by the two experimental methods agree closely with each other and with the calculations of Heine [28]. This component of the pseudopotential plays the main part in the optical properties of lead. The values of the Fourier component V_{200} obtained by the different methods lie close together, but the difference exceeds the measuring error. The reasons for this are not yet clear.

Table 39 also shows the temperature dependence of V_{200}. With increasing temperature, $|V_{200}|$ diminishes. It is difficult to be specific regarding the temperature dependence of V_{111}, but it would appear to be considerably weaker.

C. Indium

Indium has a face-centered tetragonal lattice. It was shown in Section 7 of Chapter II that the sphere of free electrons was intersected by eight {111} planes, four (200) and (020) planes, and two (002) planes. We may reasonably expect that there will be three bands of interband conductivity. Experiment reveals only two $\tilde{\sigma}_g$ bands. So far as we know there has been no detailed analysis of the results obtained by studying the de Haas–van Alphen effect, similar to that carried out for aluminum or lead. It is accordingly less easy to identify the observed bands uniquely as in the case of the other two metals. Allowing for the foregoing values of n_g, which give the number of physically equivalent Bragg planes, we may suppose that the principal maximum in the region of 1.5 eV will be associated with the {111} planes and the second with the {200}. As already indicated, we failed to observe any third maximum in the infrared region. The observed $\tilde{\sigma}_g$ maxima enable us to find $|V_{111}|$ and $|V_{200}|$. The values of the Fourier components of pseudopotential V_{111}, V_{200}, and V_{002} obtained by the different methods are presented in Table 40.

Using the optical method $|V_{111}|$ and $|V_{200}|$ were determined at helium and room temperatures by means of the results presented in Table 29. For the coefficient t_g we took the values

TABLE 39. Determination of the Fourier Components of the Pseudopotential for Lead (in eV)

T, °K	From optical measurements			From the de Haas–van Alphen effect		Calculation [28]	
	$\|V_{111}\|$		$\|V_{200}\|$	V_{111}	V_{200}	V_{111}	V_{200}
	from $\omega_{1\,max}$	from ω_0					
4.2	1.00	1.11*	0.70	−1.14	−0.53	−1.16	−0.44
78	1.02	1.13	0.65	—	—		
293	1.06	1.17	0.51	—	—		

* This value of V_{111} is obtained using the value of V_{111} = 1.13 relating to T = 78°K and allowing for the temperature dependence of V_{111} determined from the previous column.

TABLE 40. Determination of the Fourier Components of the Pseudopotential of Indium (in eV)

T, °K	From optical measurements			From cyclotron resonance [70]			Calculation [143]		
	$\|V_{111}\|$	$\|V_{200}\|$	$\|V_{002}\|$	$\|V_{111}\|$	$\|V_{200}\|$	$\|V_{002}\|$	$\|V_{111}\|$	$\|V_{200}\|$	$\|V_{002}\|$
4.2	0.70	0.28	—	0.31	0.25	<0.07	0.36	0.46	0.10
295	0.59	0.27	—	—	—	—	0.36	0.46	0.10

corresponding to the values of γ'_g given in the same table. The error was 1.5% at helium and 2% at room temperatures for $|V_{111}|$, and 2 and 7%, respectively, for $|V_{200}|$.

The calculation of V_{111}, V_{200}, and V_{002} from measurements of cyclotron resonance was carried out by Mina and Khaikin [70], the measurements being made at helium temperatures. The errors for $|V_{111}|$ and $|V_{200}|$ were 30 and 20%, respectively. For $|V_{002}|$ only the upper limit was obtained.

The results of the theoretical calculation are taken from Animalu and Heine [143]. The error was 40 and 30% for V_{111} and V_{200}, respectively; for V_{002} it was over 100%.

We see from Table 40 that the values of $|V_{200}|$ obtained from the two experiments agree with each other. The difference is considerable for $|V_{111}|$, far greater than the measuring error. The reason for the difference is not altogether clear.

The table also clearly shows the temperature dependence of $|V_{111}|$. With increasing temperature, $|V_{111}|$ falls. It is hard to be specific regarding the temperature dependence of $|V_{200}|$, since the changes in this quantity are smaller than the measuring errors.

D. Tin

Tin has a complex tetragonal lattice. According to Table 4 the sphere of free electrons is intersected by four (200) and (020) planes, eight (101) and (011) planes, four (220) planes, and sixteen (211) and (121) planes. These may all be divided into two groups with approximately the same distances from the center of the zone Γ. The first group, with a ratio of $p_g/p_F^0 \approx 0.67$, contains the (200), (020), (101), and (011) (twelve planes in all). The second group, with a ratio $p_g/p_F^0 \approx 0.94$, contains the (220), (211), and (121) (twenty planes in all). We may therefore expect that there will be two $\tilde{\sigma}_g$ bands with a complex structure. Experiment in fact reveals two $\tilde{\sigma}_g$ bands. It is difficult to say anything specific regarding the structure of the bands, as the measurements were made with too large a step in λ. A detailed analysis of the de Haas–van Alphen effect similar to that carried out for aluminum or lead has (so far as we know) never been attempted in the case of tin. Our identification of the $\tilde{\sigma}_g$ bands observed is therefore not entirely unambiguous. Some idea as to the interpretation of the bands may be obtained from measurements of the concentration of the conduction electrons N; this will be considered in Section 5. Here we denote the potentials found from the $\tilde{\sigma}_g$ maxima as V_1 and V_2.

The results of the determination of $|V_1|$ and $|V_2|$ at helium, nitrogen, and room temperatures are presented in Table 41. In calculating $|V_1|$ and $|V_2|$ the data presented in Table 36 were employed. The coefficient t_1 for the $\tilde{\sigma}_1$ band was taken in accordance with the measured values of γ_1'. For the band $\tilde{\sigma}_2$ the difference between t_2 and unity could be neglected (it is inappropriate here to allow for the difference $t_2 - 1$ in view of the large error associated with $\omega_{2\,max}$). The error relating to $|V_1|$ was 8-9% at all temperatures, and that relating to $|V_2|$, ~ 20-25%.

TABLE 41. Determination of the Fourier Components of the Pseudopotential for Tin from Optical Measurements

T, °K	$\lvert V_1 \rvert$	$\lvert V_2 \rvert$
4.2	0.61	0.30
78	0.60	0.30
293	0.57	0.22

In regard to the temperature dependence of $|V_1|$ and $|V_2|$ it is hard to say anything. More detailed measurements are required.

In summarizing this section, we should note the fair degree of agreement between the Fourier components of the pseudopotential derived from optical and de Haas—van Alphen measurements, respectively.

The optical method is the more direct; it enables us to determine the Fourier components of the pseudopotential over a wide temperature range.

The results also confirm our assumption to the effect that the structures of the bands of interband conductivity characterizing polyvalent metals in the visible and near-infrared parts of the spectrum are mainly associated with the Bragg energy splitting.

§ 3. Shape of the Bands of Interband Conductivity and Absolute Values of $\tilde{\sigma}_g$

The shape of the bands of interband conductivity is easiest to follow in the case of aluminum, since its principal $\tilde{\sigma}_{200}$ band is fairly well separated from the other $\tilde{\sigma}_{111}$ band and lies in a spectral region in which the contribution of the conduction electrons to σ is small. In addition to this, no structure associated with spin-orbital interaction should appear in the aluminum bands.

In Fig. 61, the $\tilde{\sigma}_{200}$ band is indicated by a broken line. The parameter $\gamma'_{200} = 0.09$. We may compare the shape of this band with the shape of the band calculated in Section 9 of Chapter II for $\gamma' = 0.1$ and shown in Fig. 11. Comparison of the two figures indicates that the shapes of the two bands agree reasonably closely with each other.

Let us compare the experimental and theoretical* absolute values of the bands $\tilde{\sigma}_g$. In order to calculate $\tilde{\sigma}_g$ we must know the relative broadening of the energy levels $\gamma'_g = \gamma_g/(\hbar\omega_g)$, which we determined from the shape of the long-wave edge of the experimental $\tilde{\sigma}_g(\omega)$ curve by the method indicated earlier.

The results of our comparison of the theoretical and experimental maximum values of the $\tilde{\sigma}_g$ bands for aluminum, indium, and lead are presented in Table 42. The same table gives the values of n_g, p_g, and γ'_g used in the calculation.

The number of Bragg planes and the values of the ratios $p_g/(2\pi\hbar/a)$ for aluminum and lead were taken from Table 3. For indium we assumed $n_{200} = 6$. This is equivalent to the as-

*In [20] and [35] and in the original Dissertation, an error was committed in the numerical determination of the coefficient $e^2/(12\pi^2\hbar^2)$ in Eq. (II.99). For this reason the calculated values of $\tilde{\sigma}_g$ given in these publications should be multiplied by four.

TABLE 42. Comparison of the Theoretical and Experimental Maximum Values of $\tilde{\sigma}_{111}$ and $\tilde{\sigma}_{200}$

Metal	T, °K	Band {111}					Band {200}				
		n_{111}	$p_{111}\cdot 10^{19}$, g-cm/sec	γ'_{111}	$\tilde{\sigma}_{max}\cdot 10^{-14}$, cgse exptl.	$\tilde{\sigma}_{max}\cdot 10^{-14}$, cgse theor.	n_{200}	$p_{200}\cdot 10^{19}$, g-cm/sec	γ'_{200}	$\tilde{\sigma}_{max}\cdot 10^{-14}$, cgse exptl.	$\tilde{\sigma}_{max}\cdot 10^{-14}$, cgse theor.
Al	295	8	—	—	—	—	6	1.64	0.09	46.8	96
In	4.2	8	1.22	0.16	34.3	69	6	1.44	0.25	30.1	47
	295	8	1.22	0.22	25.7	57	6	1.44	0.36	22.4	38
Pb	4.2	8	1.17	0.10	32.3	86	6	1.35	0.20	12.9	51
	78	8	1.165	0.15	29.4	68	6	1.345	0.20	12.1	50
	293	8	1.16	0.21	28.5	56	6	1.34	0.27	11.6	42

sumption that V_{200} and V_{002} lie close together and are not resolved in the optical experiments. The basis for this assumption is, first, that we found no V_{002} band in the infrared region and, second, that the $\tilde{\sigma}_{200}$ band in indium has a slight structure, as mentioned in Chapter VI. The ratio p_{200}/p_F^0 for indium was taken from Eq. (II.67).

The lattice constants a at room temperature were taken from the Ormont handbook [67]. For lead we allowed for the change in lattice constant with changing temperature.

The experimental values of $\tilde{\sigma}_{max}$ and γ' for aluminum, indium, and lead were taken from Tables 32, 29, and 34, respectively. These tables give the errors for both $\tilde{\sigma}_{max}$ and γ'.

We see from Table 42 that the experimental values for the principal bands lie approximately a factor of two below the calculated values.* This difference is apparently associated, first, with the fact that the ΔE in Eq. (II.79) varies slightly along the plane of the ring M_1M_2 (Fig. 8) and, second, with the fact that Eq. (II.79) is not satisfied in the region of intersection of several Bragg planes.

On the whole we may consider that the results of this section, like those of the previous one, support our own point of view as to the structure of the bands of interband conductivity exhibited by polyvalent metals in the visible and near-infrared parts of the spectrum.

§ 4. Effective Collision Frequencies of the Conduction Electrons and Electrons Taking Part in Interband Transitions

Let us compare the effective collision frequencies of the conduction electrons ν with the effective collision frequency of the electrons taking part in the interband transitions ν_g. We estimate the latter from the indeterminacy relation $\Delta E \Delta \tau \approx 2\pi\hbar$. In the present case, $\Delta E \approx 2\gamma_g$, $\Delta\tau \approx 2\pi/\nu_g$, so that

$$\nu_g \approx 4\gamma'_g \mid V_g \mid /\hbar. \tag{VII.1}$$

The values of ν_g and ν obtained from the results of Chapters V and VI are given in Table 43. We see from the table that $\nu_g > \nu$. The temperature dependence of ν_g is much smaller than that of ν; correspondingly, at low temperatures, $\nu_g \gg \nu$.

*The theories based on the special role of the high-symmetry points give values 5-10 times smaller than experimental for $\tilde{\sigma}_g$ [75].

TABLE 43. Collision Frequencies of the Conduction Electrons and the Electrons Taking Part in Interband Transitions

Metal	T, °K	Interband transitions		Conduction electrons
		$\nu_{111}\cdot 10^{-14}$, sec^{-1}	$\nu_{200}\cdot 10^{-14}$, sec^{-1}	$\nu\cdot 10^{-14}$, sec^{-1}
Al	295	—	2	1.2
In	4.2	7	4	0.58
	295	8	6	2.24
Pb	4.2	7	8	0.41
	78	10	8	0.87
	293	15	8	3.07
Sn	4.2	4*	—	0.53
	78	6*	—	0.77
	293	9*	—	2.27

* For tin the values given relate to the first band, i.e., $\nu_{111} = \nu_1$.

The short lifetimes of the excited states may be associated with strong interelectron interaction. This interaction is greater than the corresponding interaction for electrons on the Fermi surface, since the energies of the ground and excited states differ from the Fermi energy by an amount much greater than kT. Furthermore, owing to the fact that the isoenergy bands are parallel to one another, the phase volume into which an electron may be scattered with energy conservation is very great.

The interelectron interaction may lead to the collective Auger effect mentioned by Phillips and Nozières [73, 144], which also reduces the lifetime of the excited state.

Finally, there is yet another factor reducing the lifetime; this is the interaction of the electrons with phonons. It is well known that the interaction of the electrons with the static lattice potential near a Bragg plane is so great, in view of the corresponding phase relationships, that these electrons cannot propagate through the crystal. We may therefore expect that the interaction of these electrons with phonons will be considerably greater than that of electrons lying a long way from the Bragg plane.

§ 5. Comparison of the Values of N Obtained from Measurements in the Long-Wave Region with Those Calculated from the Fourier Components of the Pseudopotential

It was shown in Section 3 of Chapter II that the Fourier components of the pseudopotential led to the inequality $N < N_{val}$. Let us compare the results of the section in question with experiment. Table 44 shows the values of N/N_a derived from measurements in the long-wave part of the spectrum and also those calculated from the values of $|V_g|$ determined by optical measurements in the short-wave region.

In the calculation we used (II.42) and (II.43) for aluminum and indium (neglecting the slight tetragonality of indium), (II.45), and (II.46) for lead, and (II.74a) and (II.75a) for tin. The values of the Fourier components of the pseudopotential were taken from Tables 38-41. In the case of tin, we assumed that the main band of interband transitions $\tilde{\sigma}_1$ was associated with the Fourier components $V_{101} \approx V_{200}$, and the second band $\tilde{\sigma}_2$ with the Fourier components $V_{220} \approx V_{211}$. If the reverse were the case, the experimental results would be contradicted.

TABLE 44. Concentration of Conduction Electrons Obtained from Measurements in the Long-Wave Region and from the Fourier Components of the Pseudopotential

Metal	T, °K	N/N_a		N_{val}/N_a
		from measurements in the long-wave region	from $\|V_g\|$	
Al	78	1.25	—	3
	295	1.31	2.02	3
In	4.2	1.42	1.52	3
	295	1.76	1.70	3
Pb	4.2	1.11	1.13	4
	78	1.15	1.13	4
	293	1.23	1.25	4
Sn	4.2	1.12	0.96	4
	78	1.12	0.99	4
	293	1.30	1.41	4

The values of E_F^0 for lead at 4.2, 78, and 293°K, respectively, equalled 9.62, 9.52, and 9.46 eV. The temperature dependence of E_F^0 was slight and had hardly any effect on the results of the calculation. Hence, in the case of the other metals, no allowance was made for this factor. For aluminum, indium, and tin we used the following values of E_F^0: 11.6, 8.63, and 10.2 eV. The error in determining N/N_a from measurements in the long-wave region was 1-4%. The error in calculating N from $|V_g|$ was 10-30%.

We see from Table 44 that in indium, lead, and tin at all temperatures the concentration of the conduction electrons N obtained from measurements in the long-wave region practically coincides with the value of N calculated from the Fourier components of the pseudopotential. Thus, for these metals, the difference $N_{val} - N$ is mainly determined by the periodic lattice potential. In aluminum the N value obtained from the long-wave measurements is smaller than the N calculated from $|V_g|$. The question as to the reason for these differences requires further consideration. It is quite possible that the reason lies principally in the effect of interactions between the electrons discussed in [145].* However, even for aluminum the difference $N_{val} - N$ is mainly determined by the periodic lattice potential.

Thus the theory relating N to $|V_g|/E_F^0$ developed in Chapter II is confirmed. For tin the identification of the experimentally determined Fourier component V_1 with the Fourier components $V_{101} \approx V_{200}$ becomes still more probable.

*In papers by Silin [146], Pitaevskii [47], and Gurzhi [48] it was shown that the application of the Landau theory of the Fermi fluid to the optical properties of metals led to the following expression for N in cubic crystals:

$$\frac{N}{m} = \frac{2}{3(2\pi\hbar)^3} \oint \frac{v_F V_F}{v_F} dS_F,$$

$$V_F = v - \int \Phi(\mathbf{p}, \mathbf{p}')(\partial f_0/\partial \mathbf{p}') \frac{2d^3p'}{(2\pi\hbar)^3}$$

Here f_0 is the equilibrium distribution function, $\Phi(\mathbf{p}, \mathbf{p}')$ is the second variational derivative of the energy of the system with respect to the distribution function. These relations indicate yet another reason for the difference between N and N_{val}. However, for the majority of metals, the Fermi fluid effects are less important than those of the Fourier components of the pseudopotential.

We note that, as indicated by Eq. (II.31), the effect of Bragg planes on N is the greater, the greater V_g and n_g (n_g is the number of physically equivalent Bragg planes). For polyvalent metals, $E_F^0 \sim 10$ eV, $n_g \sim 10$ eV, so that if $V_g \leqslant 0.1$ eV the effect of the corresponding Bragg planes on N is only slight. This means that only the bands of interband conductivity situated in the visible and near-infrared parts of the spectrum have an appreciable influence on N.

§ 6. Temperature Dependence of the Concentration of Conduction Electrons in Metals

It was shown in Section 10 of Chapter II that thermal vibrations, in disrupting the periodicity of the lattice, reduced the Fourier components of pseudopotential and increased the concentration of the conduction electrons. The temperature dependence of N became apparent in every case which we studied, N tending to increase with rising temperature, as may clearly be seen from Fig. 71 and Table 44. The increase in N on passing from nitrogen to room temperature is greater than on passing from helium to nitrogen temperature. For indium and aluminum the measurements were only made at two temperatures, so the corresponding points in Fig. 71 are joined by straight lines.

It follows from Table 44 that the temperature dependence of N for indium, lead, and tin is determined by the temperature dependence of $|V_g|$. For aluminum we have no data regarding the temperature dependence of V_{200} and therefore cannot estimate the change in N associated with the change in $|V_g|$.

However, for aluminum and lead we do have some x-ray measurements of the Debye–Waller factor, which we may compare with the optical measurements of the temperature dependence of N [33]. Both metals are cubic, and the approximate estimate of the temperature factor presented in Section 10 of Chapter II may be applied to them with a greater credibility than to metals with a noncubic lattice.

In experiments by Chipman [147] relating to the temperature dependence of the intensity of x-ray diffraction maxima, the Debye–Waller factor was determined over the whole range between nitrogen and room temperature.* Taking the same Debye model as ourselves, Chipman used the value of W to find the corresponding Debye temperature θ_x, given in Table 45. By using these values of θ_x for aluminum and lead together with Eqs. (II.119) to (II.121), we may calculate the change in the concentration of conduction electrons on passing from nitrogen to room temperature. The results of the calculation are presented in Table 45, where T_K = 293°K, T_N = 78°K. In applying Eq. (II.121) we used the value of $N_{val}/N(T_N) - 1$ obtained in the optical experiments. As mean value of $\overline{W}_g(T)$ we took a value corresponding to the index g for which the effect of V_g on N was greatest, i.e., for aluminum, $\overline{W}_g \approx W_{200}$; for lead, $\overline{W}_g \approx W_{111}$.

The last column of Table 45 gives the temperature dependence of N based on optical experiments. The temperature range is the same as for the x-ray experiments.

We see from Table 45 that the changes in the conduction-electron concentration deduced from the optical and x-ray experiments agree with each other. The optical measurements show that the temperature dependence of the N of tin and indium is several times greater than that of aluminum and lead. The two metals in question are tetragonal. Unfortunately, we know of no x-ray data for the temperature factor of these metals.

The change in N(T) should be much greater at high temperatures than at low. This conclusion is qualitatively supported by the determination of the optical constants of tungsten at

*The temperature factor of the intensity of the x-ray maxima equals $\exp[-2M(T)]$, where M(T) may be obtained from the formulas for W on replacing p_g by $2\pi\hbar(\sin\theta)/\lambda$.

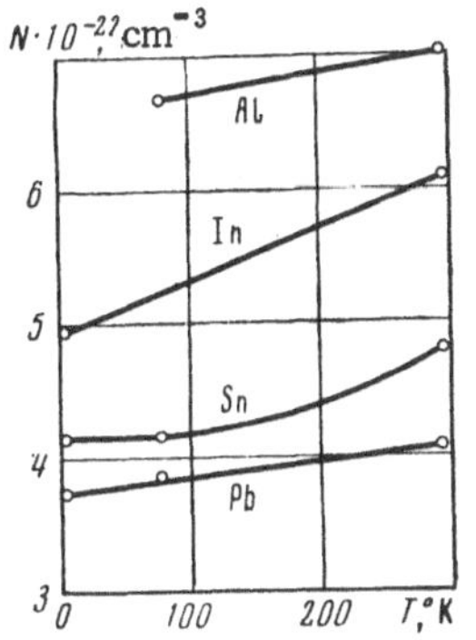

Fig. 71. Temperature dependence of the concentration of conduction electrons based on measurements in the long-wave part of the spectrum.

TABLE 45. The N(T) Relationship for Al and Pb Obtained by Optical and X-Ray Methods

Metal	θ_X, K°	$N(T_K)/N(T_N)$	
		x	optical
Al	405	1.04	1.05
Pb	89	1.11	1.07

high temperatures presented in [148]. Unfortunately, in this paper, the optical constants were only determined in the near-infrared, and the contribution from interband transitions may be considerable in this region.

Thus experiment confirms the theory developed in Section 10 of Chapter II. With increasing temperature the concentration of the conduction electrons in metals becomes greater. The rise in N is associated with the temperature dependence of the Fourier component of the pseudopotential.

§ 7. Area of the Fermi Surface and Mean Velocity of the Electrons on the Latter

The results obtained enable us to determine the area of the Fermi surface S_F and the average velocity of the electrons on the surface $\langle v_F \rangle$. These quantities may be found by the optical method if we use the values of $|V_g|$ obtained from measurements in the short-wave region and N from measurements in the long-wave region. In addition to this, S_F may be obtained from measurements of the anomalous skin effect in the radio-frequency range. The results of the determination of S_F and $\langle v_F \rangle$ by the different methods are given in Table 46.

The calculation of S_F and $\langle v_F \rangle$ using the values of $|V_g|$ may be based* on either (II.31) and (II.32) or the approximate Eq. (II.33a). The accuracy of the former equations will be considered shortly. The accuracy of the approximation Eq. (II.33a) is related to the difference between the average values of $\langle 1/v_F \rangle$ and $1/\langle v_F \rangle$ on the Fermi surface. It is difficult to estimate this accuracy a priori; however, we see from Table 46 that the results obtained on determining $\langle v_F \rangle$ and S_F from the approximate equation differ little from the corresponding results based on the more accurate equations. The difference is 2-10%. Thus we may use the

*For tin the calculations were not based on (II.31) and (II.32), as the identification of the bands V_1 and V_2 was uncertain.

TABLE 46. Area of the Fermi Surface S_F and Average Velocity of the Electrons on the Surface $\langle v_F \rangle$

Metal	T, °K	S_F/S_F^0			
		from $\lvert V_g \rvert$		from measure., long-range region (II.33a)	from measure., radio-frequency region
		(II.31)	(II.33a)		
Al	78	—	—	0.64	1.0
	295	0.81	0.82	0.66	
In	4.2	0.72	0.71	0.69	
	295	0.77	0.75	0.77	—
Pb	4.2	0.61	0.53	0.53	0.55
	78	0.61	0.53	0.54	
	293	0.62	0.56	0.56	
Sn	4.2	—	0.49	0.53	0.55
	78	—	0.50	0.53	
	293	—	0.59	0.57	

Metal	T, °K	$\langle v_F \rangle / v_F^0$			$S_F^0 \cdot 10^{37}$, g^2-cm^2/sec^2	$v_F^0 \cdot 10^{-8}$ cm/sec
		from $\lvert V_g \rvert$		from measure., long-range reg. (II.33a)		
		(II.32)	(II.33a)			
Al	78	—	—	0.64	—	—
	295	0.83	0.82	0.66	4.25	2.02
In	4.2	0.70	0.71	0.69	—	—
	295	0.74	0.75	0.77	3.16	1.74
Pb	4.2	0.46	0.53	0.53	3.52	1.84
	78	0.46	0.53	0.54	3.50	1.83
	293	0.50	0.50	0.56	3.46	1.82
Sn	4.2	—	—	0.53	—	—
	78	—	—	0.53	—	—
	293	—	—	0.57	3.73	1.89

approximate Eq. (II.33a) in analyzing experimental results in the case of the weakly anomalous skin effect.

In order to estimate the accuracy of Eq. (II.31), we may compare the ratio S_F/S_F^0 calculated from this equation (based on the solution of the second-order secular equation) with the ratio S_F/S_F^0 based on the solution of the fourth-order secular equation. The latter calculation was performed for lead by Anderson and Gold [62]. Using the values of V_{111} and V_{200} obtained in de Haas–van Alphen experiments (Table 39), Anderson and Gold obtained $S_F/S_F^0 = 0.589$. Using the same values of V_{111} and V_{200}, we obtained $S_F/S_F^0 = 0.638$ from (II.31). The difference is 8%. Thus, Eq. (II.31) gives the chief part of the change in S_F associated with the periodic lattice potential. The accuracy of (II.32) agrees with that of (II.31).

The calculation of S_F and $\langle v_F \rangle$ from measurements in the long-wave region was carried out by using the approximate Eq. (II.32a). We see from Table 46 that the determination of these quantities from the results of measurements in the long- and short-wave parts of the spectrum agree with each other for indium, lead, and tin. For aluminum the S_F and $\langle v_F \rangle$ obtained from $\lvert V_g \rvert$ are greater than the corresponding quantities obtained from measurements in the long-wave region. This reflects the difference in the values of N determined in these two parts of the spectrum.

Fawcett [149] and Aubrey [150] determined S_F from measurements of the anomalous skin effect in the radio-frequency range to an accuracy of 10%.

We see from Table 46 that, for lead and tin, the results obtained on determining the area of the Fermi surface by the three methods agree with each other, subject to the limits of experimental error. For aluminum there is a difference which exceeds the measuring errors. The reason for this is not clear.

The results presented in Table 46 show that, in polyvalent metals, the total area of the Fermi surface and the average velocity of the electrons on it are considerably smaller than the values relating to free electrons for concentrations equal to the valence concentration. Since $S_F/S_F^0 \approx \langle v_F \rangle / v_F^0$, we consider that there is little point in introducing the concept of the optical effective mass m_{opt} in metals, using the ratio $N/m = N_{val}/m_{opt}$. This definition of m_{opt} would only be appropriate if the difference between N_{val} and N were determined by the difference between $\langle v_F \rangle$ and v_F^0.

§ 8. Density of States of the Electrons on the Fermi Surface

It was shown in Section 3 of Chapter II that to a first approximation in $|V_g|/E_F^0$ the periodic potential of the lattice had no effect on the density of states dY/dE, as indicated by Eq. (II.16). A more accurate calculation based on a solution of the fourth-order secular equation was carried out by Anderson and Gold for lead [62], obtaining $(dY/dE)_F/(dY/dE)_{0F} = 0.87$. Here $(dY/dE)_F$ is the density of states of the electrons on the Fermi surface; $(dY/dE)_{0F}$ is the analogous quantity for free electrons.

We see that the foregoing ratio differs little from unity. We may thus consider that Eq. (II.16) is satisfied to an accuracy of ≈10%.

Experimentally the density of states on the Fermi surface is found from the electron specific heat $c_e = \gamma t$. The value of γ is proportional to $(dY/dE)_F$. The experiments of Phillips, Lambert, and Gardner [151] and Decker, Mapother, and Shaw [152] give $\gamma/\gamma_0 = 1.93$ for lead (the subscript 0 relates to free electrons). The difference between this and the calculated value is very great. The reason apparently lies in the phonon overgrowth of the electrons on the Fermi surface. This effect was first considered by Migdal [153] and was developed in detail in a book by Abrikosov, Gor'kov, and Dzyaloshinskii [154]. The same effect is apparently responsible for the fact that the cyclotron masses measured experimentally by Khaikin and Mina [155] for lead were roughly a factor of 2.2 greater than the calculated values [62].

However, the phonon overgrowth in question influences the static characteristics, or the characteristics relating to frequencies smaller than the Debye frequency. In optics the phonon effect plays no serious part. This was shown by Kogan in [142].

§ 9. Temperature Dependence of the Electron – Phonon Collision Frequency

The electron–phonon collision frequency ν_{ep} determining the optical properties of metals depends on the temperature much less than the analogous frequency ν_{ep}^{cl} entering into the static conductivity. The reason lies in the quantum character of the interaction between light and the electron. This problem was discussed theoretically by Gurzhi [11, 12] and Holstein [13].

At high temperatures, for which $T > \theta_R$ and there are many phonons, the interaction between the electrons and the phonons leads to the same collision frequency in both optics and statics, i.e., $\nu_{ep} \approx \nu_{ep}^{cl}$ for $T > \theta_R$. Here, θ_R is the characteristic temperature obtained from the temperature dependence of the resistance, being close to the Debye temperature θ_D (Chap-

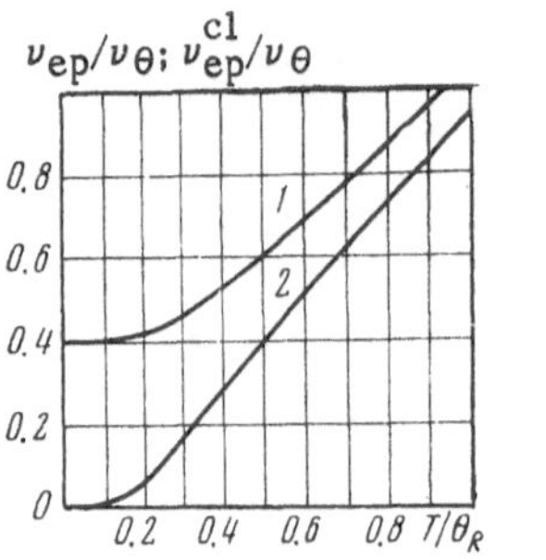

Fig. 72. Theoretical dependence of the electron–phonon collision frequency on reduced temperature. 1) Dependence of ν_{ep}/ν_θ on T/θ_R; 2) dependence of ν_{ep}^{cl}/ν_θ on T/θ_R.

ter I). At low temperatures $T \ll \theta_R$ there are few phonons, particularly few phonons of high energies. Hence, in static phenomena, an electron may hardly absorb any high-energy phonons at all. The electron may also fail to emit large phonons, since it has insufficient energy to do this. As a result of this we find that $\nu_{ep}^{cl}(T) \sim T^5$ for $T \ll \theta_R$. In optics an electron which has absorbed a light quantum is thrown a long way beyond the level of confusion on the Fermi surface; it cannot absorb large phonons, of which there are none, but it is able to emit a whole phonon spectrum. This leads to a high collision frequency. According to [11-13], $\nu_{ep} \to 0.4\nu_\theta$ for $T \ll \theta_R$, where $\nu_\theta = 0.94\nu_{ep}(\theta_R)$.

The expression for the function $\nu_{ep}(T)$ obtained by Gurzhi was given in Chapter I [see Eqs. (I.6) and (I.6a)]. Figure 72 shows the dependence of ν_{ep}/ν_θ and ν_{ep}^{cl}/ν_θ on T/θ_R. The figure demonstrates the large discrepancy between these curves for T values close to zero.

We carried out an experimental verification of the conclusions of the Gurzhi–Holstein theory [17-20, 22, 23, 50, 51]. For this purpose we required measurements of the optical constants n and $\varkappa$ over a wide temperature range, including helium temperatures. The subjects for study had to be metals in which the absorption was determined by the electron–phonon interaction at all temperatures, including helium.* In view of this requirement, we chose to study polyvalent metals: indium, lead, and tin. The detailed measurement of the optical constants of these metals in the infrared part of the spectrum at helium, nitrogen, and room temperatures enabled us to determine the character of the skin effect in these metals and take account of surface losses. Measurements of the static conductivity of the same samples carried out between helium and room temperatures enabled us to allow for the collision frequency of electrons with impurities and defects. As a result of this we were able to separate out the collision frequency of electrons with phonons ν_{ep} at the three temperatures in question and verify the theoretical temperature dependence of this frequency.

It should be noted that the corrections due to the anomalous nature of the skin effect were not very great. Even at helium temperatures these amounted to 20-25% (see Chapter V). Hence the slight inaccuracy associated with the fact that the equations for the corrections were derived for a spherical Fermi surface makes no difference to the result. The corrections associated with the presence of impurities and defects are also small, amounting to 2-10%.

*In papers by Biondi [10, 156] and Rayne [157], an attempt was made to separate out the absorption associated with the electron–phonon interaction in silver and copper at helium temperature. Only one quantity was measured: the absorption. Neither the character of the skin effect nor the relationship between volume and surface losses could therefore be estimated experimentally. The metals selected by these authors were not very suitable for the problem in hand, as the losses in them at helium temperatures were largely associated with the anomalous character of the skin effect. The resultant data cannot therefore be used to verify the Gurzhi–Holstein theory.

TABLE 47. Temperature Dependence of the Collision Frequency of Electrons with Phonons

Metal	Experiment: $\nu_{ep}\cdot 10^{-14}$, sec^{-1}, T_K	T_N	T_{He}	$\nu_{ep}(T_N)/\nu_{ep}(T_K)$	$\nu_{ep}(T_{He})/\nu_{ep}(T_K)$	θ_R, °K
In	2.19	—	0.54	—	0.25	190
Pb	3.06	0.86	0.40	0.28	0.13	85
Sn	2.21	0.72	0.48	0.33	0.22	187

Metal	Theory: $\nu_{ep}(T_N)/\nu_{ep}(T_K)$	Theory: $\nu_{ep}(T_{He})/\nu_{ep}(T_K)$	$\nu_{ep}(T_{He})/\nu_\theta$: exptl.	theor.	ν_{ep}/ν_{ep}^{cl}: T_K	T_{He}
In	—	0.25	0.41	0.40	1.39	10^6
Pb	0.29	0.12	0.48	0.40	1.23	$4\cdot 10^3$
Sn	0.34	0.25	0.36	0.40	1.34	10^6

$T_K = 295°$ K for In; 293° K for Pb, Sn; $T_N = 78°$ K, $T_{He} = 4.2°$ K.

Hence the slight error due to the application of the Matthiessen rule, allowing for ν_{ed}, also has no serious influence.

The experimental temperature dependences of $\nu_{ep}(T)$ for In, Pb, and Sn are presented in Table 47 (data taken from Chapter V). The fundamental qualitative effect is immediately obvious. The value of ν_{ep}, even at helium temperatures, remains finite and large, equal to (4-5) $\cdot 10^{13}$ sec^{-1}. For comparison we may mention that for gold at room temperature $\nu_{ep} = 4 \cdot 10^{13}$ sec^{-1}.

In order to compare theory with experiment, it is best to consider the ratio of the frequencies ν_{ep} relating to different temperatures. In this way all factors which are only determined to a low accuracy in the theory fall out. The experimental and calculated frequency ratios are shown in Table 47. In the calculation we used the value of the temperature θ_R measured independently, by reference to the temperature dependence of the electrical resistance under dc conditions. These values of θ_R are also given in the table. *

It follows from Table 47 that the theory gives the correct temperature dependence $\nu_{ep}(T)$. This may also clearly be seen from Fig. 73, in which the theoretical and experimental dependence of ν_{ep}/ν_θ on T/θ_R is presented. The value of ν_θ for indium, lead, and tin was determined by comparing the experimental values relating to room temperature with the theoretical curve. It follows from the figure that the values of $\nu_{ep}(T_N)$ and $\nu_{ep}(T_{He})$ lie excellently on the theoretical curve for all three metals. This once again supports the validity of the theoretical $\nu_{ep}(T)$ relationship.

Table 47 gives the experimental and theoretical values of the ratios $\nu_{ep}(T_{He})/\nu_\theta$. These values may be regarded as limiting, i.e., relating to T = 0. The agreement between theory and experiment is excellent. The average value of the ratio for the three metals is 0.42. The theoretical value is 0.40.

*We took no account of the dependence of θ_R on T, and used the value corresponding to room temperature. Allowance for the $\theta_R(T)$ dependence would increase the accuracy of the theoretical equation.

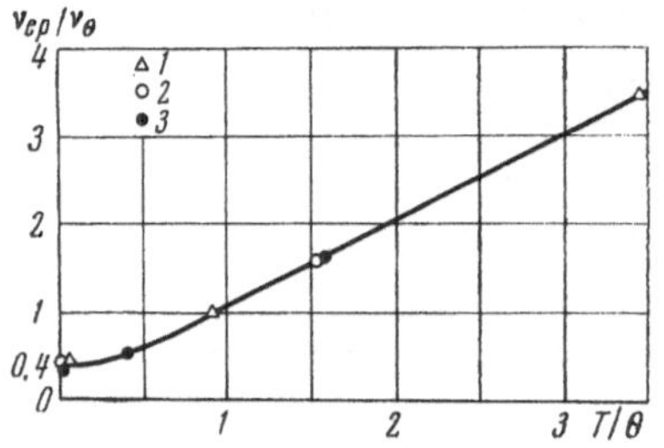

Fig. 73. Dependence of the electron–phonon collision frequency for indium, lead, and tin on the reduced temperature. Curve: theoretical; 1) lead; 2) indium; 3) tin.

The experimental data obtained in Chapter V also enable us to compare the experimental and theoretical absolute values of the electron–phonon collision frequency. According to theory, for $T > \theta_R$, $\nu_{ep} \approx \nu_{ep}^{cl}$. For indium, tin, and lead $T_K > \theta_R$; however, $\nu_{ep} > \nu_{ep}^{cl}$. The ratio ν_{ep}/ν_{ep}^{cl} for room temperature is given in the penultimate column of Table 47. It follows from the table that ν_{ep} is approximately 30% greater than ν_{ep}^{cl}. Thus the absolute values of ν_{ep} obtained from the experiments are approximately 30% greater than the theoretical. This may be associated with the imprecise allowance made by the theory for the high-frequency acoustic and optical oscillations.

The last column of Table 47 gives the ratio ν_{ep}/ν_{ep}^{cl} for helium temperature. We see that this ratio equals 10^4-10^6. Thus, at low temperatures, the frequency of the electron–phonon collisions in optics is many orders of magnitude greater than the classical static frequency.

§ 10. Indirect Interband Transitions of Electrons in Lead

In Section 5 of Chapter VI we noted that, apart from the $\tilde{\sigma}_{111}$ and $\tilde{\sigma}_{200}$ bands of interband transitions, lead exhibited a slight additional absorption in the range $\hbar\omega < 0.9$ eV. We believe that this may be associated with the indirect transitions of electrons lying close to the Fermi surface, which pass into an unfilled band in the neighborhood of the W point. The energy difference between the closest unfilled band at the point W and the Fermi level is 0.2-0.3 eV (according to Anderson and Gold [62] and Loucks [158]). This energy difference agrees with the threshold of the additional absorption estimated from our experimental data.

Another argument in favor of this assumption is the strong dependence of this absorption on temperature. It follows from the investigations of Moss [159] that the absorption associated with indirect transitions should be proportional to the quantity

$$\zeta = \frac{[\hbar\omega - (\Delta E)_{min} + k\theta]^2}{\exp(\theta/T) - 1} + \frac{[\hbar\omega - (\Delta E)_{min} - k\theta]^2}{1 - \exp(-\theta/T)}. \qquad \text{(VII.2)}$$

Here $(\Delta E)_{min}$ is the minimum width of the forbidden band, k is Boltzmann's constant, and θ is the Debye temperature. For $[\hbar\omega - (\Delta E)_{min}] \gg k\theta$, the formula simplifies:

$$\zeta \sim 1/[\exp(\theta/T) - 1] + 1/[1 - \exp(-\theta/T)]. \qquad \text{(VII.2a)}$$

For lead ($\theta \approx 86$°K) the ratio of the values of this quantity at room, nitrogen, and helium temperatures equals 7 : 2 : 1. The experimental values of $\tilde{\sigma}$ at the same temperatures in the range $\hbar\omega < 0.9$ eV yield the ratio 5 : 2 : 1, in fair agreement with the theoretical estimate.

§ 11. Effect of the Periodic Lattice Potential on the Optical Properties and Hall Constant of Gold

The principal subject for study in the present investigation has been a set of polyvalent metals. Information relating to the electron properties of these derived from optical measure-

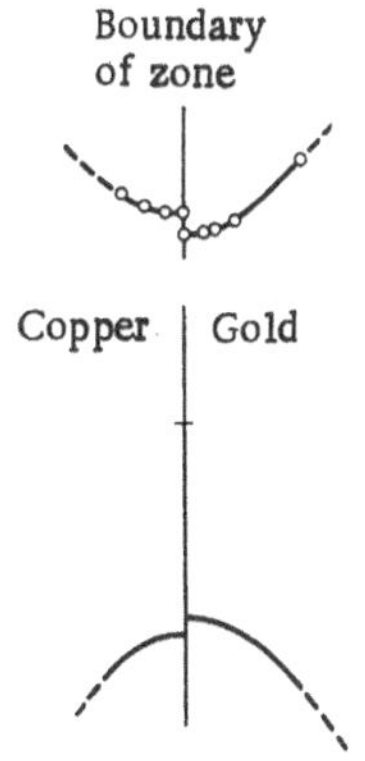

Fig. 74. Fermi surface of gold and copper close to the point of contact with the {111} Bragg plane. Circles indicate experimental points [160].

ments has been discussed in the foregoing sections. In this section we shall consider the case of a monovalent metal – gold.

Optical investigations in the infrared region yielded the concentration of conduction electrons, which proved to be 0.95 electron/atom. Thus the effect of the periodic lattice potential on this quantity is slight. However, the Fourier components of the pseudopotential of gold are relatively large. In particular, the ratio $|V_{111}|/E_F^0 \approx 0.3$, as we shall see later. The slight influence of V_g on N is associated with the fact that the Bragg planes do not intersect the sphere of free electrons for a concentration of valence electrons equal to 1 electron/atom. Gold has a face-centered cubic lattice. The {111} are the nearest Bragg planes to the sphere of free electrons. For a weak potential these would have practically no effect on the Fermi surface, nor on the velocity of the electrons on the latter. For a strong potential, however, the Fermi surface distorts, and the value of v_F changes as well. The Fermi surface "grows out" to the {111} planes. Thus the real Fermi surface constitutes a sphere, with eight "necks" extending in the {111} directions. Figure 74 shows the region of the Fermi surface close to its contact with the {111} Bragg plane. The figure is taken from Morse [160], in which the Fermi surface of gold, silver, and copper was studied by analyzing the damping of ultrasound in a magnetic field. For the radius of the neck in gold a value of $r = 0.22 \cdot 10^{-19}$ g-cm/sec was obtained. The radius of the sphere of free electrons $p_F^0 = 1.27 \cdot 10^{-19}$ g-cm/sec. The distance from the center of the zone Γ to the {111} plane $p_{111} = 1.41 \cdot 10^{-19}$ g-cm/sec. If we neglect the difference between the Fermi energy E_F and the energy of the free* electrons, we may obtain the value of $|V_{111}|$ from the relation

$$\frac{p_{111}^2}{2m} + \frac{r^2}{2m} - |V_{111}| = \frac{(p_F^0)^2}{2m}, \tag{VII.3}$$

whence

$$|V_{111}|/E_F^0 = 0.26.$$

The value of $|V_{111}|$ may also be obtained from optical measurements in the ultraviolet part of the spectrum. These measurements are given by Cooper, Ehrenreich, and Philipp [100], in the review by Suffczynski [161], in a paper by Canfield, Haas, and Hunter [162], and also by Beaglehole [163]. We used the Beaglehole data, since, in this case, both optical constants were measured in the region of the principal maximum of the imaginary part of the complex dielectric constant. According to this investigation, the maximum in question lies at approximately 3.3 eV (its position cannot be specified more precisely, since the results are presented in the

*This means that we neglect the volume of the eight necks by comparison with that of the sphere of free electrons.

form of a graph relating ε_2 to $\hbar\omega$ with a small scale along the $\hbar\omega$ axis). Identifying the maximum with the {111} band,* we obtain $|V_{111}| \approx 1.6$ eV. For gold $E_F^0 = 5.5$ eV, whence $|V_{111}|/E_F^0 = 0.29$.

Thus the results of the determination of $|V_{111}|$ by the two methods agree closely with each other. Since the value of $|V_{111}|/E_F^0$ is relatively large, allowance must be made for the quadratic terms in any theories using an expression with respect to this "small" parameter. This means that the equations of Section 3 in Chapter II cannot be employed. A generalization of the equations to allow for terms of the second order is not particularly difficult, but it lies beyond the scope of the present investigation.

Apart from the optical constants of gold, the Hall effect of gold films prepared at the same time as those for the optical measurements was also determined. Measurement of the Hall constant R_X enables us to determine the quantity $N_X = 1/(R_X ec)$, where e is the charge on the electron and c is the velocity of light. Sometimes N_X is taken as the concentration of conduction electrons. However, this should only be done for a spherical Fermi surface, as may clearly be seen from the example of gold.

For gold $N_a = 5.9 \cdot 10^{22}$ cm^{-3}, $N = 5.6 \cdot 10^{22}$ cm^{-3}, $N_X = 8.7 \cdot 10^{22}$ cm^{-3}. We see that slight deviations of the Fermi surface from spherical form have little effect on N but a great influence on N_X. This is quite natural. All parts of the Fermi surface make a contribution of the same sign to N. The contribution of an electron in a neck is smaller than that of an electron on the main part of the surface, owing to the smaller velocity; the number of such electrons, however, is not very great, so that the resultant value of N differs little from N_a. Different parts of the Fermi surface make contributions of different signs to the Hall emf, and this reduces R_X. The contribution of an electron in a neck is greater than that of an electron in the main part of the Fermi surface, owing to the fact that the effective mass of the former is considerably smaller (we use the definition of the reciprocal effective mass in terms of the second derivative of the energy with respect to momentum). Hence R_X falls substantially, and this is equivalent to a corresponding increase in N_X. An analogous situation should always arise when "bubbles" associated with the approach of the Fermi surface to the boundaries of the Brillouin zone appear on an almost spherical Fermi surface. A combined study of the optical constants and the Hall effect of the same samples (with a known Fermi surface) enables us to determine the mean velocities of the electrons and the mean effective masses for various parts of the Fermi surface. However, this does not form part of the current problem.

In concluding this section, we may note that the information as to the electron structure of a metal obtained from all the effects associated with the use of a steady magnetic field differs from the information derived from optics or specific heat. In experiments involving magnetic fields, small groups of electrons lying in the bubbles of the Fermi surface are chiefly involved. The characteristics of these groups may be obtained to a high accuracy. The characteristics of the main body of the electrons are much less accurately determined. In optical experiments, on the other hand, it is the main body of the electrons which are particularly concerned. The effect of the small groups of electrons lying in the bubbles of the Fermi surface is slight.

CONCLUSION

1. We have obtained an analytical solution to the kinetic equation for the weakly anomalous skin effect in metal optics. We have established experimentally that for polyvalent metals

*The possibility of this interpretation was also indicated by Beaglehole [163].

the weakly anomalous skin effect holds at all temperatures; for monovalent metals it holds at relatively high (room) temperatures.

2. We have made a theoretical study of the effect of the Fourier components of the pseudopotential V_g on the concentration of conduction electrons N, on the total area of the Fermi surface S_F, and on the average velocity of the electrons on the Fermi surface $\langle v_F \rangle$.

3. We have established a relationship between the optical constants of the metal in the infrared region and its fundamental electron characteristics.

4. We have calculated the concentration of the conduction electrons for metals with a face-centered cubic lattice, a body-centered cubic lattice, and a cubic lattice of the diamond type, and also for indium and white tin, metals with a tetragonal lattice.

5. We have made a theoretical study of interband transitions associated with the Bragg energy splitting, and have shown that it is the Bragg reflection of electrons which largely determines the structure of the bands of interband conductivity in the visible and near-infrared parts of the spectrum.

6. We have proposed and justified a method of determining the Fourier components of the pseudopotential from optical measurements.

7. We have theoretically considered the temperature dependence of the concentration of the conduction electrons in metals, showing that the thermal vibrations of the lattice reduce the Fourier components of the pseudopotential, leading to an increase in the concentration of the conduction electrons on raising the temperature. The temperature dependence of N is determined by the Debye–Waller factor, which also gives the temperature dependence of the x-ray diffraction maxima.

8. We have developed a polarization method of measuring the optical constants of metals in the infrared part of the spectrum, giving the optical constants of metals to a very high accuracy, and have developed a method of measuring the optical constants of metals at low temperatures.

9. We have set up two forms of apparatus for measuring the optical constants of metals by the polarization method in the infrared and visible parts of the spectrum over a wide range of temperatures, from 4.2 to 295°K. The accuracy of the determination of optical constants n and $\varkappa$ in these forms of apparatus is 1-2%.

10. We have developed a method of obtaining specular metallic layers with properties similar to or coinciding with those of the bulk metal.

11. We have measured the optical constants n and $\varkappa$ of the metals indicated in Table 48. The table also shows the spectral ranges of the investigation and the temperatures at which the measurements were made.

A study of a number of static characteristics of the samples in question showed that indium, lead, tin, and gold layers had properties coinciding very nearly with the properties of the bulk metal; the properties of aluminum layers differed somewhat from the bulk metal, but this difference was not very great.

TABLE 48

Metal	Spectral range, μ	T, °K	Metal	Spectral range, μ	T, °K
Indium	0.55–12	295	Tin	0.73–12	293
	0.55–12	4.2		0.73–12	78
Aluminum	0.8–9	295		0.93–12	4.2
	0.8–7	78	Gold	1–12	293
Lead	0.45–12	293			
	0.45–12	78			
	0.54–12	4.2			

12. We have developed schemes for analyzing the experimental values of the n and $\varkappa$ of metals in the infrared and visible parts of the spectrum, yielding the principal electron characteristics of the metals.

13. By optical measurements in the infrared part of the spectrum, we have determined the principal microcharacteristics of the conduction electrons for all the metals studied: the concentration of the conduction electrons, the area of the Fermi surface, the average velocity of the electrons on the Fermi surface, the electron—phonon collision frequency, and a number of other characteristics.

14. By optical measurements in the visible and near-infrared parts of the spectrum, we have obtained the conductivity and the real part of the dielectric constant for interband transitions. We have made an experimental study of the structure of the bands of interband conductivity for indium, aluminum, lead, and tin, as well as its temperature dependence. The results confirm the theory developed in the dissertation.

15. We have determined the Fourier components of the pseudopotential of aluminum, indium, lead, tin, and gold by the optical method. In cases in which reliable data relating to the calculation of $|V_g|$ from other effects are available, the values obtained by the optical method agree closely with the other results.

16. From the resultant Fourier components of the pseudopotential we have calculated the concentration of the conduction electrons N, the total area of the Fermi surface S_F, and the average velocity of the electrons on the Fermi surface $\langle v_F \rangle$ for indium, aluminum, lead, and tin. Agreement between the principal electron characteristics obtained from measurements in the long-wave region and those calculated from the values of $|V_g|$ confirms the theory relating the optical properties of metals to their principal electron characteristics, as developed in this dissertation.

17. We have established the temperature dependence of the Fourier components of the pseudopotential experimentally.

18. We have experimentally established the temperature dependence of the concentration of the conduction electrons in metals N(T). The observed rise in N with increasing temperature is much greater than the changes which might be expected from the direct influence of the thermal motion of the electrons on this quantity. We have shown that the effect in question is associated with the thermal vibrations of the lattice and agrees with the theory here developed.

19. We have made an experimental study of the effective collision frequencies for electrons on the Fermi surface and for electrons close to the Bragg planes in the presence of photons, and also the temperature dependence of these frequencies. The results obtained for the electrons near the Bragg planes show that the corresponding effective collision frequency is greater than the effective collision frequency for electrons on the Fermi surface, and has a weaker temperature dependence. This question requires special theoretical consideration.

20. Using gold as an example, we have compared the effect of the periodic lattice potential on the optical properties and the Hall constant. We have shown that the optical properties determined by the average values of the electron characteristics are considerably less sensitive to deviations of the Fermi surface from spherical form than the Hall effect.

The foregoing investigations confirm the picture of metal-optical phenomena developed in the first two chapters of the dissertation. In order to describe the optical properties of nontransition metals in the infrared and visible parts of the spectrum the weak-coupling scheme may be successfully used. The difference between the characteristics of the conduction electrons determining the optical properties in the infrared part of the spectrum and the free-electron characteristics is mainly due to the Fourier components of the pseudopotential to which the Bragg planes intersecting the sphere of free electrons correspond. The optical properties in the visible and near-infrared parts of the spectrum are mainly determined by interband transitions associated with the Bragg energy splitting. The electron characteristics obtained from measurements in both these spectral regions agree closely with one another.

The study of metal optics enables us to obtain a number of fundamental electron characteristics of metals: the concentration of the conduction electrons, the mean velocity of the electrons on the Fermi surface, the Fourier components of the pseudopotential, and the effective electron collision frequencies. The results obtained by the optical method agree closely with the analogous results obtained from other effects. However, metal optics give a wider set of microcharacteristics than the other effects. Furthermore, at the present time only optics give the temperature dependence of the characteristics in question. It is to be hoped that the subsequent development of metal-optical investigations will still further enhance their role in the development of metal physics.

In conclusion, the author wishes to express his sincere thanks to colleagues in the Physics Institute of the Academy of Sciences who have cooperated in the execution of this work.

The author particularly remembers the counsel of G. S. Landsberg, on whose initiative this work was started. The labors of V. L. Ginzburg served as an impetus for the development of metal-optical investigations in the Optical Laboratory of the Physics Institute of the Academy of Sciences, and his constant interest promoted their successful execution. The author is deeply grateful to A. A. Shubin and A. I. Golovashkin for help in the work, and L. V. Keldysh and V. P. Silin for discussing the results.

LITERATURE CITED

1. P. Drude, Wied. Ann., 32:584 (1887); 36:532 (1889); 39:481, 507 (1890); Lehrbuch der Optik, Leipzig (1900); R. S. Minor, Ann. der Phys., 10:581 (1903); W. Meier, Ann. der Phys., 31:1017 (1910); C. Zener, Nature, 132:968 (1933).
2. V. L. Ginzburg and G. P. Motulevich, Uspekhi Fiz. Nauk, 55:469 (1955); Fortschr. der Phys., 3:309 (1955).
3. R. B. Dingle, Physica, 19:312, 348, 729 (1953).
4. V. L. Ginzburg, Dokl. Akad. Nauk SSSR, 97:999 (1954).
5. V. L. Ginzburg and G. P. Motulevich, Izv. Akad. Nauk SSSR, Ser. Fiz., 18:631 (1954).
6. A. B. Pippard, Proc. Roy. Soc., A191:385 (1947).
7. G. E. Reuter and E. H. Sondheimer, Proc. Roy. Soc., A195:336 (1949).
8. R. G. Chambers, Nature, 165:239 (1950); Proc. Roy. Soc., 215:481 (1952).
9. T. Holstein, Phys. Rev., 88:1427 (1952).
10. M. A. Biondi, Phys. Rev., 96:534 (1954).
11. R. N. Gurzhi, Zh. Éksp. Teor. Fiz., 33:451, 660 (1957).
12. R. N. Gurzhi, Candidate's Dissertation, Khar'kov (1958).
13. T. Holstein, Phys. Rev., 96:535 (1954).
14. G. P. Motulevich and A. A. Shubin, Optika i Spektroskopiya, 2:633 (1957).
15. G. P. Motulevich and A. A. Shubin, Transactions of the Tenth Conference on Spectroscopy, Vol. 1 [in Russian], Izd. L'vov Univ. (1957), p. 95.
16. A. I. Golovashkin, G. P. Motulevich, and A. A. Shubin, Pribory i Tekh. Éksperim., No. 5, 74 (1960).
17. A. I. Golovashkin and G. P. Motulevich, Zh. Éksp. Teor. Fiz., 47:64 (1964).

18. A. I. Golovashkin and G. P. Motulevich, Work of the Commission on Spectroscopy, Academy of Sciences of the USSR, Transactions of the Fifteenth Conference on Spectroscopy [in Russian], Vol. 3, No. 1, Minsk (1965), p. 437.
19. A. I. Golovashkin, I. S. Levchenko, G. P. Motulevich, and A. A. Shubin, Zh. Éksp. Teor. Fiz., 51 :1622 (1966).
20. A. I. Golovashkin and G. P. Motulevich, Zh. Éksp. Teor. Fiz., 53 :1526 (1967).
21. G. P. Motulevich and A. A. Shubin, Zh. Éksp. Teor. Fiz., 44 :48 (1963).
22. A. I. Golovashkin and G. P. Motulevich, Zh. Éksp. Teor. Fiz., 44 :398 (1963).
23. A. I. Golovashkin and G. P. Motulevich, Zh. Éksp. Teor. Fiz., 46 :460 (1964).
24. G. P. Motulevich and A. A. Shubin, Zh. Éksp. Teor. Fiz., 47 :840 (1964).
25. I. M. Lifshits and M. I. Kaganov, Uspekhi Fiz. Nauk, 78 :411 (1962).
26. W. A. Harrison, Phys. Rev., 118 :1182, 1190 (1960); 126 :497 (1962); 129 :2503, 2512 (1963).
27. V. Heine, Low-Temperature Physics LT-9, Part B, Plenum Press, New York (1964), p. 698.
28. V. Heine, Optical Properties and Electronic Structure of Metals and Alloys, Proc. Internat. Colloq., Paris (1965), p. 16.
29. J. Ziman, Electrons and Phonons, Oxford University Press (1960).
30. J. M. Ziman, Contemporary Physics, 3 :401 (1962).
31. J. M. Ziman, Adv. Physics, 13 :89 (1964).
32. J. Ziman, Principles of the Theory of Solids, Cambridge University Press (1964).
33. G. P. Motulevich, Zh. Éksp. Teor. Fiz., 51 :1918 (1966).
34. A. I. Golovashkin, A. I. Kopeliovich, and G. P. Motulevich, ZhÉTF, Pis. Red., 6 :651 (1967).
35. A. I. Golovashkin, A. I. Kopeliovich, and G. P. Motulevich, Zh. Éksp. Teor. Fiz., 53 : 2053 (1967).
36. A. I. Golovashkin, G. P. Motulevich, and A. A. Shubin, Zh. Éksp. Teor. Fiz., 38 :51 (1960).
37. G. P. Motulevich, Zh. Éksp. Teor. Fiz., 46 :287 (1964).
38. G. P. Motulevich, A. A. Shubin, and O. F. Shustova, Zh. Éksp. Teor. Fiz., 49 :1431 (1965).
39. G. P. Motulevich, Zh. Éksp. Teor. Fiz., 37 :1770 (1959).
40. H. Bethe and A. Sommerfeld, Electron Theory of Metals [Russian translation], ONTI, Moscow-Leningrad (1938).
41. A. Wilson, Quantum Theory of Metals [Russian translation], Gostekhizdat, Moscow (1941).
42. E. H. Sondheimer, Adv. Physics, 1 :1 (1952).
43. M. Ya. Azbel' and É. A. Kaner, Zh. Éksp. Teor. Fiz., 32 :896 (1957).
44. N. N. Bogolyubov and K. P. Gurov, Zh. Éksp. Teor. Fiz., 17 :614 (1947).
45. Yu. P. Klimontovich and V. P. Silin, Zh. Éksp. Teor. Fiz., 23 :151 (1952).
46. V. L. Ginzburg and V. P. Silin, Zh. Éksp. Teor. Fiz., 29 :64 (1955).
47. L. P. Pitaevskii, Zh. Éksp. Teor. Fiz., 34 :942 (1958).
48. R. N. Gurzhi, Zh. Éksp. Teor. Fiz., 35 :965 (1958).
49. G. P. Motulevich and A. A. Shubin, Zh. Éksp. Teor. Fiz., 34 :757 (1958).
50. A. I. Golovashkin, Zh. Éksp. Teor. Fiz., 48 :825 (1965).
51. A. I. Golovashkin, Trudy Fiz. Inst. Akad. Nauk, 39 :91 (1967).
52. A. I. Shal'nikova (editor), Low-Temperature Physics [Russian translation], IL, Moscow (1959).
53. R. B. Dingle, Appl. Sci. Res., B3 :69 (1953).
54. A. N. Gordon and E. H. Sondheimer, Appl. Sci. Res., B3 :297 (1953).
55. A. V. Sokolov, Optical Properties of Metals [in Russian], Fizmatgiz, Moscow (1961).
56. Ya. L. Al'pert, V. L. Ginzburg, and E. L. Feinberg, Propagation of Radio Waves [in Russian], Gostekhizdat, Moscow (1953).
57. S. V. Vonsovskii, Izv. Akad. Nauk SSSR, Ser. Fiz., 12 :337 (1948); Uspekhi Fiz. Nauk, 48 :289 (1952).

58. G. P. Motulevich, Work of the Commission on Spectroscopy, Academy of Sciences of the USSR [in Russian], Vol. 3, No. 1, Minsk (1965), p. 428.
59. M. I. Kaganov and V. V. Slezov, Zh. Éksp. Teor. Fiz., 32:1696 (1957).
60. R. N. Gurzhi and G. P. Motulevich, Zh. Éksp. Teor. Fiz., 51:1220 (1966).
61. N. W. Ashcroft, Phil. Mag., 8:2055 (1963).
62. J. R. Anderson and A. V. Gold, Phys. Rev., 139:A1459 (1965).
63. A. I. Ansel'm, Introduction to the Theory of Semiconductors [in Russian], Izd. Nauka, Moscow (1965).
64. E. Jahnke and F. Emde, Tables of Functions [Russian translation], Gostekhizdat, Moscow-Leningrad (1948).
65. D. Pines, The Many-Body Problem, Benjamin, New York (1961).
66. P. Nozières, Le problème à les corps, Dunod, Paris (1963).
67. B. F. Ormont, Structures of Inorganic Materials [in Russian], Gostekhteorizdat, Moscow-Leningrad (1950).
68. N. E. Mott and J. Jones, The Theory of the Properties of Metals and Alloys, Oxford (1936).
69. W. A. Harrison, Phys. Rev., 147:467 (1966).
70. R. T. Mina and M. S. Khaikin, Zh. Éksp. Teor. Fiz., 51:62 (1966); R. T. Mina, Summaries of Contributions to LT-10, 254 (1966).
71. M. I. Sergeiev and M. G. Tchernikovsky, Phys. Z. Sowjetunion, 5:106 (1934).
72. Van Hove, Phys. Rev., 89:1189 (1953).
73. J. C. Phillips, Phys. Rev., 104:1263 (1956).
74. J. C. Phillips, J. Phys. Chem. Solids, 12:208 (1960).
75. J. C. Phillips, Optical Properties and Electronic Structure of Metals and Alloys, Proc. Internat. Colloq., Paris (1965), p. 22.
76. D. Brust, J. C. Phillips, and F. Bassani, Phys. Rev. Lett., 9:94 (1962).
77. W. Kohn, Optical Properties and Electronic Structure of Metals and Alloys, Proc. Internat. Colloq., Paris (1965), p. 1.
78. H. Ehrenreich, H. R. Philipp, and B. Segall, Phys. Rev., 132:1918 (1963).
79. I. N. Shklyarevskii and R. G. Yarovaya, Optika i Spektroskopiya, 16:85 (1964).
80. L. D. Landau and E. M. Lifshits, Quantum Mechanics [in Russian], Fizmatgiz, Moscow (1963); W. Heitler, Quantum Theory of Radiation, Oxford University Press (1954).
81. M. I. Kaganov and I. M. Lifshits, Zh. Éksp. Teor. Fiz., 45:948 (1963).
82. I. M. Ryzhik and I. S. Gradshtein, Tables of Integrals, Sums, Series, and Products [in Russian], Gostekhteorizdat, Moscow (1951).
83. R. James, Optical Principles of the Diffraction of X-Rays, Cornell University Press, Ithaca (1948).
84. P. Debye, Ann. der Phys., 43:49 (1914); I. Waller, Z. Phys., 17:398 (1923).
85. L. I. Mirkind, Handbook on the X-Ray Structural Analysis of Polycrystalline Aggregates [in Russian], Fizmatgiz, Moscow (1961).
86. M. Born, Optics [Russian translation], ONTI, Moscow (1937).
87. I. M. Shklyarevskii and V. K. Miloslavskii, Optika i Spektroskopiya, 3:361 (1957).
88. I. N. Shklyarevskii, N. G. Starunov, and V. G. Padalka, Optika i Spektroskopiya, 4:792 (1958).
89. J. R. Beattie, Phil. Mag., 46:235 (1955); Physica, 23:898 (1957).
90. M. M. Noskov and B. A. Charikov, Optika i Spektroskopiya, 1:1007 (1956).
91. H. Bode, Network Analysis and Feedback Amplifier Design, New Jersey (1945).
92. D. E. Thomas, Bell System Techn. J., 26:870 (1947).
93. T. S. Robinson, Proc. Phys. Soc. (London), 65:910 (1952).
94. F. C. Jahoda, Phys. Rev., 107:1261 (1957).
95. H. R. Philipp and E. A. Taft, Phys. Rev., 113:1002 (1959).

96. F. Stern, Solid State Physics, 15:299 (1963).
97. L. D. Landau and E. M. Lifshits, Electrodynamics of Continuous Media [in Russian], Gostekhteorizdat, Moscow (1957).
98. E. A. Taft and H. R. Philipp, Phys. Rev., 121:1100 (1961).
99. H. Ehrenreich and H. R. Philipp, Phys. Rev., 128:1622 (1962).
100. B. R. Cooper, H. Ehrenreich, and H. R. Philipp, Phys. Rev., 138:A494 (1965).
101. M. M. Kirillova and B. A. Charikov, Fiz. Met. Metallov., 19:495 (1965).
102. M. P. Givens, Solid State Physics, 6:313 (1957).
103. L. G. Schulz, Adv. Phys., 6:102 (1957).
104. G. Pfestorf, Ann. der Phys., 81:906 (1926).
105. G. K. T. Conn and G. K. Eaton, J. Opt. Soc. Amer., 44:484 (1954).
106. A. Elliott, E. J. Ambrose, and R. Temple, J. Opt. Soc. Amer., 38:212 (1948).
107. L. Holland, Vacuum Deposition of Thin Films, Chapman and Hall, Ltd., London (1956).
108. K. V. Bol'shova, I. S. Levchenko, and A. A. Shubin, Pribory i Tekh. Éksperim., No. 2, 228 (1968).
109. M. M. Kirillova, M. M. Noskov, and B. A. Charikov, Fiz. Met. Metallov., 13:798 (1962).
110. E. Grüneisen, Ann. der Phys., 16:530 (1933).
111. Handbook of Chemistry and Physics, 33rd edition, Cleveland, Ohio (1951).
112. C. J. Smithells, Metals Reference Book, Butterworths Sci. Pub., London (1955).
113. J. G. Dorfman and S. E. Frisch (editors), Handbook of Physical Constants [Russian translation], ONTI, Moscow (1937).
114. E. A. Lynton, Superconductivity, John Wiley, New York (1962).
115. B. T. Matthias, T. H. Geballe, and V. B. Compton, Rev. Mod. Phys., 35:1 (1963).
116. W. Meissner and B. Voigt, Ann. der Phys., 7:761, 892 (1930).
117. W. Meissner, H. Franz, and H. Westerhoff, Ann. der Phys., 13:555 (1932).
118. R. G. Yarovaya, Candidate's Dissertation, Khar'kov (1965).
119. A. P. Lenham and D. M. Trehern, Proc. Phys. Soc., 85:167 (1965).
120. I. N. Shklyarevskii and R. G. Yarovaya, Optika i Spektroskopiya, 14:252 (1963).
121. G. Hass and J. E. Waylonis, J. Opt. Soc. Amer., 51:719 (1961).
122. H. M. O'Bryan, J. Opt. Soc. Amer., 26:122 (1936).
123. L. G. Schulz, J. Opt. Soc. Amer., 44:357 (1954).
124. L. G. Schulz and F. K. Tangherlini, J. Opt. Soc. Amer., 44:362 (1954).
125. J. N. Hodgson, Proc. Phys. Soc., B68:593 (1955).
126. P. H. Berning, G. Hass, and R. P. Madden, J. Opt. Soc. Amer., 50:586 (1960).
127. H. E. Bennett, M. Silver, and E. J. Ashley, J. Opt. Soc. Amer., 53:1089 (1963).
128. R. P. Madden, L. R. Canfield, and G. Hass, J. Opt. Soc. Amer., 53:620 (1963).
129. K. N. Onnes and W. Tuyn, Com. Leid. Suppl., No. 58, 1 (1926).
130. K. G. Ramanathan, Proc. Phys. Soc., A65:532 (1952).
131. P. Drude, Wied. Ann., 39:481 (1890).
132. J. Trompette, Compt. Rend. Acad. Sci., 248:207 (1959).
133. W. H. G. Childs, Physical Constants, John Wiley, New York (1958).
134. G. W. C. Kaye and T. H. Laby, Tables of Physical and Chemical Constants, John Wiley, New York (1959).
135. V. N. Kachinskii, Dokl. Akad. Nauk SSSR, 135:818 (1960); Zh. Éksp. Teor. Fiz., 43:1158 (1962).
136. G. A. Bolotin, A. N. Voloshinskii, M. M. Kirillova, M. M. Noskov, A. V. Sokolov, and B. A. Charikov, Fiz. Met. Metallov., 13:823 (1962).
137. V. G. Padalka and I. N. Shklyarevskii, Optika i Spektroskopiya, 11:527 (1961).
138. K. Försterling and V. Fréedericksz, Ann. der Phys., 40:201 (1913).
139. I. N. Shklyarevskii and R. G. Yarovaya, Optika i Spektroskopiya, 21:197 (1966).
140. G. K. White, Phil. Mag., 7:271 (1962).

141. J. N. Hodgson, Phil. Mag., 4:183 (1959); 5:272 (1960); 6:509 (1961); 7:229 (1962).
142. Sh. M. Kogan, Fiz. Tverd. Tela, 9:1510 (1967).
143. A. C. E. Animalu and V. Heine, Phil. Mag., 12:1249 (1965).
144. P. Nozières, Optical Properties and Electronic Structure of Metals and Alloys, Proceedings of the Internat. Colloq., Paris (1965), p. 363.
145. V. L. Ginzburg, G. P. Motulevich, and L. P. Pitaevskii, Dokl. Akad. Nauk SSSR, 163:1352 (1965).
146. V. P. Silin, Zh. Éksp. Teor. Fiz., 33:1282 (1957); 34:707 (1958).
147. D. R. Chipman, J. Appl. Phys., 31:2012 (1960).
148. W. S. Martin, E. H. Duchane, and H. H. Blay, J. Opt. Soc. Amer., 55:1623 (1965).
149. E. Fawcett, The Fermi Surface, Proc. Internat. Conf., New Jersey (1960), p. 197.
150. J. E. Aubrey, Phil. Mag., 5:1001 (1960).
151. N. E. Phillips, M. H. Lambert, and W. R. Gardner, Rev. Mod. Phys., 36:131 (1964).
152. D. L. Decker, D. E. Mapother, and R. W. Shaw, Phys. Rev., 112:1888 (1958).
153. A. B. Migdal, Zh. Éksp. Teor. Fiz., 34:1438 (1958).
154. A. A. Abrikosov, L. P. Gor'kov, and I. E. Dzyaloshinskii, Methods of Quantum Field Theory in Statistical Physics [in Russian], Fizmatgiz, Moscow (1962).
155. M. S. Khaikin and R. T. Mina, Zh. Éksp. Teor. Fiz., 42:35 (1962); R. T. Mina and M. S. Khaikin, Zh. Éksp. Teor. Fiz., 45:1304 (1963).
156. M. Biondi, Phys. Rev., 102:964 (1956).
157. J. A. Rayne, Phys. Rev. Lett., 3:512 (1959).
158. T. L. Loucks, Phys. Rev. Lett., 14:1072 (1965).
159. T. Moss, Optical Properties of Semiconductors [Russian translation], IL, Moscow (1961).
160. R. W. Morse, The Fermi Surface, Proc. Internat. Conf. on the Fermi Surface, Cooperstown (1960), p. 214.
161. M. Suffczynski, Phys. Stat. Sol., 4:3 (1964).
162. L. R. Canfield, G. Haas, and W. R. Hunter, J. Physique, 25:124 (1964).
163. D. Beaglehole, Proc. Phys. Soc., 85:1007 (1965); 87:461 (1966).

Dedicated to the memory of G. S. Landsberg

EXPERIMENTAL STUDIES OF INTERMOLECULAR FORCES BY SPECTROSCOPIC METHODS AND THE DEVELOPMENT OF SPECTRAL APPARATUS *

V. I. Malyshev

INTRODUCTION

The study of intermolecular forces by spectroscopic methods is based on an analysis of the changes taking place in the spectra of molecules as a result of changes in the state of aggregation, pressure, or temperature, changes in other physical parameters of the substances under consideration, or as a result of dissolution.

It is well known that the frequencies of the intramolecular vibrations and the rotational frequencies observed in Raman and infrared absorption spectra are determined by intramolecular forces, the masses of the atoms, and the structures of the molecules. Intermolecular forces may exert a considerable perturbation on the interacting molecules (by perturbing them and changing the value of the intramolecular forces), thus leading to the creation of various molecular complexes, quasi-crystalline structures, and so on.

On changing the pressure or the state of aggregation, or on dissolving the substance in question, the magnitude and character of the intermolecular forces are modified, and this in turn leads to a change in the perturbation of the molecules and hence their molecular spectra. These changes in the spectrum may be extremely variegated; there may be a displacement of the band maxima, a change in the width, shape, or intensity of the bands, splitting of the bands into several components, the appearance of new bands, and so on.

A study of these changes in the molecular spectra and a comparison with the microscopic characteristics of the molecules and the macroscopic characteristics of the material as a whole is of particular interest from the point of view of analyzing the nature of the condensed state.

Furthermore, a study of the molecular spectra on changing the concentration of a solution or the nature of the solvent, and (in the case of gases) on changing the pressure, is of particular importance when using molecular spectra for quantitative spectral analysis. If, for example, there is any deviation from the linear concentration dependence of the Raman line intensities or the optical density, quantitative analysis is made far more complicated, and can indeed only be carried out with the help of calibration curves plotted on the basis of standard mixtures.

---

* Presented in pursuit of the degree of Doctor of Physico-Mathematical Sciences and defended in 1967.

Many theoretical and experimental investigations have been devoted to the study of intermolecular forces by spectroscopic methods. The first experimental investigations in this sphere were those carried out in 1905-1909 [1-3], in which a deviation from the Lambert–Bouguer–Beer law $I(\lambda) = I_0(\lambda)e^{-\varkappa(\lambda)pl}$ was observed during a study of the infrared spectrum of CO_2.

It was found in these investigations that the absorption of radiation depended on the pressure of the absorbing gas p, increasing with p so as to give a constant product pl. Absorption also rises on adding another, nonabsorbent gas to the absorbing gas. On the basis of these experiments it was concluded that the absorption of radiation was determined, not only by the number of absorbing molecules, but also by the distance between these molecules. Later the number of investigations increased rapidly and extended to many other substances; the introduction of better apparatus facilitated the quantitative measurement of individual parameters of the molecular spectra [4].

A large number of experimental investigations were carried out in relation to the changes taking place in the spectra on changing the state of aggregation. Thus the infrared absorption spectra of a large number of substances were compared [5] in the solid and liquid states. It was shown that in the liquid state the bands were usually wider than in the solid state, and on passing into the solid state a number of bands split into several components; there was also a displacement of the bands.

Still sharper changes in the spectra appear on passing from the gas phase to the liquid, or on substantially increasing the pressure of gases, or dissolving gases in liquids. The absorption spectra of molecular gases exhibit a clear rotational–vibration structure at low pressures, but this vanishes on passing to the liquid state [6], or on greatly raising the gas pressure (to hundreds of atmospheres) it is transformed into wide bands.

Among all these investigations, those of greatest interest are those relating to the changes which take place in the molecular spectra of gases on varying the pressure of the absorbing gas and various gaseous impurities [7, 8] (see also reviews [9, 10]). In addition to the broadening of individual rotational components of the P and R branches of the vibrational–rotational bands of frequencies active in the infrared part of the spectrum, at high pressures the inactive frequencies belonging to a number of gases (H_2, O_2, N_2, CO_2) also start exhibiting absorption (this being absent at low pressures). The appearance of this induced absorption is associated with a change in the transition probabilities, resulting from the action of intermolecular forces. Mixtures of absorbing and nonabsorbing gases ($CO_2 + O_2$, $CO_2 + N_2$, $CO_2 + H_2$) also exhibited absorption bands at high pressures [11], the frequencies of these being equal to the sum of the frequencies of the active (CO_2) and inactive (O_2, N_2, H_2) vibrations. This phenomenon is associated with the occurrence of simultaneous transitions in two different molecules, resulting from intermolecular interaction.

However, the greatest number of experimental investigations has been devoted to the so-called dissolution effect, i.e., to the changes taking place in molecular spectra when substances are dissolved in various solvents [12-16]. Experiments revealed a displacement of the band maxima and a change in the shape, width, and intensity of the bands. The first papers devoted to this subject were mainly concerned with the displacement of the band maxima in the infrared absorption spectrum as a function of the nature of the solvent, since large apparatus distortions made it difficult to measure the other parameters of these bands. With the appearance of infrared spectrometers of high resolving power, investigations started into the changes taking place in the intensity and other parameters of the absorption bands.

A vast amount of experimental material relating to the dissolution effect has now been accumulated.

Apart from experimental work, a large number of theoretical papers have been published; these have attempted to explain the observed changes in the molecular spectra and to compare them with the microscopic and macroscopic characteristics of the test materials [9, 10, 16, 17]. A particularly large number of papers has been concerned with the dissolution effect. Several attempts have been made at relating the observed displacement of the infrared absorption bands on dissolution to the macroscopic characteristics, such as the dielectric constant [18, 19] or the refractive indices [13] of the solvent in the form

$$\frac{\nu_v - \nu_s}{\nu_v} = k \frac{\varepsilon - 1}{2\varepsilon + 1},$$

where ν_v is the vibration frequency of an isolated molecule in the vapor, ν_s is the frequency in the solution, and ε is the dielectric constant of the solvent. (In [13], ε is replaced by n^2.) However, experimental verification showed [16, 20, 21] that this "Kirkwood–Bauer–Magat" relation was only approximately satisfied for nonpolar solvents, and not satisfied at all for polar solvents and benzene. Furthermore, in a number of cases, $\nu_s > \nu_v$ rather than $\nu_v > \nu_s$, as would follow from the theoretical relation. For a number of materials the frequency of the bands hardly changed at all on passing into solution [22].

Later a number of other macroscopic theories were proposed [23]; these took a more rigorous account of the interaction between the absorbing molecule and the surrounding medium, yet none of them led to an entirely satisfactory agreement between theory and experiment. Strictly speaking, no rigorous agreement between the macroscopic theory and experiment is really to be expected since, in the interaction between the molecules, a considerable contribution may arise from microscopic parameters of the molecules such as the molecular structure, the dipole moments of the molecules as a whole and of individual bonds, the presence of specific atoms or groups of atoms in the molecules, and so on.

As a result of this, specific forms of intermolecular interaction may appear: dipole–dipole interaction, the local interaction of individual atomic groups in the molecules, and resonance interaction may lead to the dimeric and polymeric association of the molecules of the dissolved substance with one another and with the molecules of the solvent. The character and magnitude of these specific interactions may vary considerably with the nature of the molecules, which will impede the construction of any general theory aimed at explaining the spectroscopic effects of dissolution. Nevertheless, attempts have been made at creating theories of various types to allow for the effects of both macroscopic and specific interaction [12, 14, 24, 25].

In order to study the nature of the specific interaction, a large number of experiments relating to solutions of one of several substances in various solvents possessing widely varying physical and chemical properties [24, 25], have been undertaken.

A still more complicated problem is the creation of a theory to explain the observed changes in the width and intensity of the bands of molecular spectra which take place when the state of aggregation is changed or the test substance suffers dissolution. The difficulties which arise in creating such a theory are due to the fact that a considerable proportion of the experimental data have been obtained in instruments with a relatively poor resolving power, in which these effects are masked by serious apparatus distortions, although in principle the intensity of the bands is a more sensitive parameter than their displacement. A review of theoretical and experimental investigations devoted to this problem was presented in [16].

There is no doubt that, in order to create a rigorous and consistent theory so as to explain all the observed changes in the molecular spectra associated with changes in the state of aggregation or dissolution, experimental research involving a wide variation in experimental conditions and providing reliable quantitative experimental data is of particular importance.

The author's own experimental research was of this kind; it included the following:

1. Study of the hydrogen bond by spectroscopic methods.
2. Study of the effect of pressure and the nature of foreign gases on the width of the components of the P, Q, and R branches.
3. Study of the changes taking place in the infrared absorption spectra on passing through the critical point.
4. Study of the effect of viscosity on the width of the infrared absorption bands, and so on.

In order to obtain reliable experimental data containing the minimum apparatus distortion, new designs of apparatus were formulated and produced.

The work on interpreting intermolecular forces by spectroscopic methods was started and executed by the author under the direction of Academician G. S. Landsberg, who until the end of his life afforded inestimable help both in setting up the experiments and in analyzing the results.

The work was carried out first in the Optical Laboratory of the Institute of Physics of Moscow State University and later in the Optical Laboratory of the Physical Institute of the Academy of Sciences.

It should be mentioned that G. S. Landsberg laid very great stress on the study of intermolecular forces by spectroscopy [26]; work which he started in 1932 continued, after his death, in the hands of his colleagues and the author.

CHAPTER I

STUDY OF THE HYDROGEN BOND BY SPECTROSCOPIC METHODS

The hydrogen bond is one of the particular forms of the specific interaction of molecules containing O−H, N−H, S−H, and sometimes C−H groups. The hydrogen bond is also formed between different molecules if one of these contains one of the groups indicated and the other molecule contains an oxygen, nitrogen, or sometimes chlorine atom. In addition to this, the hydrogen bond is formed between HF, HCl, HBr, and HI molecules. The hydrogen bond may be formed not only between molecules but also within a single molecule. The energy of the hydrogen bond is fairly high (4-8 kcal/mole); it constitutes, as it were, an intermediate step between the energies of intramolecular chemical bonds (about 100 kcal/mole) and the energy of the ordinary intermolecular interaction of the Van de Waals forces (1-2 kcal/mole). As a result of this, substances containing molecules capable of forming a hydrogen bond possess a number of anomalous physicochemical properties, which are usually associated with the existence of fairly strong complexes.

Great attention is now being paid to the study of the hydrogen bond, as it plays an important part both in physics (determining the properties of a number of substances in the condensed phase) and also in chemistry and biology (determining the structure of organic crystals, proteins, and so on). Interest in the hydrogen bond has, in particular, recently developed in view of the assertion that it plays a major part in determining the structure of proteins and nucleic acids. The protein molecule constitutes a long chain built of α amino acids incorporating N−H, O−H, NH_2, COOH, C=O, and other groups, i.e., those groups of atoms and radicals which are capable of forming intermolecular and intramolecular bonds. Nucleic acids such as DNA (desoxyribonucleic acid) also contain N−H, NH_2, C=O, C−N, and other groups capable of forming a hydrogen bond of the type N−H...O, N−H...N. According to [27], it is because of the formation of a hydrogen bond that the DNA molecule takes the form of double spiral cylinders. The authors of [27] also suggest that the hydrogen bond plays a major part in determining the structure of genes.

It is therefore no chance matter that the study of the hydrogen bond has attracted a large number of authors [15, 28, 29] of both chemical and physical (including spectroscopic) persuasions. We may mention by way of example that the monograph [29] devoted to the hydrogen bond contains 2584 bibliographical references.

Spectroscopic methods have proved especially fruitful in studying the hydrogen bond, since the formation of a hydrogen bond is usually accompanied by considerable changes in the molecular spectra. Some very interesting results were obtained in [30], in which the infrared absorption spectrum of the intrinsic hydrogen bond of water (i.e., the vibrations of the complexes of water molecules connected by hydrogen bonds) was studied in the far-infrared part of the spectrum.

The author's Candidate's dissertation [31] also concerned the hydrogen bond, and presented the first results of a systematic examination of the hydrogen bond in water and alcohols and their solutions in various solvents by the Raman technique. It was shown quite convincingly [31-34] that, for small concentrations of methyl alcohol in neutral organic solvents, both alcohol molecules associated into complexes and also isolated, single alcohol molecules, surrounded by solvent molecules only, existed at the same time.

The foregoing results are illustrated by Fig. 1, which presents the Raman spectra of pure methyl alcohol and its solutions in CCl_4. In the spectrum of pure alcohol (Fig. 1a) there is a wide band with a maximum at ν_m = 3370 cm^{-1}, characterizing the perturbed vibration of the O−H group of the alcohol molecule in complexes. The spectrum of a 1% solution of alcohol in CCl_4 (Fig. 1b) contains a narrow, displaced line with a frequency ν_l = 3647 cm^{-1} instead of the wide band; this characterizes isolated, single alcohol molecules (one strong and two weaker lines in the spectrum correspond to the three exciting lines of the mercury lamp ν_c = 27,388, 27,353, and 27,290 cm^{-1}). In this solution single alcohol molecules exist almost exclusively. Figure 1c shows the Raman spectrum of a 5% solution of alcohol, which shows both the narrow line ν_l and the wide band ν_m. In this solution we thus find both single alcohol molecules and

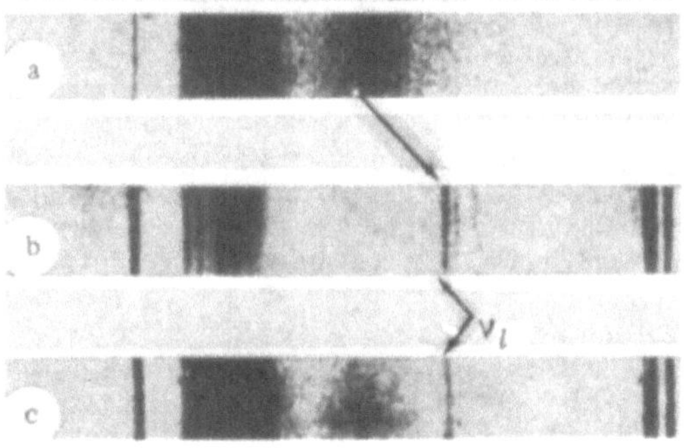

Fig. 1. Raman spectra of pure methyl alcohol and solutions of the latter in CCl_4: a) 100% CH_3OH; b) 1% CH_3OH; c) 5% CH_3OH.

also alcohol molecules associated into complexes at the same time. The relative number of the single and associated molecules depends on the concentration of the solution. Single molecules have also been observed in very concentrated solutions (about 50%), when the number of alcohol molecules is greater than the number of solvent molecules and the alcohol molecules cannot be separated from one another by the solvent.

This is only possible if we assume that a hydrogen bond, accompanied by the perturbation of the O−H vibrations, is only formed for a specific mutual orientation of the interacting molecules, and is not simply determined by the distance between them.

The directional character of the hydrogen bond is also indicated by the results obtained when studying solutions of alcohols in substances containing molecules with a dipole moment [33, 34]. Single alcohol molecules with unperturbed O−H vibrations are observed in these solutions for alcohol concentrations higher than in solutions of dipole-free solvents, and for equal concentrations of alcohol the relative number of single molecules is considerably greater in solutions of dipolar solvents. By way of illustration, Fig. 2 shows the Raman spectra of 10% solutions of methyl alcohol in CCl_4 (dipole moment $\mu = 0$) and in chloroform ($\mu = 1.15$). We readily appreciate the difference in the relative intensities of the ν_l and ν_m lines. Thus the dipole moment of the solvent molecules, interacting with the dipole moment of the alcohol molecule, impedes the mutual orientation of the latter which would be favorable for the formation of hydrogen bonds. In addition to this, the dipole interaction changes the relative number of alcohol complexes with different forms of association, since the spectra of solutions of alcohols in dipolar solvents also exhibit [34] a change in the shape of the band corresponding to the perturbed O−H vibrations.

Some interesting results were obtained when studying ternary solutions; two different substances were introduced into a neutral solvent (usually CCl_4) and the interaction of the different molecules in the solutions was examined. This method was used to study the formation of a hydrogen bond between the molecules of methyl alcohol and various other molecules containing oxygen or nitrogen atoms [33, 34]. It was found that the band of the perturbed O−H vibration observed in the spectrum of ternary mixtures of alcohol and oxygen-containing substances with CCl_4 differed in width and position from the band observed in the spectrum of pure alcohol or solutions of alcohol in CCl_4. The band observed in the spectrum of the ternary mixtures has a smaller width and is less displaced ($\nu = 3530\ cm^{-1}$) than that in the spectrum of pure alcohol ($\nu = 3370\ cm^{-1}$). Thus the character of the interaction between an alcohol molecule and a molecule containing an oxygen atom differs from that of the interaction between two alcohol molecules.

Analogous results were obtained when studying the Raman spectra of ternary mixtures of alcohol and nitrogen-containing molecules in CCl_4; the band of the perturbed O−H vibration observed here has a maximum at $\nu = 3400\ cm^{-1}$. If we estimate the interaction energy from

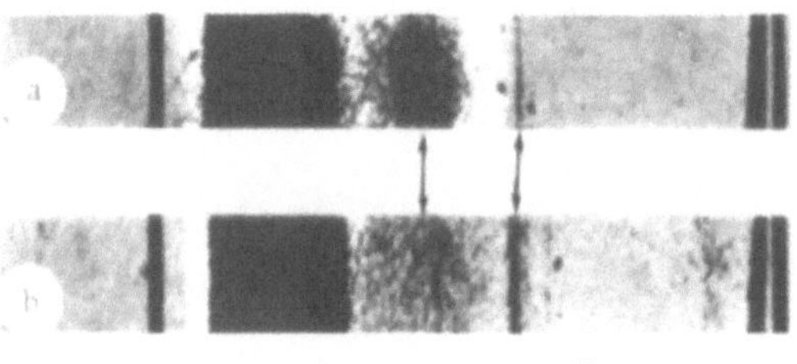

Fig. 2. Raman spectra of 10% solutions of methyl alcohol in CCl_4 (a) and chloroform $CHCl_3$ (b).

the frequency displacement of the perturbed O−H vibration relative to the O−H vibration of the monomers, we may conclude from the results presented that the interaction energy between the O−H group of the alcohol molecule and the oxygen atom of the oxygen-containing molecules (acetone, ether, dioxane, etc.) is smaller than the interaction energy between the O−H group and the nitrogen atom, and this latter, in turn, is smaller than the interaction energy between two alcohol molecules.

§ 1. Determination of the Energy of the Hydrogen Bond by the Spectroscopic Method

It was pointed out in the author's Candidate's dissertation [31] that the relative intensity of the line (characterizing the isolated molecules) and the band (characterizing the associated molecules) depends on the temperature, owing to the displacement of the thermodynamic equilibrium between the number of isolated and associated molecules (Fig. 3).

A spectroscopic method of determining the energy of the hydrogen bond was later proposed on the basis of these considerations [35]. The method is based on measuring the temperature dependence of the line intensity of the vibrations of the O−H group, proportional to the number of isolated molecules, and the line intensity of the vibrations of the C−H group, proportional to the total number of alcohol (CH_3OH) molecules in the solution (CCl_4).

As a first approximation we may consider that, before dissolution, the alcohol molecules were associated in the form of similar complexes, each comprising m molecules. Then equilibrium is established in the solution between the single molecules and the remaining complexes:

$$(CH_3OH)_m \rightleftarrows (CH_3OH)_1 + (CH_3OH)_{m-1}.$$

If α is the degree of dissociation at temperature T, then on the basis of the law of mass action we may write down the following expression for the equilibrium constant:

$$K = \frac{n_0 \alpha^2}{V(1-\alpha)},$$

where V is the volume and n_0 is the number of initial complexes.

If we then suppose that the energy of the hydrogen bond remains constant for small changes in temperature, then, integrating the equation of the isochore

$$\left(\frac{\partial \ln K}{\partial T}\right)_v = \frac{W}{RT^2},$$

we obtain the following for the energy of the hydrogen bond W:

$$W = 4.57 \frac{T_1 T_2}{T_2 - T_1} \log \frac{K_2}{K_1}; \quad \frac{K_2}{K_1} = \left(\frac{\alpha_2}{\alpha_1}\right)^2 \frac{(1-\alpha_1)}{(1-\alpha_2)},$$

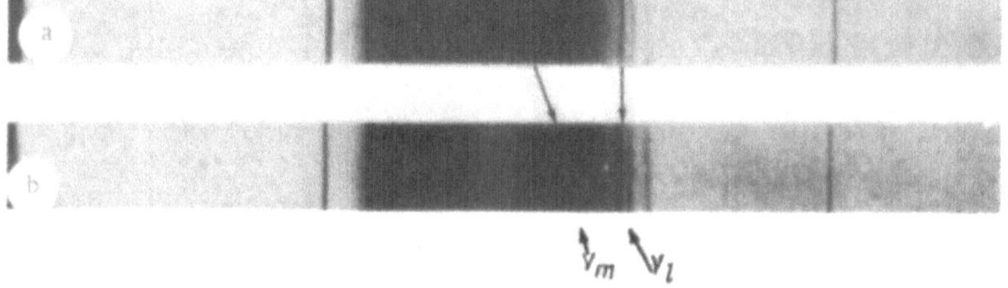

Fig. 3. Raman spectrum of a 25% solution of methyl alcohol in CCl_4 at: a) 6°C; b) 65°C.

i.e., in order to determine the energy of the bond we must know the degree of dissociation α at two temperatures T_1 and T_2. If we choose a concentration of the solution such as to make the degree of dissociation small, $\alpha \ll 1$, then $\frac{K_2}{K_1} \approx \left(\frac{\alpha_2}{\alpha_1}\right)^2$. Since the intensity of the Raman lines is proportional to the number of scattered molecules, $I(O-H) = a\alpha n_0$ and $I(C-H) = bn_0 m$ (considering that the intensity of the line of frequency $C-H$ remains unaltered on formation of the complex). Then the relative intensity of these lines $\delta = \frac{I(O-H)}{I(C-H)} = \alpha \frac{a}{b} \frac{1}{m} = f\alpha$, i.e., $\delta \sim \alpha$, and hence, $\frac{K_2}{K_1} = \left(\frac{\alpha_2}{\alpha_1}\right)^2 = \left(\frac{\delta_2}{\delta_1}\right)^2$, which enables us to determine W.

Measurements were carried out for a 25 vol.% solution of CH_3OH in CCl_4, for which $I(O-H) \ll I(C-H)$, and hence $\alpha \ll 1$, so that the foregoing approximate relationships remain valid. The Raman spectra were measured at 9-65°C by photographic photometry; the value of δ was determined from the results, and a curve relating δ to temperature was plotted. For various parts of this curve $(T_2 - T_1 \approx 10°)$ the energy of the hydrogen bond was calculated. The following values were obtained: $W(9°-15°) = 13.6$ kcal/mole; $W(25°-35°) = 12.3$ kcal/mole; $W(35°-45°) = 12.3$ kcal/mole; $W(55°-65°) = 13.3$ kcal/mole; $\overline{W} = 13.0$ kcal/mole.

In these measurements the accuracy was not particularly high, since the Raman spectra contained a fairly strong background, and the line of the $O-H$ vibration was superimposed on the edge of the $O-H$ band, so that these had to be separated graphically. Nevertheless, the value obtained for the hydrogen bond energy (W = 13 kcal/mole) of the methyl alcohol molecules in CCl_4 solution was quite close to that of simple carboxylic acids (acetic, formic), in which W = 13.5-14 kcal/mole in the vapor phase [36] and in CCl_4 solutions. At the same time it is well known that the molecules of these acids are associated into dimers, and their dissociation is accompanied by the rupture of two hydrogen bonds:

$$R-C\underset{O-H\cdots\cdots O}{\overset{O\cdots\cdots H-O}{\big\langle\quad\big\rangle}}C-R.$$

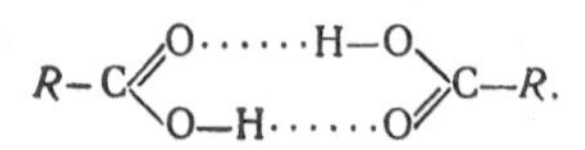

The similarity between the hydrogen-bond energies obtained for the alcohol and acid molecules suggests that two hydrogen bonds are broken in the dissociation of the alcohol molecules also, and that, therefore, the majority of the alcohol molecules in CCl_4 solution are associated into dimers.

This conclusion supports the author's earlier [31] assertion, that in CCl_4 solutions of alcohols (at relatively low concentrations) the alcohol molecules are mainly associated into cyclic dimers with the following structure:

$$\underset{\substack{H-O \\ \diagdown R}}{\overset{\substack{R \\ \diagup O-H}}{\vdots\qquad\vdots}}$$

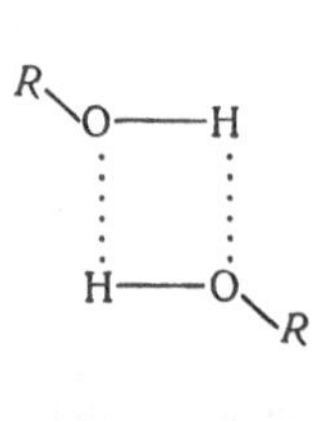

This conclusion is based on a comparison between the frequencies of the bands of the $O-H$ group in the Raman spectra of solutions of alcohol and solutions of alcohol and acetone, in which the formation of cyclic complexes is eliminated.

An analogous proposition was later advanced by other authors. Thus the intensity of the infrared absorption band of the monomers $(\nu = 3630 \text{ cm}^{-1})$ was measured as a function of temperature in [37] for solutions of methyl alcohol in CCl_4, and the hydrogen-bond energy was deduced as W = 9.2 ± 2.5 kcal/mole; it was accordingly suggested that the majority of the alcohol molecules were associated into cyclic dimers.

The infrared absorption spectra of methyl alcohol dissolved in solid nitrogen at T = 20°K was studied in [38]; the results again indicated the possibility that the alcohol molecules were associated into cyclic dimers. Finally, the same suggestion was made in [39] on the basis of a study of solutions of ethyl alcohol in CCl_4 by the NPR method.

In addition to the formation of cyclic dimers of alcohols, in which the two hydroxyl groups interact simultaneously, we may also encounter the formation of the so-called open dimers [29, 31, 34], in which the hydroxyl group of one molecule interacts with the oxygen atom of the other.

The open dimer of an alcohol molecule is

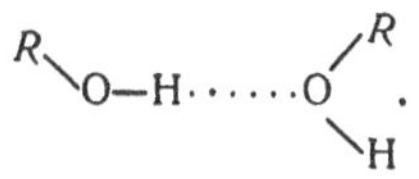

These dimers are distinguished by the bond energy and, correspondingly, the frequency displacement: ν = 3370 cm^{-1} for cyclic dimers and ν = 3500 cm^{-1} for open ones (this is similar to the frequency of the O−H group for solutions of alcohols in oxygen-containing solvents, in which only this type of bond is possible).

There is no doubt that, in pure alcohols and concentrated solutions of alcohols in neutral solvents, more complicated polymeric complexes may also be formed

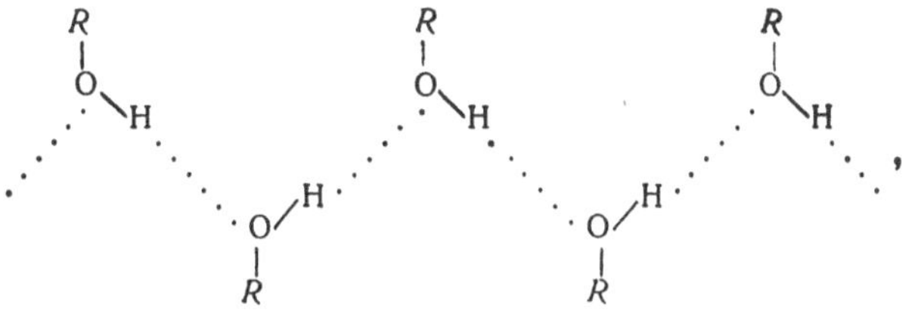

the dimers (distinguished by the character of their interaction) may enter into these complexes as constituent parts.

Since the dimers in question differ considerably in structure, we may thus speak of two forms of hydrogen bond: the true hydrogen bond, in which the O−H group of one molecule interacts with the oxygen atom of the other, and the "hydroxyl bond" in which the two hydroxyl groups interact with each other. This assumption (as to the existence of two forms of hydrogen bond) was confirmed in the author's later investigations [40] on the infrared absorption spectra of alcohols and a number of other compounds containing the O−H group in the region of the fundamental tone and the first overtone of the O−H vibration frequency. In the majority of the pure substances studied, the absorption band of the O−H group in the region of the first overtone, relating to associated molecules, has two maxima (Table 1). The relative intensity of these maxima depends on the temperature, the intensity of the more displaced maximum ν_2' falling with rising temperature, while that of the less displaced maximum ν_1' increases.

In addition to these maxima, the spectra of octyl alcohol and ethylene chlorohydrine (even at room temperature) and the spectrum of ethyl alcohol (at 75°C) exhibit a narrow band ν_0' associated with the unperturbed vibration of the O−H group of the monomers.

In the region of the fundamental tone of the O−H group frequency, however, the infrared absorption spectra of the substances in question exhibit a single wide band at room temperature, characterizing the associated molecules; some substances also exhibit the narrow line of unperturbed O−H vibrations.

The existence of two maxima ν_1' and ν_2' in the band of the perturbed vibration of the O−H group in the region of the first overtone and one maximum in the region of the fundamental may

TABLE 1. Position of the Maxima of the ν_1' and ν_2' Bands of the Perturbed and the ν_0' Line of the Unperturbed O–H Vibration of Various Substances in the Region of the First Overtone (cm^{-1})

Substance	T, °C	ν_0'	ν_1'	ν_2'
Methyl alcohol	25	—	6690	6370
Ethyl alcohol	25	—	6635	6320
	75	7100	6800	6370
Octyl alcohol	25	7100	6530	6270
Ethylene glycol	25	—	6690	6320
Ethylene chlorohydrine	25	6985	6800	6350
Phenol	25	—	6970	6420
Trichloroacetic acid	25	7080	6800	5750

be explained if we suppose that, first, there are two different types of hydrogen bond, differing in the degree of perturbation of the O–H vibration (and hence in the bond energy), and, second, that the optical activity of the perturbed vibrations of one particular type is different in the region of the fundamental vibration and the overtone. The perturbed vibration characterized by the smaller displacement ν_1' is relatively more active in the region of the first overtone than in the region of the fundamental. Hence, in the absorption spectra of the majority of substances, the O–H band has two maxima in the region of the first overtone and only one in the region of the fundamental.

It should nevertheless be noted that, in the infrared absorption spectra of weak solutions of alcohols in CCl_4, the absorption band of the associated molecules has two maxima even in the region of the fundamental, $\nu_1 = 3500\ cm^{-1}$ and $\nu_2 = 3340\ cm^{-1}$ [41], which may be related to the two forms of the hydrogen bond.

In accordance with the foregoing assumption as to the existence of two forms of hydrogen bond, we may assign the more displaced maximum ν_2' in the region of the first overtone and ν_2 in the region of the fundamental to the "hydroxyl" bonds (cyclic dimers), and the less displaced one ν_1' and ν_1 to the purely hydrogen bond (open dimers).

Confirmation of the foregoing ascription of the ν_1' and ν_2' bands to the different forms of hydrogen bond comes from the results of our own investigation into the infrared absorption spectra of acetic and trichloroacetic acids in the region of the first overtone of the O–H vibration frequency. The clearest results were obtained with trichloroacetic acid, in which the infrared absorption spectrum contains no band of the C–H frequencies, so that it is easy to observe the considerably displaced band of the perturbed O–H vibration. In the infrared absorption spectra of acetic and trichloroacetic acids, the band of the perturbed vibration of the O–H group also has a complex structure (Fig. 4), with two maxima. One of these is wide and extremely displaced, with a frequency of $\nu_2' = 5750\ cm^{-1}$; it is also very strong. The other maximum is weaker and narrower, with a frequency of $\nu_1' = 6800\ cm^{-1}$, close to the frequency of the maximum of the ν_1' band observed in alcohols (Table 1). On raising the temperature (Fig. 4) the intensity of the ν_1' maximum rises considerably, and the intensity of the $\nu_0' \approx 7080\ cm^{-1}$ band of the unperturbed vibration of the O–H groups rises at the same time. The ν_2' band, however, diminishes.

Thus, in the case of acetic and trichloroacetic acid also, we may speak of two kinds of hydrogen bond. On the other hand, it is well known [36] that the molecules of fatty acids are mainly associated into dimers involving two hydroxyl groups, and the strongly displaced ν_2' band characterizes the perturbed O–H vibration of these dimers. The ν_1' band, however, clearly relates to open dimers. Hence, if we suppose that, in forming ring dimers with the

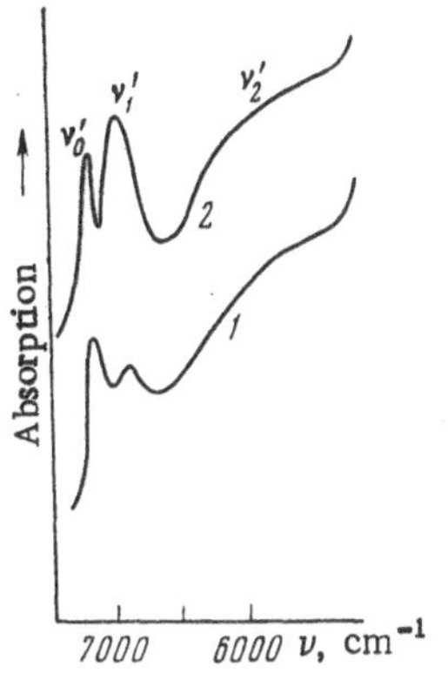

Fig. 4. Change in the absorption band of the O–H group in the spectrum of trichloroacetic acid (in the region of the first overtone) on changing the temperature. 1) 30°C; 2) 300°C.

simultaneous interaction of both hydroxyl groups, there is a considerable displacement of the frequency of the perturbed O–H vibration, then we have grounds for considering that, in the case of alcohols also, the more displaced maximum ν_2' characterizes the cyclic dimers of the hydroxyl bond. The difference in the frequencies of these maxima in the acids and alcohols is associated with the fact that the dimers of the hydroxyl bond still differ considerably from the dimers of the fatty acids in respect of the orientation of the hydroxyl groups.

§ 2. Study of the Effects of the Steric Factor

As already indicated, the hydrogen bond, as a specific form of intermolecular interaction, is formed as a result of the interaction of specific groups of atoms entering into the molecules, and it is thus able to occur for a specific mutual orientation of the interacting molecules. The existence of any factors impeding the mutual approach and the favorable orientation of these molecules should reduce the probability of the formation of a hydrogen bond and hence increase the number of isolated monomeric molecules. As already mentioned, this probably explains the increase in the relative numbers of monomers in solutions of alcohols in polar solvents ($CHCl_3$, C_6H_5Cl, etc.) by comparison with solutions in CCl_4 for the same molar concentration [31, 34]. The dipole–dipole interaction of the solvent and alcohol molecules limits the mobility of the latter in the solution and thus impedes their mutual orientation in a position favorable toward the formation of a hydrogen bond.

On the basis of the foregoing considerations we may suppose that, as a result of the operation of the steric factor, the approach and mutual orientation of the hydroxyl groups in the higher monatomic alcohols $C_nH_{2n+1}OH$ become more and more difficult the larger the hydrocarbon part of the molecules. Thus, even at comparatively low temperatures and in the pure liquid phase, the higher alcohols will contain a certain number of isolated molecules not connected by any hydrogen bond; the number of these will be the greater, the larger the hydrocarbon section of the molecule.

In order to confirm the foregoing, we studied the Raman spectra of a number of monatomic normal alcohols, from methyl (CH_3OH) to octyl ($C_8H_{17}OH$) at 5 and 75°C [34, 42, 43]. At 5°C the spectra of all the alcohols showed simply the wide band of the O–H group of the associated alcohol molecules with a maximum ν_2 between 3376 and 3333 cm^{-1} for the different alcohols. At 75°C, however, the Raman spectra of all the alcohols except methyl contained a sharp line of the O–H group as well; this had a frequency of ν_0 = 3630-3639 cm^{-1} and belonged to the monomeric molecules. The intensity of the line relative to the band was the greater, the larger the hydrocarbon part of the molecule. In order to secure a quantitative estimate of the effect, we determined the ratio of the intensity of the monomer line to the intensity of the associated-molecule band maximum (this ratio being proportional to the degree of dissociation, subject to certain simplifying assumptions) using photographic photometry. The results of the measure-

TABLE 2. Relative Intensities of the Line and Band for Normal Alcohols at T = 75°C

Alcohol	Formula	I_l/I_{bm}*	Alcohol	Formula	I_l/I_{mb}*
Methyl	CH_3OH	0	Hexyl	$C_6H_{13}OH$	0.35
Ethyl	C_2H_5OH	$\ll 1$	Heptyl	$C_7H_{15}OH$	0.39
Propyl	C_3H_7OH	0.11	Octyl	$C_8H_{17}OH$	0.54
Butyl	C_4H_9OH	0.19			

* I_l = intensity of the line; I_{bm} = intensity of band maximum.

TABLE 3. Relative Intensities of the Line and Band, and Boiling Points of Various Alcohol Isomers

Alcohol	Formula	I_l/I_{bm}	B.p., T, °C
Butyl, normal	C_4H_9OH	0.19	118
Isobutyl	$(CH_3)_2CHCH_2OH$	0.22	108
Isoamyl, primary	$(CH_3)_2CHCH_2CH_2OH$	0.30	132
Isoamyl, secondary	$(CH_3)_2CHCHOHCH_3$	0.37	113

ments are presented in Table 2. Although the I_l/I_{bm} (line/band maximum) ratio could not be measured with any great accuracy (owing to the presence of background and overlapping between the line and band), the results presented in Table 2 nevertheless give a true picture of the influence of the steric factor on the association of the alcohol molecules on passing from one homolog to another.

It is interesting to estimate the effect of the degree of branching in the hydrocarbon part of the molecule (for the same number of carbon and hydrogen atoms) and also that of the position of the hydroxyl group in this chain. With this aim in view we studied the Raman spectra of three isomeric alcohols: isobutyl and primary and secondary isoamyl.

The I_l/I_{bm} ratio measured for these alcohols is shown in Table 3. Thus the relative intensity of the line in the isobutyl alcohol is greater than in the normal and, analogously, greater in the secondary alcohol than in the primary, i.e., the branched hydrocarbon part has a greater screening effect on the hydroxyl group. These results are supported by the fall in the boiling point of alcohol isomers which is usually observed on increasing the degree of dissociation, as indicated in Table 3.

It should also be noted that, on raising the temperature of all the alcohols studied, the Raman spectra exhibited a displacement of the band maximum of the associated molecules in the high-frequency direction, extending to some 50-100 cm^{-1}; at the same time, the band width increased, but no systematic law was observed in this case.

§ 3. Study of the Intramolecular Hydrogen Bond in Substances with Molecules Containing Two Hydroxyl Groups

In pure liquids or solutions of substances having molecules capable of forming hydrogen bonds, it is always possible to bring the molecules close enough together and to ensure the right mutual orientation to promote the formation of an intermolecular hydrogen bond. In the case of crystals, however, in which the distances between the molecules and the mutual orientation of the latter are rigidly fixed, it is not always possible for a hydrogen bond to be formed. A comparison between the molecular spectra of such crystals and the distances between the interacting groups enables us to estimate the minimum distance between these groups at which the formation of a hydrogen bond becomes possible. G. S. Landsberg and F. S. Baryshanskaya [41, 44] attempted to do this by studying the Raman spectra of a number of hydroxides and crystal hydrates; they concluded that a hydrogen bond was capable of being formed if the distance between the oxygen atoms R(O−O) of the hydroxyl groups was less than 3 Å.

Subsequently a number of other authors used x-ray and neutron-diffraction methods, as well as the NPR technique to measure the R(O−O) and R(O−H) distances in crystals in which an intermolecular hydrogen bond was formed. These measurements were summarized in [29]. The mean value of R(O−O) for all the crystalline alcohols was 2.44 Å (including methyl alcohol, 2.67 Å), for carboxylic acids 2.63 Å, hydroxides 2.82 Å, crystal hydrates 2.45 Å, and so on. These later data confirmed the conclusions of [41, 44].

In view of the foregoing investigations, it is interesting to discover the nature of the interaction between two hydroxyl groups within a single molecule (instead of between two molecules), in which the distance between the O−H groups and their mutual orientation are restricted by the structure of the molecule. The author accordingly studied the infrared absorption spectra of weak solutions of substances containing two hydroxyl groups with different distances between them in a neutral solvent [40].

The infrared absorption spectra of solutions of the following substances in CCl_4 were studied:

glycols: ethylene glycol $OHCH_2CH_2OH$,
1,3-butylene glycol $OHCH_2CH_2CHOHCH_3$,
1,4-butylene glycol $OHCH_2CH_2CH_2CH_2OH$;

catechols: hydroquinone 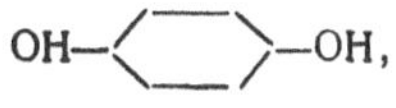

resorcinol 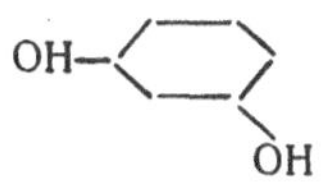

pyrocatechin 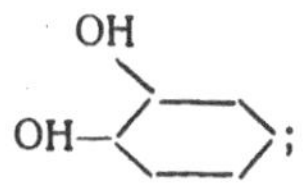

pinacol 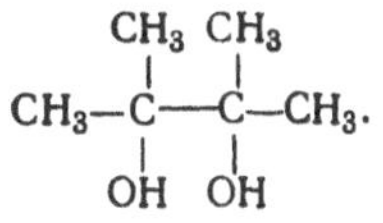

In addition to this, the absorption spectra of substances containing molecules with the same structure as those indicated but in which one of the hydroxyl groups had been replaced by Cl or O atoms or OCH_3 or NO_2 groups were also studied for purposes of comparison.

These included the following: ethylene chlorohydrin $OHCH_2CH_2Cl$, guaiacol 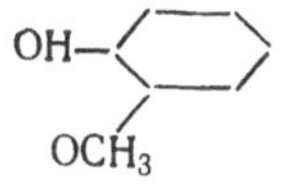

o-nitrophenol 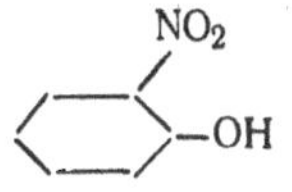, m-nitrophenol 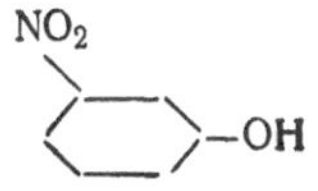, and benzoin 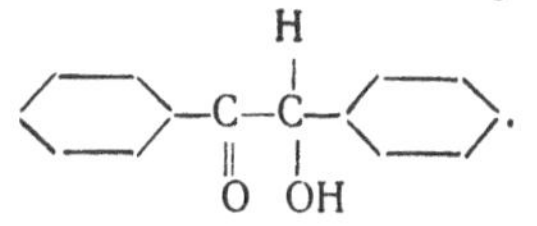.

The concentration of the solutions of these substances was taken so low (under 0.1%) that the probability of the formation of an intermolecular hydrogen bond was also small. Under these conditions, any observed change in the frequency of the O−H group can only be attributed to the formation of intramolecular hydrogen bonds, which are preserved in the solution. Since the relative values of the absorption coefficient of perturbed and unperturbed valence vibrations of the O–H group were different in the region of the fundamental and overtone [40], the infrared absorption spectra were studied in both of these regions in order to ensure more specific results. The infrared absorption spectra were obtained by means of double-beam infrared spectrophotometers constructed in the optical laboratory of the Physics Institute of the Academy of Sciences with the participation of the author (Chapter III).

The results of this analysis of the spectra of dissolved substances containing two hydroxyl groups were as follows: In the spectra derived from solutions of the glycols, pyrocatechin, and pinacol, two comparatively narrow absorption bands appeared both in the region of the fundamental and in the region of the overtone. The frequency of one band ν_{10} was approximately the same in all three cases ($\nu_{10}' \approx 7100$ cm^{-1} in the region of the overtone and $\nu_{10} \approx 3635$ cm^{-1} in that of the fundamental), while the frequency of the other ν_{20} was displaced in the low-frequency direction, its intensity, width, and displacement being different in the different molecules. The width of the band was greater, the greater the displacement (Table 4).

In addition to this, the relative intensity of the displaced band is different in the region of the first overtone and fundamental, respectively, this difference again being the greater, the greater the displacement. Thus, for example, in the region of the fundamental the intensity of the displaced band is approximately equal to that of the undisplaced band in all the glycols. However, in the region of the first overtone, in ethylene glycol ($\Delta\nu = \nu_{10} - \nu_{20} = 41$ cm^{-1}) the two bands have the same intensity, while in 1,3-butylene glycol ($\Delta\nu = 83$ cm^{-1}) the intensity of the displaced band is much lower than that of the undisplaced band, and in 1,4-butylene glycol ($\Delta\nu = 166$ cm^{-1}) only one undisplaced band appears (Fig. 5).

In the spectra of hydroquinone and resorcinol, only one undisplaced band appears in both regions of the spectrum. These results may be explained in the following way. The existence of two absorption bands in the solutions of the glycols, pyrocatechin, and pinacol indicates that the molecules of these substances have two configurations with different mutual orientations of the hydroxyl groups. In one of these configurations the hydroxyl groups are isolated from one

TABLE 4. Frequencies of the Unperturbed (Free) ν_{10} and Perturbed ν_{20} Valence Vibrations of the O–H Group in the Spectra of Various Solutions in CCl_4 (in cm^{-1})

Substance	Overtone region			Fundamental region		
	ν_{10}	ν'_{20}	$\Delta\nu' = \nu'_{10} - \nu'_{20}$	ν_{10}	ν_{20}	$\Delta\nu = \nu_{10} - \nu_{20}$
Ethylene glycol	7120	7070	50	3643	3602	41
1,3-Butylene glycol	7100	6960	140	3633	3550	83
1,4-Butylene glycol	7105	—	—	3637	3471	166
Pinacol	7090	7005	85	3620	3580	40
Pyrocatechin	7070	6990	80	3602	3562	40
Resorcinol	7070	—	—	3606	—	—
Hydroquinone	7080	—	—	Spectrum not studied		
Ethylene chlorohydrin	7100	7030	70	3633	3599	34
Guaiacol	7100	6965	135	3600	3555	45
Benzoin	7100	6755	345	—	3470	—
m-Nitrophenol	7025	—	—	3599	—	—
o-Nitrophenol	—	6270	755*	—	3240	359*

* Displacement with respect to the unperturbed line observed in m-nitrophenol.

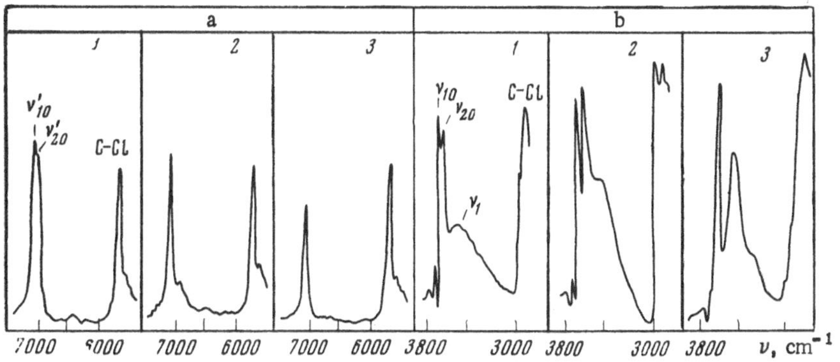

Fig. 5. Infrared absorption spectra of glycols in the region of the first overtone (a) and the fundamental (b): 1) ethylene glycol; 2) 1,3-butylene glycol; 3) 1,4-butylene glycol.

another and their vibrations are unperturbed (undisplaced bands ν_{10} and ν_{10}'), while in the other the hydroxyl groups interact with each other, causing a slight perturbation in the vibrations of the O−H groups (displaced ν_{20}, ν_{20}' bands). These configurations in the case of the glycols and pinacol constitute rotational isomers in view of the possibility of free rotation around the C−C and C−O bonds.

In the catechins such isomers only appear in pyrocatechin, in which the distance between the hydroxyl groups is small. In the case of hydroquinone and resorcinol, however, the distances between the hydroxyl groups are too large to allow them to interact, and the spectrum therefore only exhibits one band of the unperturbed vibration of the O−H group. We might expect that a similar effect would occur in the glycols, as well, as the distance between the hydroxyl groups increased. However, experiments showed that both in the case of ethylene glycol and in the case of 1,3- and 1,4-butylene glycol the spectrum of the fundamental exhibited two bands of the O−H group vibrations. Contrary to expectations, with increasing length of the chain in the glycols, and hence with increasing distance between the hydroxyls along these chains, the frequency displacement of the perturbed O−H vibration increases, and so does the width of the band (Fig. 6). We must therefore suppose that, in the molecules of 1,3- and 1,4-butylene glycol, the interaction between the hydroxyls is greater than in the ethylene glycol molecules. We may suppose that, in view of the possibility of free rotation around the C−C and C−O bonds, one of the possible configurations of the 1,3- and 1,4-butylene glycol molecules is of such a shape that the distance between the hydroxyl groups and the mutual orientation of the latter favor interaction. In the case of the 1,4-butylene glycol, this means that the molecules may not simply have the ordinary shape of a zigzag chain, but may also have the shape of a ring, which is closed by virtue of the interaction between the hydroxyl groups at the ends of the molecule (Fig. 6). This agrees with the conclusions of [45], in which the clotting of blood in the presence of glycols was studied and an annular structure was deduced for these molecules.

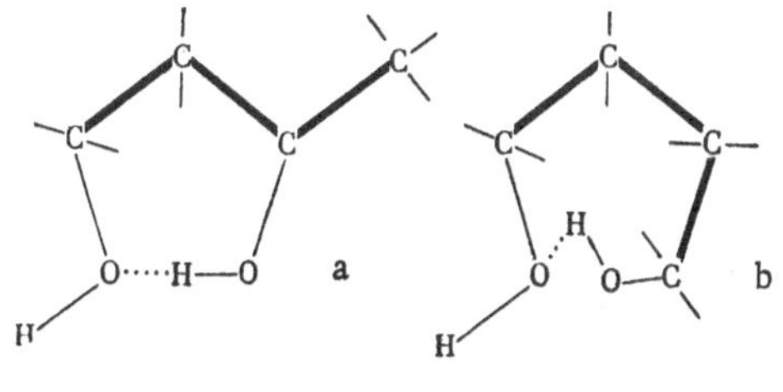

Fig. 6. Possible ring structures of the molecules of 1,3- (a) and 1,4-butylene glycol (b).

In order to compare the character of the intramolecular interaction of the hydroxyl groups with the intermolecular hydrogen bond, the infrared absorption spectra of the pure substances were studied at various temperatures. The infrared absorption spectra of trichloroacetic acid and a number of alcohols were also studied for comparison under the same conditions. It is important to remember that, for these substances also, the infrared absorption spectra in the region of the fundamental and overtone frequency of the O–H group differ considerably. In the region of the overtone the band of the perturbed valence vibration of the O–H bond in the majority of the substances (except for benzoin and o-nitrophenol) has a complex structure with two clearly expressed maxima ν_1' and ν_2', the frequencies of which are given in Table 5 (Fig. 7 [1]). The relative intensity of these maxima is different in different substances and depends on the temperature. On raising the temperature the intensity of the more displaced maximum ν_2' diminishes, that of the maximum ν_1' increases, and the frequency of both does likewise, while at fairly high temperatures the spectra of a number of substances exhibit a narrow band ν_0' corresponding to the unperturbed vibration of the O–H bond of the monomers.

In the region of the fundamental, however, in the majority of the substances studied the band of the perturbed O–H vibration has a single maximum ν_2 corresponding to the ν_2' band in the region of the first overtone. Only in pyrocatechin, resorcinol, and guaiacol are there two maxima ν_1 and ν_2 (Table 6).

The existence of two maxima on the band of the perturbed vibration of the O–H group in the region of the overtone confirms the earlier view as to the existence of two forms of intermolecular hydrogen bond, differing in respect to the bond energy. The differences in the shape of the band corresponding to the perturbed O–H vibration in the region of the fundamental indicate that these two forms of bond differ not only in the bond energy but also in the character of the bond: They correspond to open and cyclic dimers, differing in respect of the absorption

TABLE 5. Frequencies of the Unperturbed and Perturbed Vibrations ν_0' and ν' of the O–H Group of Pure Substances in the Region of the First Overtone (in cm^{-1})

Substance	T, °C	ν_0'	ν_1'	$\Delta\nu_1' = \nu_0' - \nu_1'$	ν_2'	$\Delta\nu_2' = \nu_0' - \nu_2'$
Ethylene glycol	25	7120*	6690	430	6320	800
1,3-Butylene glycol	25	7100*	6635	465	6320	780
1,4-Butylene glycol	25	7105*	6470	635	6280	825
Pinacol	25	7090*	6960	130	6500	590
Pyrocatechin	50	7070*	6890	180	6680	390
Resorcinol	100	7070*	6935	135	6670	400
Hydroquinone	150	7080*	6970	110	6690	390
Ethylene chlorohydrin	25	6985	6800	185	6350	635
Guaiacol	25	7100*	6905	195	6740	360
Benzoin	200	7025	6825	200		
m-Nitrophenol	190	7025*	6960	65	6690	335
o-Nitrophenol	150	—	—		6270	755**
Trichloroacetic acid	25	7080	6800	280	5750	1330
Phenol	25	7050	6970	80	6420	630
Methyl alcohol	25	7080*	6690	390	6370	710
Ethyl alcohol	25	7100	6635	465	6320	780
Octyl alcohol	25	7100	6530	570	6270	830

* Frequency of the O–H line of monomers of the corresponding substance in CCl_4 solution (Table 4 and [40]) at 25°C.

** Displacement with respect to the line of the monomers of m-nitrophenol in CCl_4.

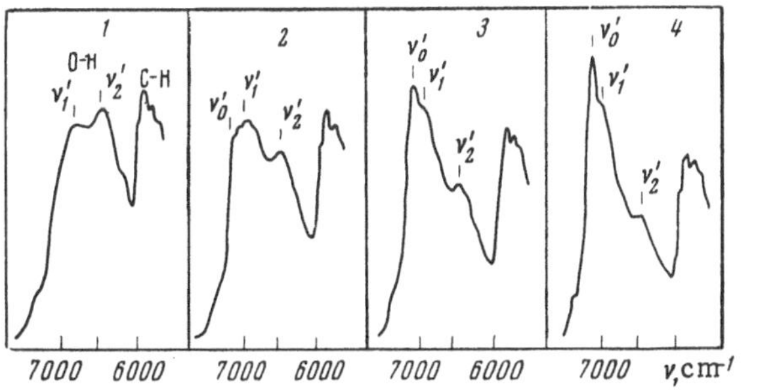

Fig. 7. Infrared absorption spectra of 1,3-butylene glycol in the region of the first overtone at various temperatures: 1) 25°C; 2) 85°C; 3) 125°C; 4) 150°C.

TABLE 6. Frequencies of the Unperturbed ν_{10} and Perturbed ν Vibrations of the O−H Group of Pure Substances in the Region of the Fundamental (in cm^{-1})

Substance	T,°C	ν_{10}, in solution	ν_1	$\Delta\nu_1 = \nu_{10} - \nu_1$	ν_2	$\Delta\nu_2 = \nu_{10} - \nu_2$
Ethylene glycol	30	3643	—	—	3345	~300
1,3-Butylene glycol	30	3633	—	—	3340	~290
1,4-Butylene glycol	30	3637	—	—	3330	~310
Pinacol	30	3620	—	—	3430	~190
Pyrocatechin	30	3602	3525	~80	3275	~330
Resorcinol	190	3606	3480	~125	3250	~355
Hydroquinone, crystalline	30	—	—	—	3205	~400*
Ethylene chlorohydrin	30	3633	—	—	3335	~300
Guaiacol	30	3600	3505	~95	3435	~165
Benzoin	30	—	—	—	3395	
o-Nitrophenol, crystalline	30	—	—	—	3245	~360**

* Displacement with respect to the line of the resorcinol monomers.

** Displacement with respect to the line of the m-nitrophenol monomers.

index of the perturbed vibration of the O−H group. The foregoing results also indicate that these two forms of intermolecular hydrogen bond occur not only in alcohols but also in glycols and catechins (with molecules containing two hydroxyl groups), and in carboxylic acids and other more complex molecules containing other groups of atoms (NO_2, Cl, O) as well as carbon and hydrogen.

It is usually considered that the energy of the hydrogen bond is the greater, the greater the frequency displacement of the O−H vibration relative to the O−H frequency of the monomers [46]. It is therefore interesting to compare the frequency change associated with the intramolecular and intermolecular interaction of the hydroxyl groups.

A comparison of the frequency displacement of the perturbed vibration of the O−H group relative to the frequency of the monomer line (Tables 4-6) shows that the frequency displacement associated with the formation of an intermolecular hydrogen bond is much greater than that associated with the intramolecular interaction of the hydroxyl groups in the materials studied. This applies to the frequencies of both maxima in the band of intermolecular interaction in the region of the overtone, and this difference is particularly great for the frequency of the maximum in the region of the fundamental.

We must therefore suppose that, in the molecules here studied, the energy of the intramolecular bond of the hydroxyl groups is much smaller than the energy of the intermolecular hydrogen bond. We may further consider that, as a result of the limited freedom afforded to the mutual orientation of the hydroxyl groups in the molecule, the interaction between these groups may differ from the interaction which exists between them in the formation of an intermolecular hydrogen bond, being, as it were, intermediate between a hydrogen bond and ordinary electrostatic interaction. In this respect it is interesting to compare the frequency displacement of the perturbed O–H vibration of the glycols and ethylene chlorohydrin (Table 4), where in the latter case one of the hydroxyl groups is replaced by a chlorine atom. We see that, in the case of the ethylene glycol molecule, the frequency displacement is approximately equal to the displacement in the molecule of ethylene chlorohydrin, and we may therefore consider that the interaction between the hydroxyl groups in the ethylene glycol molecule is roughly equal to the electrostatic interaction.

It is logical to suppose that in the interaction of the hydroxyl group with the chlorine atom a hydrogen bond is also formed; however, the energy of this bond is clearly much smaller than the energy of interaction between two O–H groups or between an O–H group and an oxygen atom. The same no doubt applies to the molecules of pinacol and pyrocatechin.

In the case of the 1,3- and particularly the 1,4-butylene glycol the displacement is considerably greater than in the ethylene chlorohydrin molecule, and hence the interaction between the hydroxyl groups in these molecules cannot be regarded as simple electrostatic; a mutual orientation of the hydroxyl groups closer to that required for the formation of a hydrogen bond may occur here.

Intermediate forms of intramolecular interaction also occur in the case of the interaction of a hydroxyl group with an oxygen atom. Thus, in the case of guaiacol the frequency displacement of the perturbed O–H vibrations is slightly greater than the displacement observed in ethylene chlorohydrin or pinacol, i.e., the interaction is almost electrostatic: the CH_3 group of the guaiacol molecule as it were screens the oxygen atom. In the case of benzoin (Table 4), in which there is no such screening, the interaction between the oxygen atom and the O–H group is similar to the interaction of a hydrogen bond. A still greater interaction occurs in the o-nitrophenol molecule, in which more favorable conditions may arise for the formation of a hydrogen bond.

The frequency of the maximum in the band of the perturbed O–H vibration in the spectrum of pure benzoin and o-nitrophenol is similar to the frequency of the perturbed O–H vibration in the spectrum of weak solutions of these substances in CCl_4 (Tables 4-6), i.e., the intramolecular interaction of the O–H group and the oxygen atom is similar to the intermolecular hydrogen bond and, hence, in the pure substance a considerable proportion of the molecules may form an intramolecular hydrogen bond.

We note that in the m-nitrophenol molecule interaction of the hydroxyl group and the nitro-group NO_2 fails to occur because of the great distance between them.

§ 4. Study of the Hydrogen Bond near the Critical Point

As already indicated, the character of the interaction and the degree of dissociation of the molecules capable of forming a hydrogen bond depend on the structure of the molecules, the state of aggregation of the material, and the temperature. It is therefore interesting to study the changes taking place in the degree of dissociation of the molecules and the character of the interaction between them on continuously passing from the liquid into the gaseous phase. We might well expect that the sharpest changes in the molecular spectrum would occur in the region of the critical point at which, according to accepted concepts, there is a particularly sharp change in the intermolecular interactions. We might also expect to discover whether the thermodynamic critical point constituted any kind of "singular" point for the hydrogen bond as a specific form of intermolecular interaction.

The first experiments in this direction were undertaken on the initiative of G. S. Landsberg in 1937 [47, 48]; the Raman spectra of water and methyl alcohol were studied over a wide temperature range, including the critical point. However, in view of the great experimental difficulties these investigations were insufficiently detailed, although they did lead to the conclusion that the transition through the critical point was not accompanied by any sharp changes in the spectrum, while the band of the associated molecules was observed even in the hypercritical region.

In order to obtain more detailed and reliable data regarding the changes taking place in the molecular spectra on passing through the critical point we undertook an investigation of the infrared absorption spectra of methyl and isobutyl alcohols over a wide temperature range, including the critical and hypercritical regions [49].

In order to carry out these investigations, we developed the construction of a special high-pressure cuvette enabling us to obtain the infrared absorption spectra of both the liquid and vapor phases and also vary the thickness of the absorbing layer during the measurement without dehermetization of the cuvette. The arrangement of the cuvette is depicted in Fig. 8a Inside the body of the cuvette 1, made of high-speed cutting steel, is an obturator 2 with a

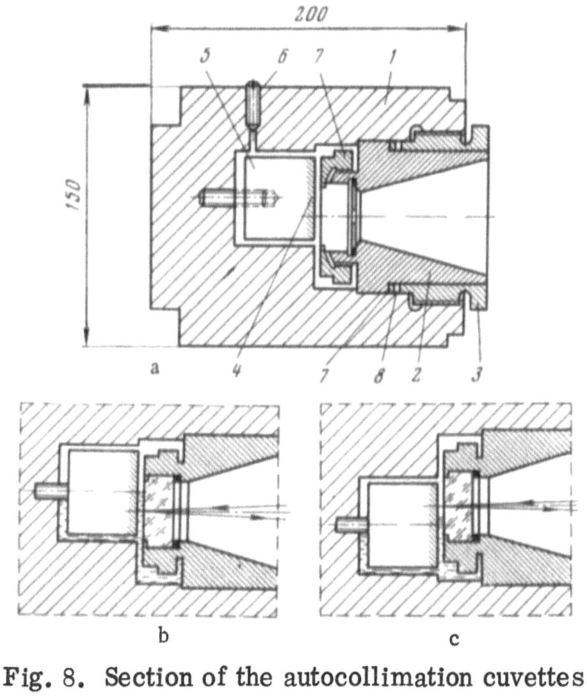

Fig. 8. Section of the autocollimation cuvettes for high-pressure work. a) General view; b,c) positions of the cuvette when studying the absorption spectrum of the liquid and gaseous phases, respectively.

quartz window, fixed by means of a nut 3. Inside the body, opposite to the window, is a chromium-covered mirror 4, fixed to the end of an unbalanced nut 5 fitted to a stationary screw. Owing to the displacement of the center of gravity of the nut relative to the axis of the screw, when the cuvette as a whole rotates the nut does not, but simply moves along the screw, changing the distance between the mirror and the window. This construction of the cuvette enables us to change the thickness of the absorbing layer of the test substances during the investigation without dehermetization, which is essential in view of the severe temperature dependence of the density. The test substance is introduced into the cuvette by means of a syringe through an opening in the side wall of the cuvette, which is then closed by means of a conical valve 6.

In this kind of investigation it is of interest to study the absorption spectrum of both the liquid and the vapor phases at the same temperature. To this end the axis of the obturator and the screw of the nut carrying the mirror were displaced with respect to the diameter of the cuvette, while the volume of the cuvette was made such that, on introducing the required amount of liquid (determined by the critical parameters of the latter), for one (horizontal) position of the cuvette a layer of liquid lay between the window and mirror (Fig. 8b) while on rotating the cuvette through 180° a layer of vapor occupied this position (Fig. 8c). This arrangement of the cuvette enabled us to secure the absorption spectrum of both the liquid and vapor phases throughout the whole temperature range, with the optimum thicknesses of the absorbing layer for both phases.

The working volume of the cuvette was sealed by means of special gaskets. For this purpose the nut 3 of the obturator presses a copper gasket 7 against the latter through a ring of quenched steel 8, which ensures a uniform pressure on the gasket and prevents its rotation. When the nut 3 is tightened the copper flows and fills the whole space between the obturator and the body of the cuvette.

Preliminary hermetization of the quartz window was achieved by means of a Teflon gasket placed between the window and the polished end of the obturator in an annular depression. The obturator was heated to about 250°C, and under a pressure of 40-50 atm the window was pressed against the obturator by means of the nut 7, using a small hydraulic press. After this the obturator was slowly reduced to room temperature and introduced into the cuvette. Under working conditions the sealing of the window and obturator was ensured by the rise in pressure within the working space of the cuvette as temperature rose.

The heating of the cuvette was achieved by means of a furnace in the shape of a hollow cylinder fitted onto the cuvette. The furnace and cuvette were placed horizontally in a special stand so as to be able to rotate around the axis.

The infrared absorption spectra were studied in the region of the first overtone of the O−H valence vibration (1.3-1.8 μ), since in obtaining the absorption spectra in this region the thickness of the liquid layer was considerably greater than in the region of the fundamental (microns and hundredths of a millimeter); for very thin layers of liquid the position of the critical point might be affected by surface phenomena at the metal−liquid−quartz interfaces. Furthermore, in the region of the overtone, the relative intensity of the O−H band of the monomers is greater than in the region of the fundamental. The absorption spectra were obtained by means of a specially developed double-beam automatic spectrophotometer (Fig. 9) with three 60° glass prisms. The light source was a tungsten strip lamp and the radiation receiver a bismuth bolometer.

The first experiments were carried out with water as test substance (T_{cr} = 374.1°C). Although the region of the critical point was reached in these experiments, the absorption spectra could not be obtained at high temperatures, since, at temperatures above 300°C, the surface of the quartz window was damaged where it came into contact with water owing to lixiviation.

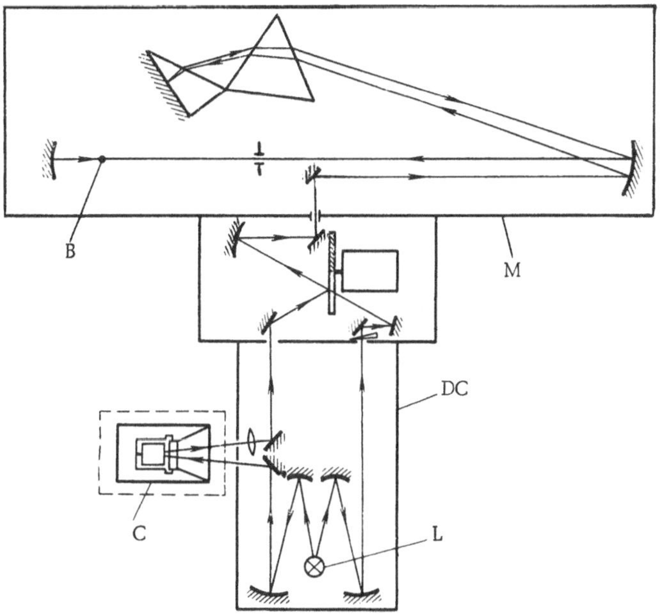

Fig. 9. Arrangement of the double-beam infrared spectrophotometer with a high-pressure cuvette (for a general view of the apparatus, see Fig. 50). M = monochromator; B = bolometer; DC = double-beam condenser; L = strip lamp; C = high-pressure cuvette.

Subsequent experiments were therefore carried out with methyl and isobutyl alcohols, for which the critical temperatures were similar (240 and 265°C) but the role of the steric factor [43] in the molecules of the two substances was quite different.

We obtained the absorption spectra of these substances in the liquid phase while varying the temperature from room temperature to critical, and in the vapor phase in the neighborhood of the critical temperatures (220-280°C).

The following results were obtained. At room temperature the infrared absorption spectrum of both alcohols showed a wide band with two maxima $\nu_1' = 6700$ cm^{-1} and $\nu_2' = 6400$ cm^{-1}, corresponding to associated molecules. On raising the temperature the intensity of the band diminishes, the intensity of the more displaced maximum $\nu_2' = 6400$ cm^{-1} doing so the more rapidly. At T ≈ 100°C for methyl alcohol and 40°C for isobutyl alcohol, the spectrum develops a narrow line at $\nu_0' = 7080$ cm^{-1}, corresponding to the unperturbed vibration of the O−H group of the monomers. On further increasing the temperature the intensity of the band of associated molecules continues decreasing, while the intensity of the monomer band rises uninterruptedly. This process continues even on passing through the critical point, which shows that associated alcohol molecules exist in both the critical and hypercritical regions.

As a quantitative characteristic of this process, we measured the temperature dependences of the transmission coefficients at the maxima of the O−H bands of the monomers ($\nu_0' =$ 7080 cm^{-1}) and the C−H group ($\nu = 5850$ cm^{-1}), characterizing the total number of alcohol molecules

$$T_m(\mathrm{O-H}) = \frac{I}{I_0} = e^{-\varkappa_m(\mathrm{O-H})} \quad \text{and} \quad T_m(\mathrm{C-H}) = \frac{I'}{I_0'} = e^{-\varkappa_m(\mathrm{C-H})},$$

where $\varkappa_m(\mathrm{O-H}) = k_m(\mathrm{O-H})\rho c l$ and $\varkappa_m(\mathrm{C-H}) = k_m(\mathrm{C-H})\rho l$. Here $k_m(\mathrm{O-H})$ and $k_m(\mathrm{C-H})$ are the absorption indices at the maxima of the O–H and C–H bands, referred to one molecule; ρ is the total number of alcohol molecules in 1 cm^3; c is the relative number of monomers in 1 cm^3; l is the thickness of the absorbing layer. If we suppose that the molecular absorption indices k_m(O–H) and k_m(C–H) of the valence vibrations of the O–H monomers and the C–H vibrations vary relatively slowly with temperature, then the ratio $\frac{\varkappa_m(\text{O–H})}{\varkappa_m(\text{C–H})} = \frac{k_m(\text{O–H})}{k_m(\text{C–H})}c$ will be proportional to the concentration of the monomers c, i.e., the degree of dissociation of the molecules α.

The experimental curves of the function $\frac{\varkappa_m(\text{O–H})}{\varkappa_m(\text{C–H})} = f(T)$ varied monotonically (Fig. 10) in the precritical and hypercritical regions, not experiencing any sharp jump at the critical point, i.e., the critical point was not distinguished in any way.

We also plotted the optical densities $\varkappa_m$ (O–H) and $\varkappa_m$(C–H) as functions of ρ/ρ_{cr} (Fig. 11), where ρ is the density of the alcohols, ρ_{cr} being its value at the critical point. We see from Fig. 11 that the quantity $\varkappa_m$(C–H) = k_m(C–H)ρl depends linearly on the densities throughout the whole density range, which indicates that k_m(C–H) is independent of temperature. The optical density $\varkappa_m$(O–H) = k_m(O–H) $\rho c l$, however, does vary, and particularly in the liquid phase at high densities (ρ/ρ_{cr} > 2), for which a considerable deviation from linearity sets in. This behavior of the $\varkappa_m$(O–H) relationship is due to the change in the degree of dissociation of the molecules with changing temperature, i.e., to the density dependence of the concentration of monomers C.

Thus the foregoing investigations show that associated complexes of alcohol molecules exist not only in the liquid phase but also in the hypercritical region. Thermodynamic equilib-

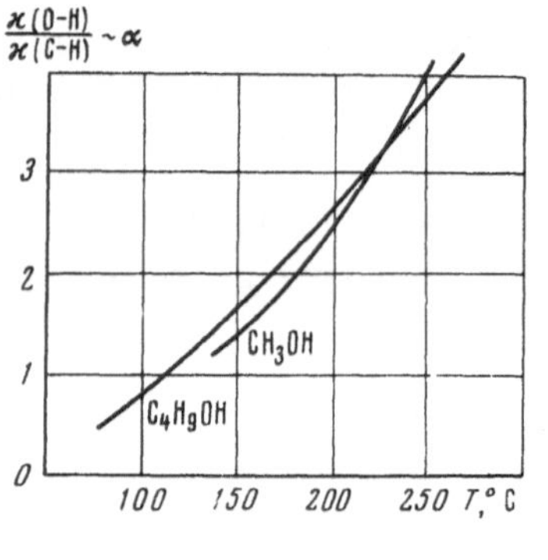

Fig. 10. Temperature dependence of the ratio of the optical densities of the O–H absorption bands of the monomers and the C–H band of the alcohol molecules.

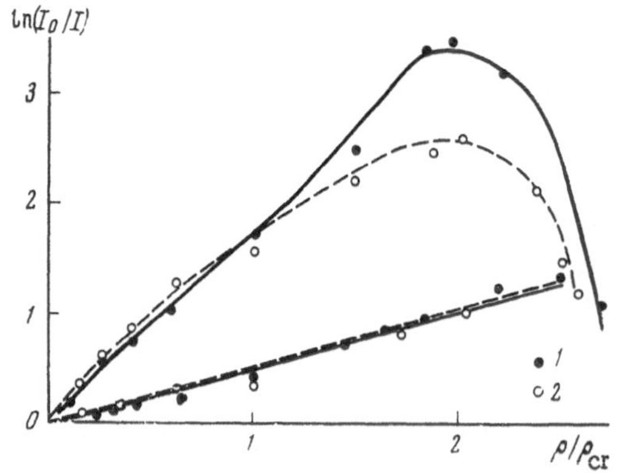

Fig. 11. Dependence of the optical densities of the O–H bands of the monomers (upper curves) and the C–H bands of the alcohol molecules (lower curves) on the ratio ρ/ρ_{cr}. 1) C_4H_9OH; 2) CH_3OH.

rium between the number of associated molecules and monomers changes continuously in the direction of an increasing number of monomers on passing through the critical point. Hence the critical point is not a singular point for the hydrogen bond; this confirms the conclusions drawn in [47, 48].

In these experiments we also studied the temperature dependence of the frequency of the O–H valence vibration of the monomers ν_0'. We found that the position of the maximum of the absorption band changed considerably in the region of the critical temperature. Thus in liquid alcohols at temperatures a long way from the critical point $\nu_0' = 7080\ cm^{-1}$; in alcohol vapor at T = 230°C the frequency rises to $\nu_0' = 7165\ cm^{-1}$, and at the critical point it falls to $\nu_0' = 7120\ cm^{-1}$.

The existence of associated complexes of alcohol in the critical and hypercritical regions was also confirmed in [50], in which the infrared absorption spectra of ethyl alcohol and its solutions in various solvents were studied in the region of the fundamental (≈ 3 μ) over a wide temperature range from room to 270°C.

§ 5. Role of the Hydrogen Bond in the Broadening of the Rotational Lines in Gas Mixtures

Many authors have examined the broadening of the rotational and vibrational–rotational lines of molecules [17]. Usually the broadening effect is simply explained as being due to the van der Waals forces; however, the experimental results cannot always be satisfactorily explained in this way. In some cases we believe that the hydrogen bond must be taken into account in addition to these forces. As already indicated, the hydrogen bond may lead to the formation of stable complexes not only in the liquid but also in the gas phase; this occurs, for example, in a number of carboxylic acids [36] and for the vapor of water and alcohols at high temperatures [47-50]. In cases in which the formation of stable complexes is impeded (low density, high temperature), the hydrogen bond may be revealed in the line broadening as the result of a brief perturbation during the collision of the molecules, which generates short-lived complexes.* We should nevertheless expect that the effectiveness of such perturbations would be fairly low, since the hydrogen bond can only develop if the interacting molecules are oriented in a particular mutual fashion. Since the broadening action of the hydrogen bond may be masked by the action of other intermolecular forces, the experimental observation of the effect in pure form presents obvious difficulties. It should nevertheless be possible to detect the effect qualitatively by comparing the line broadening in gas mixtures capable of forming hydrogen bonds between their molecules and gas mixtures in which no hydrogen bond can exist.

To this end we measured the width of the rotational lines of the infrared absorption of H_2O, HCl, and CH_4 vapor [52] for various pressures of He, Ar, O_2, and N_2. Among these mixtures, only in the case of mixtures of H_2O and HCl with N_2 and O_2 should the formation of a hydrogen bond in principle be possible; in the other mixtures the formation of a hydrogen bond is very unlikely.

The absorption spectra were measured by means of a special double-beam infrared diffraction spectrophotometer of high resolving power [53]. The gas mixtures were placed in a

* The assumption as to the existence of short-lived complexes formed by the collision of molecules in the gas phase is at present being discussed by a number of authors. For example, it is being suggested that the atmosphere contains $[O_2]_2$ complexes with a lifetime of about $2 \cdot 10^{-11}$ sec [51]. At the International Conference on Spectroscopy in January 1967 (Bombay, India), B. Vodar suggested the existence of short-lived H–Cl...Ar dimers in a mixture of HCl + Ar.

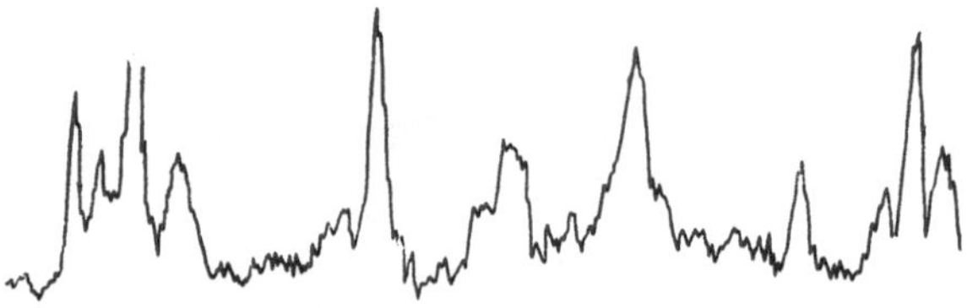

Fig. 12. Infrared absorption spectrum of water vapor in the region of 1.45 μ. Spectral slit width $\delta\nu$ = 0.4 cm^{-1}.

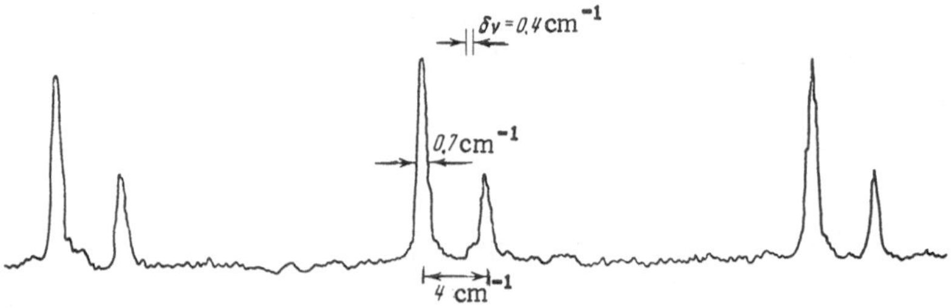

Fig. 13. Infrared absorption spectrum of hydrogen chloride in the region of 1.7 μ. Spectral slit width $\delta\nu$ = 0.4 cm^{-1}; double lines indicate isotopic splitting.

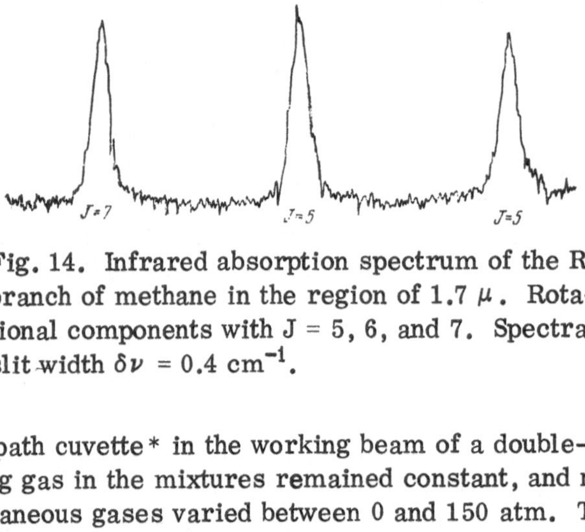

Fig. 14. Infrared absorption spectrum of the R branch of methane in the region of 1.7 μ. Rotational components with J = 5, 6, and 7. Spectral slit width $\delta\nu$ = 0.4 cm^{-1}.

high-pressure multiple-path cuvette* in the working beam of a double-beam condenser. The pressure of the absorbing gas in the mixtures remained constant, and no greater than 1 atm. The pressure of the extraneous gases varied between 0 and 150 atm. The infrared absorption spectra were measured in the region of the overtones of the O−H, C−H, and Cl−H vibrations (1.3-2.5 μ). The minimum width of the apparatus function was 0.4 cm^{-1}. The accuracy of the line-width measurement was about 10%. Figures 12-14 illustrate small sections of the spectra obtained for the vapors of water, HCl, and methane at a pressure of 1 atm.

We measured the width of various lines in the R and P branches of the rotational−vibrational bands: H_2O, λ = 1.45 and 1.8 μ ; HCl, λ = 1.7 μ, and CH_4, λ = 1.7 and 2.2 μ. For analysis we chose simply those lines which lay a reasonable way away from their neighbors. In the case of HCl we allowed for the isotopic splitting of the lines. The measurements showed that

*A description of the high-pressure cuvette and its disposition in the illuminating system are presented in Section 1 of Chapter II.

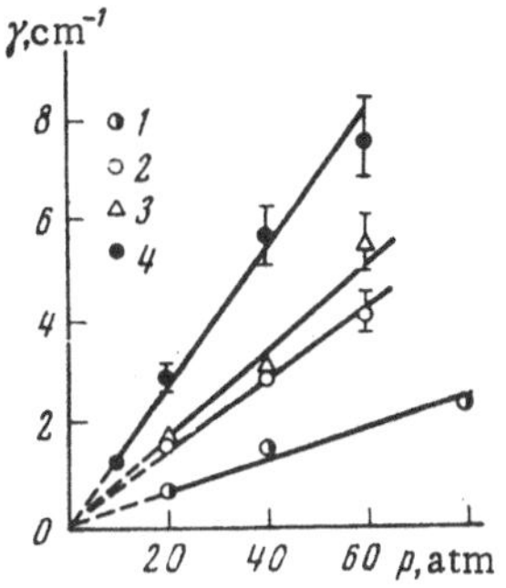

Fig. 15. Pressure dependence of the width of the rotational components of the water molecule in the region of 1.45 μ: 1) He; 2) Ar; 3) O_2; 4) N_2.

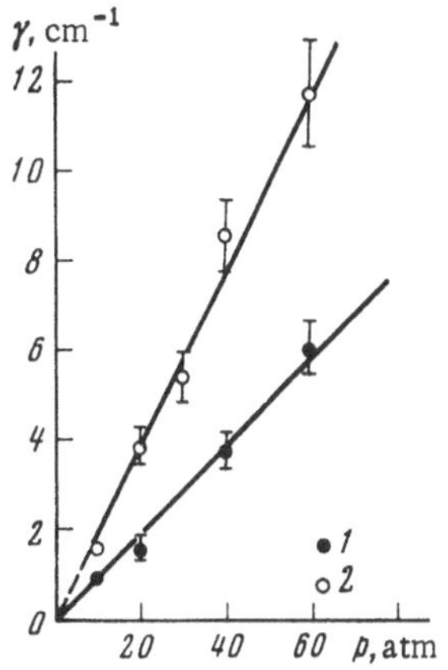

Fig. 16. Pressure dependence of the width of the rotational components of HCl molecules: 1) argon pressure; 2) nitrogen pressure.

the shape of the contour in individual bands was very close to that of the ordinary dispersion curve, while their width γ depended linearly (Figs. 15, 16) on the pressure of the foreign gases.* (Here $\gamma = \gamma_{obs} - \gamma_0$, where γ_{obs} is the observed width and γ_0 is the width of intrinsic broadening.) We may therefore consider that the impact mechanism of broadening is operative under the conditions of our experiments; using the usual relationships of the impact theory [17], we may therefore calculate the effective optical cross sections for the collisions of the absorbing molecule with the molecules of the foreign or extraneous gas. The results of these calculations are presented in Table 7. The data presented in this table and also in Figs. 15 and 16 show that the efficiency of the gases in the broadening of the H_2O ($\lambda = 1.45\mu$) and HCl lines increases in the sequence He, Ar, O_2, N_2. Here N_2 is almost twice as effective as Ar; however, they act in almost exactly the same way on the CH_4 molecule. The difference in the broadening effect of Ar and N_2 on the rotational lines of H_2O and HCl cannot be explained as being due to an inductive interaction, since the polarizabilities of Ar and N_2 are practically identical, while the dispersion interaction may be neglected [17]. The gas-kinetic collision cross sections of Ar and N_2 molecules with H_2O and HCl (and also CH_4) are practically identical as well. The results may be explained if we suppose that the effect of the hydrogen bond appears in the $H_2O + N_2$ and $HCl + N_2$ mixtures, although this conclusion is not strictly unique, since the nitrogen molecule possesses a quadrupole moment greater than that of the oxygen molecule, while the argon atom has no quadrupole moment at all [17].

* For the lines of methane the results of the measurements are presented in Section 1 of Chapter II (Figs. 36 and 37).

TABLE 7. Optical Collision Cross Sections (in 10^{-16} cm^2)

Principal substance	Band studied, μ	Extraneous gas				
		He	Ar	N_2	O_2	CH_4
H_2O	1.45	10(18)*	41(30)	85(32)	46(30)	—
H_2O	1.8	6(18)	25(30)	34(32)	—	—
HCl	1.7	—	62(50)	116(53)	—	—
CH_4	1.7 $J=6$	21(17)	50(28)	50(30)	—	70(15)
CH_4	2.2 $J=1$	17(17)	44(28)	60(30)	—	—

* The gas-kinetic cross sections appear in brackets for comparison.

The fact that O_2 has a weaker broadening effect than N_2 agrees with its lower hydrogen-bond energy [34].

The results show that in the case of the H_2O molecule the difference in the broadening effect of Ar and N_2 is more sharply expressed for the vibration $\lambda = 1.45\ \mu$ than for the vibration $\lambda = 1.8\ \mu$. This may be associated with the fact that the $\lambda = 1.45\ \mu$ is a composite, derived from the frequencies of the almost purely valence vibrations of the O–H group ($\nu_1 + \nu_3$) while the $\lambda = 1.8\mu$ band corresponds to a composite vibration ($\nu_2 + \nu_3$) in which one of the frequencies ν_2 is of the deformation type and may be less subject to the action of the hydrogen bond.

It should be noted that the $\lambda = 1.8\ \mu$ band of water broadens less than the $\lambda = 1.45\ \mu$ band under the influence of He and Ar as well as nitrogen.

As regards the $\lambda = 2.2\ \mu$ band of CH_4, for which the cross section was slightly greater in the case of broadening by nitrogen than by argon, this is probably associated with measuring error; in the case in question the cross section was determined simply from measurements of line widths at two pressures, while the width of the P_1 line itself was measured to a low degree of accuracy, since the neighboring Q branch partly overlaid the P_1 line and introduced distortions into its contour.

The broadening effect of N_2 and O_2 on the bands of the H_2O and HCl molecules (greater than that of Ar) is supported by a number of other authors. Thus, according to the results of [54], the diameter of the collision cross section of the H_2O molecule with the Ar molecule equals 4.9 Å, with the N_2 molecule, 5.5 Å; and with the CO_2 molecule, 6.8 Å. It was shown in [55] that the broadening of the HCl lines due to O_2, N_2, CO, CO_2, and N_2O was much greater than that due to Ar and Ne atoms (by a factor of four). The difference between the line broadening of HCl and CH_4 was established most clearly in [56], in which it was shown that molecules containing oxygen and nitrogen atoms (SO_2, CO_2, CO, N_2O, N_2, and O_2) produced a greater broadening of the lines of the HCl molecule relative to the lines of CH_4 than He, Ne, H_2, Ar, Kr, and Xe.

Our own experimental data therefore in no way contradict the hypothesis regarding the possible manifestation of an intermolecular hydrogen bond in the additional broadening of the rotational lines.

§ 6. Study of the True Absorption of Water in the Drop Phase (in a Water Cloud)

As already indicated, the most interesting examples of the study of intermolecular interactions by spectroscopic methods relate to transitional states: transitions between the gas and liquid phases, or between liquids and solids.

One of the problems which attracted our attention and the results of which are presented in this section was the condensation of water vapor. The condensation process is well known to be associated with the formation of nuclei and appears to take place in a number of stages intermediate between the vapor phase and the drop phase of the liquid (particularly if the medium is carefully freed from dust and ions, and the nuclei develop as fluctuation phenomena, and in particular as short-lived complexes arising from the formation of hydrogen bonds). The experimental analysis of these intermediate stages certainly offers considerable interest in connection with explaining the mechanism of condensation.

In the case of water vapor an analysis of this kind may be carried out by studying the changes in the vibration frequencies of the O−H group (for example, by recording the infrared absorption spectra), since the spectra of water and its vapor differ substantially.

In addition to this, a study of the true absorption of radiation in the drop phase is of independent interest from the point of view of studying the special characteristics of the intermolecular interaction of water molecule in a surface layer. For small drop sizes (less than 0.01 μ in diameter), in fact, the number of molecules concentrated in the surface monomolecular layer (roughly $3 \cdot 10^{-8}$ cm thick) may be comparable with the total number of molecules in the volume of the drop (for $d = 0.01\ \mu$ the number of molecules on the surface of the drop is roughly 10% of the total number), and frequencies characterizing these surface molecules will therefore appear in the absorption spectrum.

In order to solve these problems we must first of all develop a method yielding the true* absorption of the drop phase of a cloud, since (as far as we know from published data existing at the beginning of the present investigation) the absorption of radiation by the drop phase fails to appear in ordinary methods of observation [57, 58]. It is pointed out in [58] that the spectrum recorded through a cloud only exhibits the absorption band belonging to the water vapor, and also a weakening of the radiation due to scattering. "The optical properties of liquid water in drop form (associated molecules) are in no way apparent..." [58]. This assertion struck us as rather strange in view of the well-known fact that [57], in the scattering of light in drops, a considerable proportion of the light flux falling on a drop (up to 88%) passes through it and, hence, for drop diameters of a few microns (or for smaller drops but greater thicknesses of the cloud) the true absorption of the liquid phase should be substantial and easily observed, since, in a film of liquid water a few microns thick, the absorption is very considerable. We concluded that in ordinary methods of studying the transmission spectrum of a cloud the true absorption was evidently masked by some other effects.

In order to elucidate all these questions we studied the transmission spectra of water clouds with different degrees of condensation. The transmission spectra in the spectral range 2.5-15 μ were obtained under laboratory conditions using a composite double-beam prismatic automatic spectrophotometer [59]. The monochromator of the spectrophotometer was an IKS-11 monochromator with LiF and NaCl prisms.

The first experiments were carried out with a jet of steam emerging from the orifice of a boiler at various distances from the orifice, distinguished by different degrees of condensation of the steam in the air. The cross section of the jet had a rectangular shape, the width

*Here and subsequently we shall use the term "true absorption spectrum" to mean the absorption spectrum of the test substance free from physical distortion factors taking place in the actual medium under examination, such as the scattering of radiation. Even on eliminating these factors, of course, the absorption spectrum recorded by means of the spectral instrument will differ from the true spectrum as a result of apparatus distortions due to the apparatus function of the spectral instrument and the recording system of the spectrometer. The apparatus distortions are not considered in this article.

along the beam being 4-8 mm and the height 30 mm. In order to prevent the condensation of steam on the walls, the end of the jet was heated by means of a small electric furnace. Near the jet the stream of vapor was completely transparent (to the eye), indicating that there was only a slight degree of condensation. A long way from the jet the stream became opaque owing to scattering in the drops. This stream of vapor was placed in the working beam of a double-beam illuminating system and its transmission spectrum was recorded.

The experiment showed that in this case the recorded transmission spectrum of the stream of vapor at all distances from the jet exhibited a slight background due to scattering and a very strong absorption spectrum of the water vapor molecules with a characteristic vibrational–rotational structure.

The absorption of the drop phase in the stream of vapor, however, failed to appear even at considerable distances from the jet (at which the stream was opaque), as the transmission spectrum exhibited none of the broad absorption bands characterizing the perturbed vibration of the O–H group of the water molecules, analogous to the bands with maxima at $\lambda_m \approx 3$ and 6 μ observed in the transmission spectrum of a film of liquid water.

Since the double-beam spectrophotometer records the intensity ratio of the working beam and the comparison beam, the presence of strong absorption bands of water vapor in the transmission spectrum of the stream shows that the density of the vapor in the latter is much greater than the density of the vapor in the air (in the comparison beam) at room temperature, even at considerable distances from the jet.

It was suggested that the absence of the wide absorption bands of liquid water from the transmission spectrum in the drop phase was associated with the masking action of strong neighboring absorption bands of water vapor and the scattering background. In order to observe the absorption of the drop phase of the cloud we therefore undertook some experiments in which the absorption bands of the vapor were eliminated from the recorded spectrum. We were able to do this by placing cuvettes containing pure water vapor of the same density as that of the vapor in the test cloud in the comparison beam, or by reducing the vapor density in the cloud to a value equal to the density of the vapor in the air in the comparison beam. In this case the double-beam spectrometer did not, of course, record the compensated absorption bands.

Of the two compensation methods indicated, the second was in practice the simpler, and we accordingly made use of this in future experiments. In order to obtain a cloud with a vapor density similar to that in the air at room temperature we used two methods. In the first method the cloud was obtained by spraying water with the help of a pulverizer (Fig. 17), and in the second method by the intensive cooling of the water vapor in a special mixing chamber, using gaseous nitrogen at a low temperature. Figure 18 shows the arrangement for obtaining the cloud by the second method. The water vapor was obtained by means of a boiler with electric heating and was introduced into the mixing chamber. The cooled nitrogen was supplied from a metal dewar containing liquid nitrogen with a small heater let down into it.

The resultant cloud was introduced from below into a cuvette with fluorite windows placed in the working beam of the spectrophotometer. In order to prevent the condensation of the cloud on the windows, the latter were heated. In certain experiments the spectrum was obtained from a cuvette without any windows. For this purpose the cloud was sucked out of the upper side tube of the cuvette (Fig. 18) at such a velocity that a constant volume occupied by the cloud was maintained in the central section of the cuvette. The working length of the cuvette filled with the cloud was varied from 1 to 10 cm, depending on the density of the latter. The average diameter of the drops in the cloud obtained by the foregoing methods lay between 3 and 6 μ. The dimensions of the drops were determined by photographing water drops, condensed from the cloud stream onto a glass plate covered with a thin layer of oil, at high mag-

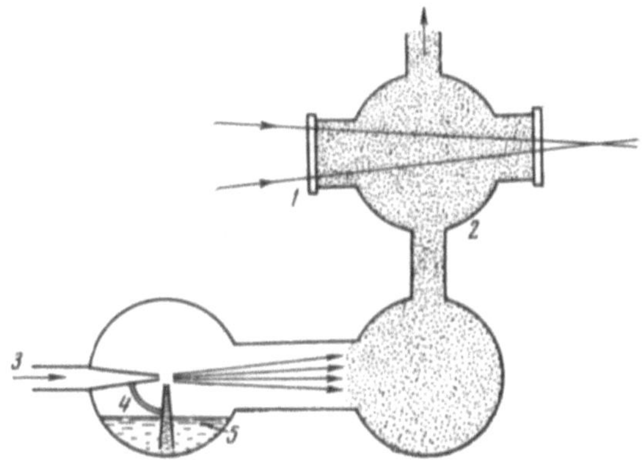

Fig. 17. Cloud "generator" with pulverizer and absorption cuvette. 1) CaF_2 windows; 2) cuvette containing cloud; 3) compressed air; 4) pulverizer; 5) water.

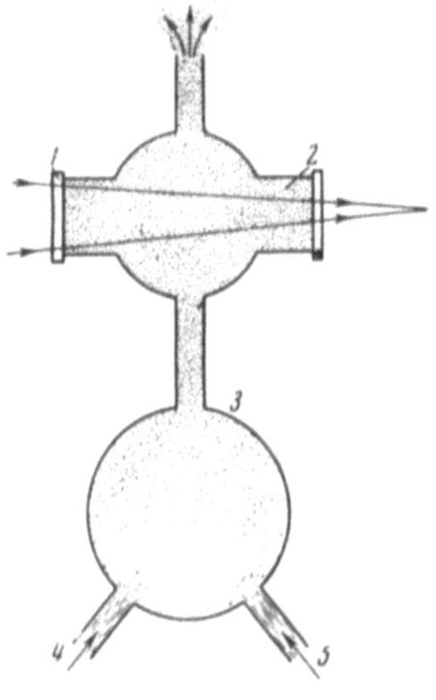

Fig. 18. Cloud "generator" with cooled water vapor and an absorption cuvette. 1) CaF_2 windows; 2) cuvette containing cloud; 3) cloud; 4) cooled nitrogen; 5) vapor from boiler.

nification (dark-field method), and directly measuring their diameters. In this way drops with diameters of over 1 μ could be measured quite reliably.

Provided that the spray or boiler and the cooling system operated stably, the supply of cloud to the cuvette was also reasonably stable, thus ensuring a constant density of the cloud in the cuvette. The constancy of the cloud density in the cuvette was monitored by recording the background due to scattering at a fixed wavelength; the level of scattering background remained constant for the time required to record the whole transmission spectrum in the range 2.5-15 μ.

We were able to choose operating conditions for the "cloud generator" such that the vapor density in the cloud occupying the cuvette was similar to the vapor density in the air in the comparison beam, so that complete compensation of the absorption bands of the vapor was achieved in the recorded spectrum of the cloud.

However, even after compensation of the absorption bands of the water vapor, the transmission spectrum of the cloud exhibited no absorption bands of liquid water. At the same time, instead of the expected absorption bands of liquid water, the part of the spectrum which ought to have exhibited the latter behaved in exactly the opposite way, yielding a selective band of transparency on the background of a general weakening of the light due to scattering in the cloud (Figs. 19 and 20). These bands of transparency lay in the region of 2.77 and 5.8 μ, not coinciding with the maxima of the absorption bands of liquid water (λ_m = 2.95 and 6.02 μ, respectively). Experiments carried out with clouds having different mean drop diameters (not extending over a very wide range) showed that the position of the bands of transparency was independent of the drop size [60].

Existing theories of the selective scattering of light by a cloud for the case in which the particle sizes are commensurable with the wavelength [61, 62] failed to explain the foregoing phenomenon. The scattering-function curve calculated in accordance with [61] for the case of a single-dispersion cloud with particle sizes similar to those obtained in our experiments ($2\rho = 6$ μ) differs considerably from that actually observed, both in the width of the minimum and also in its position (1.7 μ instead of 2.77 μ). In addition to this, the minimum in the theoretical scattering function ($\lambda_{min} = 2\pi\rho/11.2$) depends on the particle size, which disagrees with experiment.

We considered that the observed bands of transparency were due to selective scattering close to the absorption band, associated with the anomalous behavior of the refractive index in the neighborhood of the absorption band of liquid water (since the scattering coefficient associated with the cloud drops depends on the refractive index), and with the true absorption of the radiation by the water drops. One proof of this hypothesis is the fact that the minima of the bands of transparency coincide in position with the minima of the refractive index of liquid

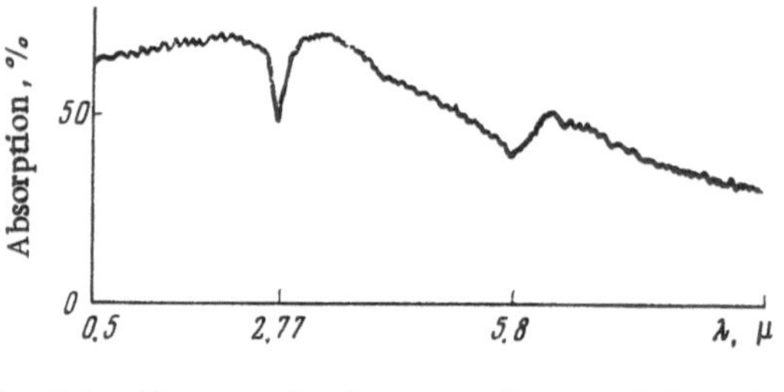

Fig. 19. Transmission spectrum of the cloud in the spectral range 0.5-8 μ. Prism made of NaCl, cloud obtained by spraying water from a pulverizer, mean drop diameter 6-7 μ.

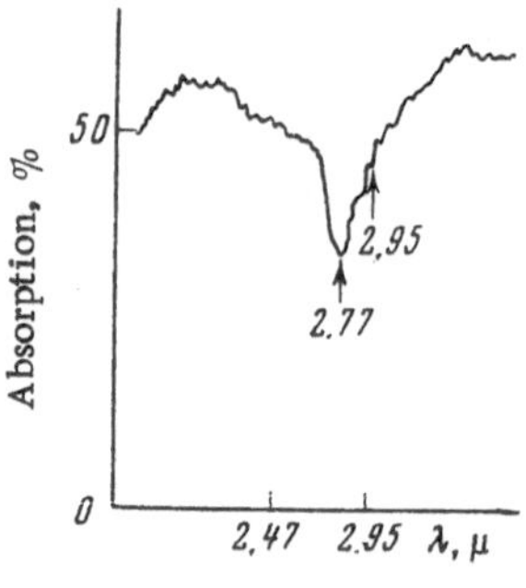

Fig. 20. Transmission spectrum of the cloud in the range 2-4 μ. Prism made of LiF, cloud obtained by cooling water vapor; mean drop diameter 4-5 μ.

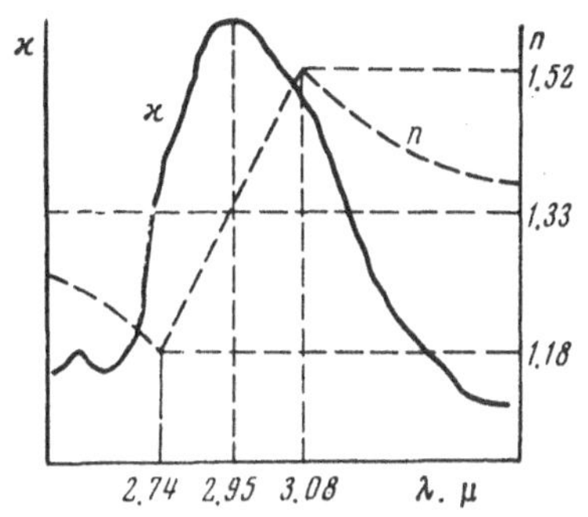

Fig. 21. Anomalous behavior of refractive index n of liquid water in region of absorption band ϰ. Data relating to n taken from [63].

water [63] (Fig. 21). In order to present a qualitative explanation, we may suppose that in the present case ($\rho \approx \lambda$, where ρ is the radius of the drops) the dependence of the scattering coefficient K on the wavelength λ has (to a first approximation) the same form as in the case of large particles $\rho > \lambda$, for which K may be represented by the formula given in [57, 64]: $K = 2\pi\rho^2\alpha^2(n-1)^2$, where n is the refractive index and $\alpha = (2\pi\rho/\lambda_0)n = \alpha_0 n$. The value of K depends on λ by way of $\alpha_0^2 \sim 1/\lambda^2$ and also by way of $n = f(\lambda)$. The first relationship gives a gradual fall in the scattering coefficient with increasing wavelength,* as actually observed in our experiments (Fig. 19); however, the function $\varphi(\lambda) = n^2(n-1)^2$ suffers very sharp changes in the region of the absorption band owing to the anomalous variation in the refractive index. According to [63], for the case of water the refractive index at $\lambda = 2.74\,\mu$ has a minimum value of $n = 1.18$, and at $\lambda = 3.08\,\mu$ a maximum value of $n = 1.52$ (Fig. 21). The value of $n^2(n-1)^2$ meanwhile changes by a factor of approximately 13. It thus follows that the scattering coefficient K should also suffer sharp changes in the region of the absorption band; its minimum value should coincide with the position of the minimum for n. Hence, on the general background of a weakening in the intensity of the radiation passing through the cloud, due to scattering, the range of wavelengths close to the K minimum should weaken considerably less and convey the effect of increased transparency.

The foregoing effect may be explained qualitatively in the following manner. If a parallel beam of light traveling in a certain direction falls on the drop, then the drop acts as a spherical lens, converging the light to a focus, beyond which it diverges at a certain angle ω (Fig. 22). This angle is inversely proportional to f^2, where $f = \rho/2(n-1)$ is the focal length of the spherical lens. Whereas in the absence of the drop the parallel beam passes completely through the aperture diaphragm of the instrument D (Fig. 22), after refraction in the drop only part of the beam will pass through the diaphragm; the spectrophotometer will record this as a reduction in the transmission coefficient – the scattering background. The transmission coefficient will be the smaller, the greater the angle ω, i.e., the smaller the distance f. In the region in which the refractive index behaves in an anomalous manner, n changes sharply, and hence so do f and ω, i.e., we encounter a special kind of chromatic aberration. For those wavelengths at which the value of n is close to unity, the value of f will be the greatest and the angle ω the smallest; hence the transmission coefficient will also be greatest in this range of wavelengths, and a band of enhanced transparency will appear.

*A more precise dependence of K on λ [without allowing for $n = f(\lambda)$] in the region of $\rho \approx \lambda$ is given in [61] for nonabsorbing particles, and in [62] for particles with a constant absorption; here we should find characteristic wide minima and maxima, which can only be observed in single-dispersion clouds (see also [65]). However, the position of these minima does not coincide with those actually observed in our experiments. In highly dispersed clouds the minima and maxima predicted by the theories of [61, 62] will be smoothed out.

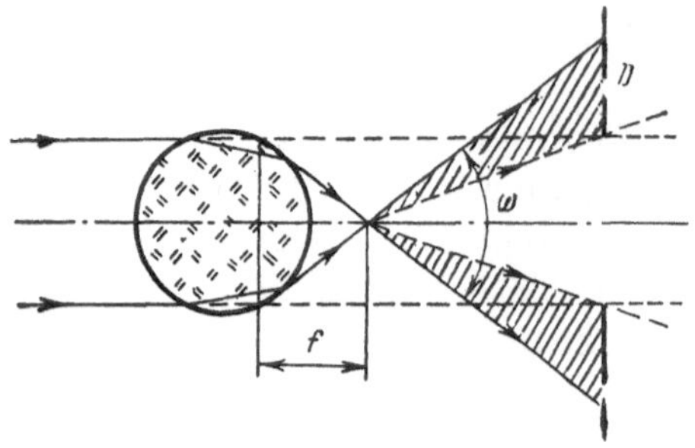

Fig. 22. Simplified scheme of change in structure of a parallel beam after passing through a drop.

It should be noted that this phenomenon is rather similar to the selective scattering of light in mercury vapor near a resonance line observed by Mandelstam and Landsberg [66], which was also associated with a sharp change in the scattering coefficient resulting from the anomalous behavior of the refractive index in the region of an atomic absorption band. The difference lies simply in the fact that in mercury vapor the scattering takes place at atoms, while in our own case it takes place at macroscopic particles consisting of a large number of molecules. However, the spectroscopic properties of the individual molecules appear in our own case also.*

There is no doubt that, in addition to the foregoing effects due to the weakening of the transmitted radiation by scattering, there should also be a weakening of the light as a result of true absorption by the molecules of water in the liquid drop phase, so that the bands of transparency observed in the recorded spectrum are the result of the simultaneous action of all these effects. The weakening of the radiation due to scattering in the cloud evidently plays the major part and masks the weakening due to true absorption. Nevertheless, the picture actually observed indicates that the spectral properties of the molecules in the drop (the dependence of the refractive index, both real and imaginary parts, on the wavelength) have a considerable effect on the optical properties of the cloud, so that the dependence of the refractive index on the wavelength λ should also manifest itself in the scattered light.

It should be noted that the reduction in the effect of true absorption in the transmission spectrum obtained under ordinary conditions of observation is probably also associated with the fact that the observations are conducted within a limited solid angle (as determined by the aperture of the monochromator), while the weakening associated with true absorption affects the whole of the scattered light, i.e., it occurs within a considerably greater solid angle, determined by the scattering indicatrix. If in some way we could measure the absorption over the whole solid angle of the scattering indicatrix, i.e., measure the integrated absorption in the whole beam of radiation passing through the drop, then the effect of true absorption would appear more clearly.

We thus have the problem of separating the weakening of the light respectively due to scattering and true absorption; without so doing, the role of the latter can never be established.

In order to determine the true absorption of liquid drops of water, we placed integrating spheres (Fig. 23) of the Ulbricht type in the optical system of the double-beam condenser of the spectrophotometer; these enabled us to measure the integrated absorption over a forward scattering angle of about $\pm 90°$ and eliminate the weakening of the beam of scattering, both within and without the region in which the refractive index behaved anomalously. The diameter of the integrating spheres was 80 mm, the inner surface being mat and covered with a layer of silver. In order to increase the integration angle, we placed the cuvette containing the cloud

*The influence of anomalous dispersion on the attenuation factor of clouds near the absorption band is briefly considered in [65].

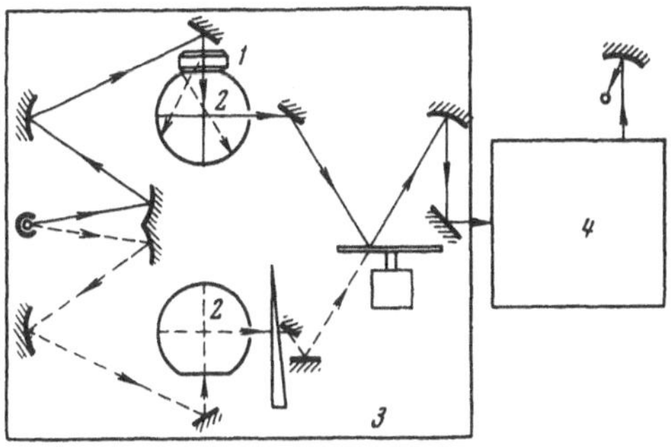

Fig. 23. Arrangement of the double-beam illuminating system incorporating integrating spheres. 1) Cuvette containing the cloud; 2) spheres; 3) double-beam illuminating system (condenser); 4) monochromator.

directly at the exit aperture of the sphere, and made the sides of the cuvette reflecting. The CaF_2 windows of the cuvette were slightly heated in order to prevent drops from condensing on them. The cooled cloud under examination was passed continuously through the cuvette from the cloud generator, which ensured a uniform feed. The thickness of the cloud layer was 20 mm.

The transmission spectrum of the cloud obtained in the system incorporating the integrating spheres differs considerably from that obtained by the ordinary arrangement. No weakening of the light due to scattering appears, nor does the band of transparency, but there is a selective weakening of the light, reminiscent of the absorption band of a film of liquid water (Fig. 24). This band lies in the same range of wavelengths as the band of liquid water obtained under the same conditions (using spheres and the same slit widths) on passing light through a thin film of liquid water between fluorite plates (Fig. 25). However, the band of weakening in the cloud differs somewhat from the absorption band of the film, being slightly wider, having a different shape, and having its maximum displaced by about 100 cm^{-1} in the long-wave direction. This difference in the shape of the bands may be due to some incompletely eliminated scattering effects arising from the limited nature of the capture angle over which the sphere integrates.

With the apparatus in question, the scattered light was only integrated forward and partly to the side; the back-scattered light was not integrated. It is quite possible that the role of the surface water molecules mentioned earlier enters in this connection. In order to elucidate this, we made a number of experiments with a special mirror cuvette and an integrating sphere

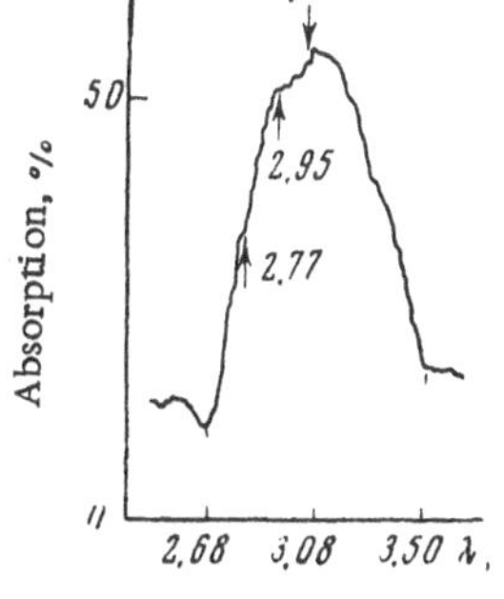

Fig. 24. Absorption bands of cloud obtained with integrating spheres. Prism made of LiF. Cloud obtained by cooling water vapor. Mean drop diameter 4-5 μ. Arrows indicate the wavelengths at which the refractive index of water has a minimum (λ = 2.77 μ) and a maximum (λ = 3.08 μ). The wavelength of λ = 2.95 μ corresponds to absorption-band maximum of a film of liquid water.

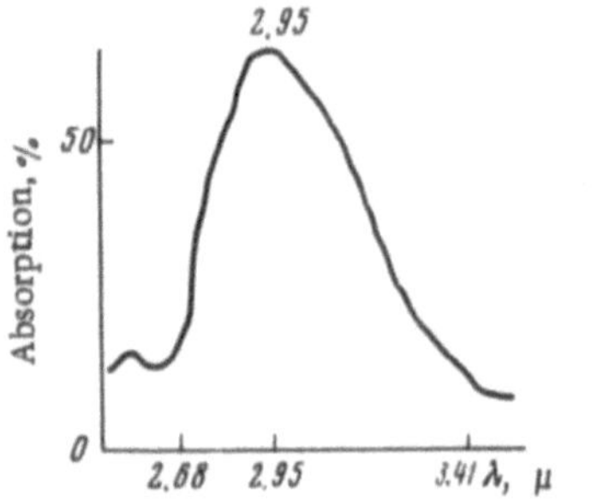

Fig. 25. Absorption spectrum of a film of liquid water between fluorite plates. Prism made of LiF.

(Fig. 26) which enabled us to integrate nearly all the scattered light with a capture angle of almost 360°. With this arrangement the light scattered by the cloud drops, both forward, backward, and laterally, after one or many reflections from the mirror walls of the cuvette fell into the integrating sphere, and after being averaged in the latter passed into the monochromator. A similar device but with an empty cuvette was placed in the comparison beam of the double-beam illuminating system to ensure complete compensation of the optical path. The transmission spectrum obtained under these conditions shows an absorption band which, in position, shape, and width agrees closely with that of a film of liquid water (Fig. 27; see also Fig. 25). However, the maxima is still slightly displaced (by about 50 cm^{-1}) in the long-wave direction, and its long-wave edge is also rather broader.

It may well be that this difference is not purely an apparatus effect and is not associated with scattering phenomena, but is simply due to the absorption of the surface molecules in the drops of water. As already indicated, the relative role of this effect should be small for drops a few microns in diameter, which corresponds to the mean size of the drops (measured by means of a microscope) in our experiments. However, since the cloud under examination was not of a single particle size but contained many small drops not visible under the microscope, it is in principle not impossible that the observed displacement of the band in the long-wave direction may have been due to the absorption of the surface water molecules in these drops.

Fig. 26. Mirror absorption cuvette (1) with an integrating sphere (2). Integration angle almost 360°.

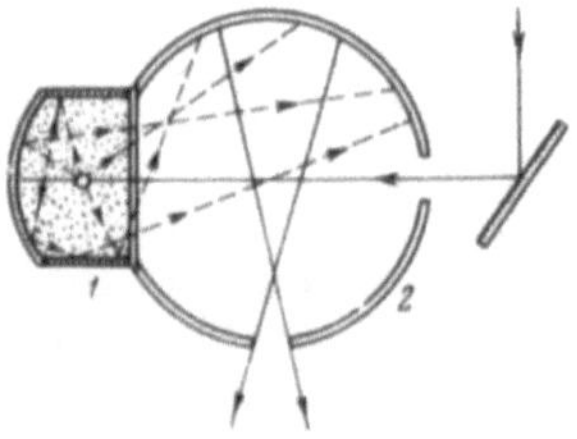

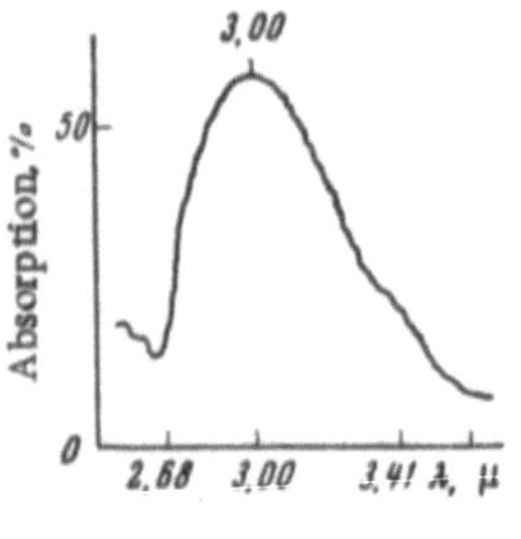

Fig. 27. Absorption spectrum of a cloud obtained from a mirror cuvette and integrating sphere. Integration angle about 360°. Prism made of LiF, cloud obtained by cooling water vapor; drop diameter 4-5 μ.

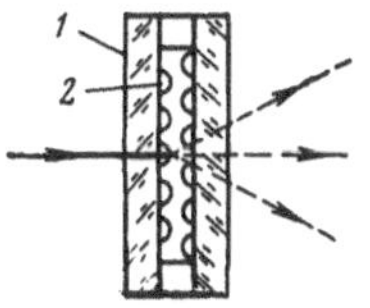

Fig. 28. Arrangement of a cuvette with a "two-dimensional cloud." 1) Cuvette walls; 2) drops of water.

On the basis of the foregoing experiments we may assert that, on observing the spectrum with integrating spheres, it is the true absorption of the water molecules in the drop phase which is mainly apparent; this is not observed in ordinary methods of studying the transmission spectrum of a cloud.

One confirmation of this conclusion lies in some of our auxiliary experiments relating to the transmission of a "two-dimensional" cloud obtained by placing drops of water on the inner surface of cuvette windows made of various materials (CaF_2, KRS-5, quartz) (Fig. 28). We were thus able to vary the size of the drops over a wide range from 1 to 100 μ. The transmission spectrum of a two-dimensional cloud under ordinary conditions of observation (in a transmitted beam) is similar to that of an ordinary cloud. For drop sizes commensurable with the wavelength ($\rho \leq \lambda$) the spectrum appeared to be intermediate between the absorption band of a film of water and the anomalous refractive index curve, while for $\rho > \lambda$ a typical band of transparency appeared, as in the case of a cloud (Fig. 29). In the spectra of the same samples obtained with the aid of integrating spheres we found an absorption band similar in shape and width to the band observed in a film of liquid water. On varying the distance between the cuvette and the entrance aperture of the sphere (thus varying the integration angle of the scattered light), we obtain all the intermediate pictures, from the absorption band to the band of transparency (Fig. 30). Thus the method of obtaining the spectrum has a major effect on the form of the spectrum observed. Indications as to effects of this kind when studying the

Fig. 29. Absorption spectrum of water drops on a fluorite plate obtained without integrating spheres. Prisms made of LiF; a) mean drop size 2 μ; b) 30 μ.

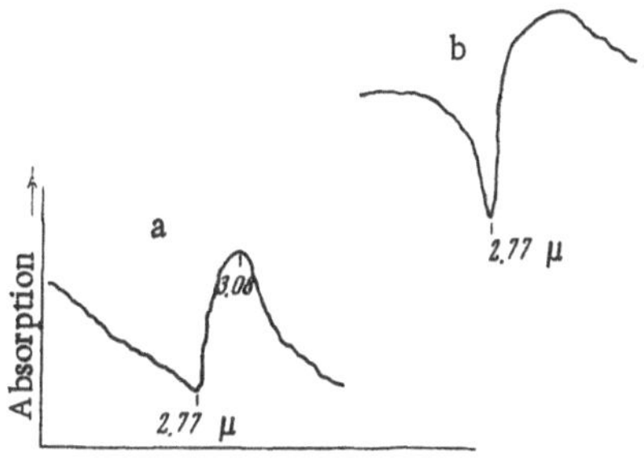

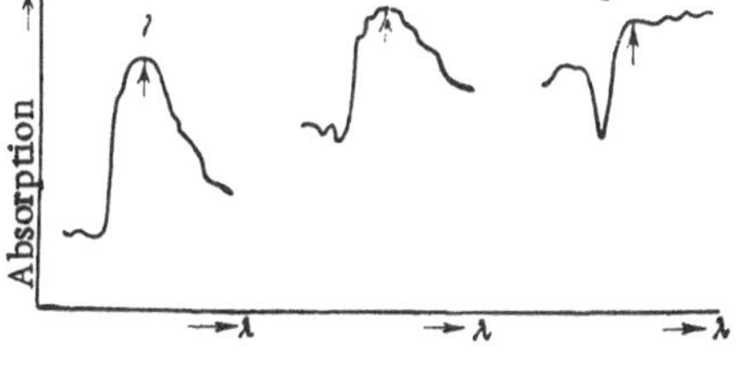

Fig. 30. Absorption spectra of water drops on a fluorite plate for various distances of the cuvette from the entrance window of the integrating sphere. Arrows indicate the same wavelength. Prism made of LiF; 1) cuvette at the window of the sphere; 2) intermediate case; 3) cuvette a long way from the window.

absorption of cloudy and dispersed media appeared in earlier publications [67, 68]. It was noted in the latter reference that when studying the absorption spectra of suspensions the form of the spectrum depended on the relation between the particle size and the wavelength as well as on the surrounding medium.

We also studied the transmission spectra of a two-dimensional cloud in the case in which the drops on the surface of the cuvette were surrounded not by air but by an immersion medium with $n > 1$, for example, CCl_4 ($n_D = 1.46$). For this purpose a cuvette with fine drops of water on its inner walls was filled with CCl_4 and placed in the working beam of the spectrophotometer. Under these conditions the scattering effects associated with the difference between the refractive indices of the drop and the surrounding medium were so weakened that even under ordinary conditions of obtaining the transmission spectra (without integrating spheres) the absorption band of liquid water duly appeared (Fig. 31).

Finally we studied the transmission spectrum of a water emulsion in CCl_4 (water drop diameter 1-2 μ), which constitutes, as it were, a "three-dimensional cloud" in CCl_4. The emulsion was obtained by rapidly cooling a heated and vigorously stirred mixture of CCl_4 with a small quantity of water. In this type of emulsion the scattering effects are also weakened, and the spectrum exhibits the wide absorption band of liquid water in drop form. In addition to this, the transmission spectrum of the emulsion also exhibits a narrow displaced absorption band of the vibration of the O−H group belonging to isolated water molecules (monomers) dissolved in the CCl_4.

It should be noted that the results obtained when studying the true absorption of the drop phase of a cloud are also of wider interest in connection with problems associated with the absorption spectra of dispersed media (emulsions, suspensions, powders, etc.). The effects of scattering in these media may seriously distort the true absorption spectrum of the substance. This applies in particular to the infrared absorption spectra of various solids, which are usually introduced in the form of fine powders into paraffin oil or pressed into a matrix of potassium bromide. In order to obtain absorption spectra as close as possible to the true versions, we must minimize the effects of scattering by carefully crushing the particles to sizes smaller than the wavelength and duly choosing the appropriate immersion medium.

On the basis of the foregoing investigations into the transmission spectrum of a cloud, we may regard it as convincingly proved that, on passing through a cloud, infrared radiation is weakened not only by scattering but also by the true absorption of the water molecules in the drop phase. The true absorption in the cloud must therefore be taken into account when calculating the range of action of various types of apparatus used in the infrared part of the spectrum. It is also to be expected that monochromatic radiation (for example, from a laser)

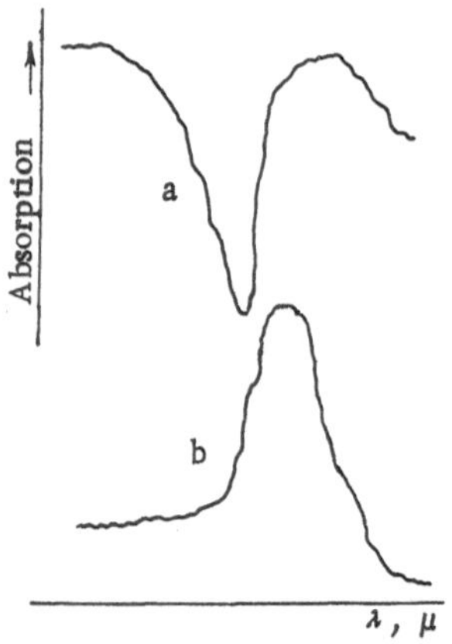

Fig. 31. Absorption spectra of water drops on a fluorite plate obtained on the same spectrogram. Prism made of LiF; a) drops surrounded by air; b) drops surrounded by immersion in CCl4; mean drop size, 6–18 μ.

with a wavelength coinciding with the minimum of the scattering coefficient (transparency band $\lambda = 2.77\ \mu$) will be weakened much less by scattering when propagating in clouds or dense mist. In order to ensure that the radiation should at the same time not be weakened by absorption in the water vapor, its wavelength should not coincide with any of the vibrational–rotational absorption bands of water molecules in the vapor phase.

§ 7. Study of the Hydrogen Bond in Hydrogen Halides

A number of authors [69-72] have studied the infrared absorption spectra of solutions of hydrogen chloride in various solvents including some containing oxygen atoms. In the latter case the spectrum exhibits a wide band severely displaced (by 400-450 cm^{-1}) in the long-wave direction with respect to the Q branch of the HCl band in the gas phase. An analogous picture occurs in the Raman spectrum [73] of such solutions. These results suggest that a hydrogen bond may be formed between the hydrogen chloride molecule and the oxygen atom in the solvent molecule.

However, the experiments in question were carried out for binary mixtures at fairly high concentrations, so that extensive complexing might have occurred, thus perturbing the H–Cl vibrations to various extents.

In order to obtain more reliable results when studying the hydrogen bond of the HCl molecule with oxygen-containing molecules, we used the method of ternary mixtures [74], successfully used earlier when studying the hydrogen bond of alcohols [31, 34]. For this purpose we introduced the test hydrogen halide into a carefully dried neutral solvent in such quantities as to ensure that the absorption spectrum of the solution should only contain the band of the unperturbed vibration of the Hal–H monomers. Then a small quantity of oxygen-containing substance R–O was introduced into the solution, so that the mixture most probably formed dimer complexes of the Hal–H...OR type and no other. The spectroscopic manifestation of the hydrogen bond between the Hal–H molecule and the OR is the appearance of a wide band considerably displaced from the band of the monomers in the long-wave direction, together with a reduction in the intensity of the monomer band. On changing the temperature of the mixture there is a change in the relative intensity of the monomer and dimer bands as a result of a displacement in the thermodynamic equilibrium between the numbers of dimers and monomers.

We used the method of ternary mixtures not only to study the hydrogen bond of the HCl molecules but also to study that of HBr and HI. We were the first to study the latter two compounds [75]. As oxygen-containing substances we took: acetone, dioxane, and ethyl ether; as neutral solvent we took CCl_4. In these experiments the careful drying of the CCl_4 is very important, since traces of water, which always exist in CCl_4 in the dissolved state, may greatly complicate the character of the interaction between the components of the solution as a result of the formation of a hydrogen bond between the water molecules and those of the hydrogen halide and the oxygen-containing substance.

Experiments showed that the preliminary drying of CCl_4 in any arbitrary environment failed to give the desired result, since while the cuvette was being filled the CCl_4 picked up water from the surrounding atmosphere, as indicated by the presence of the absorption band of monomeric water molecules in the spectrum. In order to eliminate this effect it is essential to fill the cuvette with dried CCl_4 in a dry hermetic box, or else to dry the CCl_4 directly in the working cuvette, while testing spectroscopically for the presence of water in the solution.

Experience showed that the second method was the better. For this purpose the cuvette was furnished with two side tubes at the top and bottom (Fig. 32), having ground-glass stoppers at their ends. The CCl_4 was dried by passing a jet of dry air through the lower tube; the air bubbles passing through the CCl_4 expelled the dissolved water. During the drying of the

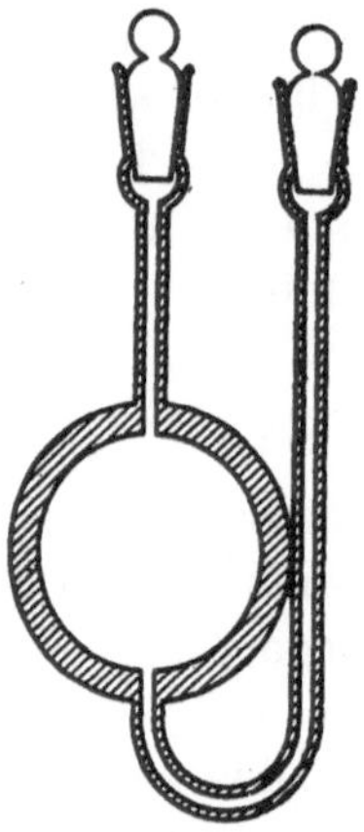

Fig. 32. Cuvette for studying the infrared absorption spectra of solutions of hydrogen halides in CCl_4.

CCl_4 the cuvette was placed in the working beam of the spectrophotometer, the wavelength drum of the latter being set to the maximum of the absorption band of the dissolved water. The curve so recorded enabled us to judge the reduction in the concentration of dissolved water during the course of the drying until completely water-free CCl_4 was left.

The hydrogen halides and oxygen-containing substances under consideration were also introduced into the cuvette through the aforementioned side tubes, the substances in question being first carefully purified by chemical methods and dried. The amount of dissolved hydrogen halide was monitored spectroscopically by reference to the intensity of the monomer band. The amount of oxygen-containing substances was also monitored spectroscopically by reference to the intensity of the absorption band of the perturbed vibration of the hydrogen halide in the dimer.

The thickness of the working cuvette was 4 mm. The absorption spectra were obtained by means of a double-beam infrared spectrophotometer [59] using an LiF prism. The spectral slit width was 2-4 cm^{-1}.

In order to eliminate interfering bands belonging to CCl_4 and in a number of cases the C−H bands of the oxygen-containing substances from the recorded spectrum, a compensating cuvette filled with the corresponding substance was placed in the comparison beam of the spectrophotometer.

As already indicated, the spectrum of a ternary solution of CCl_4 + (Hal−H) + OR exhibits the line of the Hal−H monomers, and a wide displaced band of associated complexes: Hal−H...OR dimers. It should furthermore be noted that, in the ternary CCl_4 + HCl + OR solution, the frequency of the band maximum of the perturbed H−Cl vibration obtained in the case of the dimer (in which the perturbation arises from one single oxygen atom) differs a great deal from the frequency of the band observed in the spectra of binary solutions in the same solvent [69-73]. In some cases the difference amounts to 100-200 cm^{-1} (Table 8). In addition to this, the displacement of the band depends on the nature of the oxygen-containing molecule; in the case of the oxygen-containing substances used in our experiments, the difference amounts to about 100 cm^{-1}. The results of our study of the spectra of ternary CCl_4 + HCl + OR solutions are presented in Table 8. A number of other results describing the spectra of binary solutions are also presented for comparison.

We also used the method of ternary mixtures to study the hydrogen bonds of HBr and HI molecules with oxygen atoms (acetone molecule). The results of an analysis of all the spectra appear in Table 9. The same table contains data relating to the spectra of HF solutions obtained in [76]. The data presented enable us to follow the relation between the perturbation of

TABLE 8. Frequency of the Band Maximum of the Perturbed Vibration of H−Cl in Various Solutions

Test mixture	Conc. of HCl, C_{HCl}, mole/liter	ν_m, cm^{-1}	Lit. cit.
$CCl_4+HCl+(CH_3)_2CO$	0.15	2506	[75]
$CCl_4+HCl+(C_2H_5)_2O$	0.15	2413	[75]
$CCl_4+HCl+C_4H_8O_2$	0.15	2456	[75]
$HCl+(C_2H_5)_2O$	0.82	2451	[69]
$HCl+C_4H_8O_2$	1.1	2469	[69]
$HCl+(CH_3)_2CO$	—	2381	[71]
$HCl+(C_2H_5)_2O$	—	2361	[71]
HCl+C [illegible]	—	2240	[71]
$HCl+(C_2H_5)_2O$	0.35	2393	[70]

TABLE 9. Frequencies of the Hal−H Vibrations in the Gas Phase ν_g (Q Branch) (in cm^{-1})

Hydrogen halide	ν_g	ν_l	Oxygen-containing substance	ν_m	Band width γ	$\Delta\nu = \nu_l - \nu_m$	$\frac{\Delta\nu}{\nu_l}$, %	Energy of hydrogen bond ε, kcal/mole
HF	3961	3854	Acetone	3290	—	564	14.6	25
			Ethyl ether	3210	—	644	16.7	29
HCl	2886	2827	Acetone	2506	500	321	11.3	12
			Dioxane	2456	500	371	13.1	14
			Ethyl ether	2413	—	414	14.6	16
HBr	2558	2515	Acetone	2237	400	278	11.0	10
HI	2230	2203	Acetone	2083	200	120	5.4	4

Note: ν_l = monomers in solution in CCl_4; ν_m = maxima of the perturbed vibration band in the ternary mixture (dimers Hal−H...OR).

the Hal−H vibration by the oxygen atom of the acetone molecule and the atomic number of the halogen. The relative displacement $(\nu_l - \nu_m)/(\nu_l)$ diminishes with increasing atomic number of the halogen, reaching a maximum value of 14.6% for HF and a minimum of 5.4% for HI.

It is interesting to note that the displacement $\Delta\nu = \nu_l - \nu_m$ depends linearly on the dipole moment μ of the hydrogen halide molecule (Fig. 33), where μ is given in Debye units, the straight line on the graph passing through the origin of coordinates. In the case of one particular hydrogen halide, the value of $\Delta\nu / \nu_l$ depends on the chemical composition of the molecule containing the oxygen atom. Thus, in the case of hydrogen chloride, the value of $\Delta\nu / \nu_l$ increases in the sequence acetone, dioxane, ethyl ether.

The band width of the perturbed Hal−H vibration is the greater, the greater the displacement of $\Delta\nu$, as in the case of the vibration of the O−H group.

We attempted to estimate the energy of the hydrogen bond in Hal−H...OR by using the Sokolov relation [28] based on the approximate quantum-mechanical theory of the hydrogen bond and the empirical relation of Badger [77]. According to the Sokolov theory, the energy of the hydrogen bond ε in the bridge of A−H...B is related to the elongation of the bond r(A−H)

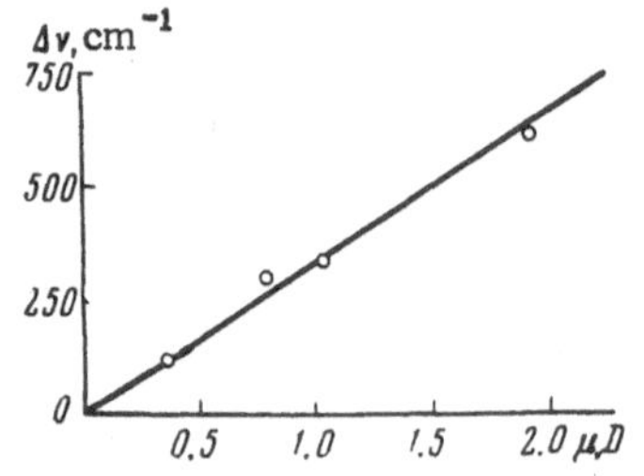

Fig. 33. Frequency displacement of perturbed vibration of hydrogen halide in ternary mixture CCl_4 + Hal · H + $(CH_3)_2CO$ as a function of the dipole moment of hydrogen halide molecule.

by the expression $\varepsilon = \frac{2a^2D}{b}\Delta r$, where D is the energy of dissociation of an isolated A−H bond, Δr is the elongation of the bond, a is the constant in the Morse function, and b is a constant equal to 3.78 $Å^{-1}$.

Using the Badger relation $k(r - d_{ij})^3 = 1.86 \cdot 10^5$, where k is the force constant of the A−H bond, and d_{ij} is a constant depending on which rows of the periodic system of the elements are occupied by the A and H atoms, we find

$$\Delta r = \left(\frac{1.86 \cdot 10^5}{4\pi^2 c^2 m}\right)^{1/3} (\nu_m^{-2/3} - \nu_l^{-2/3}),$$

where m is the reduced mass of the hydrogen halide molecule. Substituting the measured values of ν_m and ν_l (Table 9), we may determine the value of Δr, and then, by using the formula for ε calculate ε for the various hydrogen halides. The results of these calculations are presented in the last column of Table 9. Despite the fact that this method of calculation is extremely approximate, the values obtained for ε appear perfectly reasonable. The correct tendency for the value of ε to fall on passing from HF to HCl, HBr, and HI is, in particular, obtained. In absolute value, the energy of the hydrogen bond in Hal−H...OR (except for the case of HI) is several times greater than in R_1O−H...OR_2; qualitatively, this also accords with experimental data.

CHAPTER II

MANIFESTATION OF ORDINARY INTERMOLECULAR INTERACTIONS − VAN DER WAALS FORCES

In addition to our study of the hydrogen bond, we also carried out some experimental research into the manner in which ordinary van der Waals forces made their appearance. We shall now briefly set out the results of these investigations, paying special attention to methodical questions.

§ 1. Dependence of the Width of the Rotational Components of the P, R, and Q Branches of the Methane Molecule on the Pressure and the Nature of the Extraneous Gas

A number of mechanisms may give rise to the broadening of the rotational, and particularly the vibrational−rotational lines when collisions occur between complex molecules, each of which possesses a large number of internal degrees of freedom. In elastic collisions we may encounter perturbation of the vibrational and rotational levels of the molecule, and in inelastic collisions there may be a change in the vibrational and rotational energy. It is a very difficult matter to take proper account of all these factors, and as yet no theory has been developed to give an adequate explanation for all the experimental facts. The accumulation of reliable quantitative experimental data in relation to the line broadening of molecular spectra is therefore of first-rate importance.

Investigations of this kind include [78] and [79], which are concerned with the pressure dependence of the width of the rotational lines and the Q-branch components of oxygen and nitrogen. These investigations were based on the Raman spectra for gas pressures of 10-150 atm. It was found that the widths of purely rotational lines increase linearly with rising pressure, while the components of the Q branch remain practically unaltered. Analogous work [80] was later carried out at lower pressures (1-10 atm) and the results agreed with [78, 79].

We set ourselves the task of carrying out analogous investigations with the infrared absorption spectra, varying the pressure of not only the absorbing but also the extraneous gas.

As absorbing gas we took methane and as extraneous gases He, Ar, and N_2. The infrared absorption spectra were obtained with a double-beam infrared spectrophotometer of our own construction, using a diffraction grating of high resolving power [53].

The gas mixtures studied were placed in a multiple-pass cuvette of the high-pressure type enabling the pressure of the gas mixture components and the optical length of the path traveled to be varied over a wide range. The arrangement of the cuvette and the double-beam illuminating system of the spectrophotometer is shown in Fig. 34. Here M_2 is the mirror of the illuminating system, which, in the absence of the high-pressure cuvette, reflects the source of radiation onto the entrance diaphragm O and fills the diffraction grating of the monochromator; L_1' is a lens imaging the convergent lens K_2 onto the entrance diaphragm O with a magnification of 1 : 1, conserving the solid angle of the beam passing into the monochromator. The radius of curvature of the mirrors Z_1, Z_2, and Z_3 of the multiple-pass cuvette is 100 mm; the reflecting coating is gold.

Figure 35 shows a general view of the high-pressure cuvette.

In the majority of the investigations the mirrors of the multiple-pass cuvette were placed in a position corresponding to a fourfold traverse of the beam (Fig. 34), i.e., the total length of the optical path in the cuvette was 400 mm. The pressure of the absorbing gas in the mixtures was kept constant at no more than 1 atm. The pressure of the extraneous gases was varied from 0 to 150 atm. The gases were fed in from cylinders through a reducing valve, the pressure being measured with a standard manometer. The glass window of the cuvette limited the working range of the spectrum; hence all the work was carried out in the region of the first overtone of the C−H vibration, λ = 1.7-2.2 μ. The length of the optical path in the cuvette, the pressure of the absorbing gas, the rate of scanning, and other conditions governing the recording of the spectra were chosen in such a way as to enable the true line widths to be measured

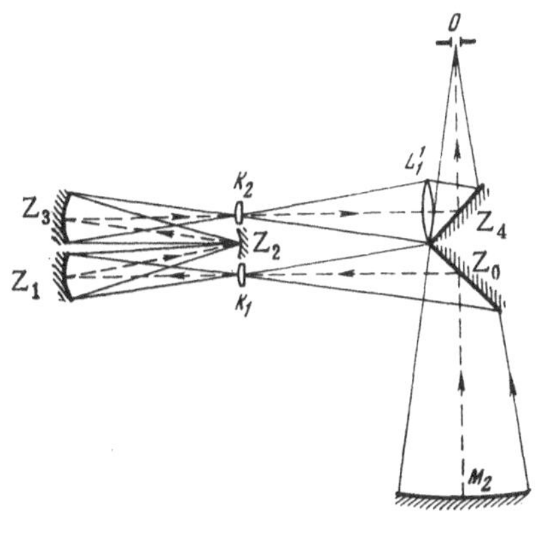

Fig. 34. Arrangement of the multiple-pass cuvette of the high-pressure type and the double-beam illuminating system. Z_0 and Z_4 are plane mirrors, K_1 and K_2 convergent lenses; Z_1, Z_2, Z_3 concave mirrors of the multiple-pass cuvette placed in a high-pressure bomb.

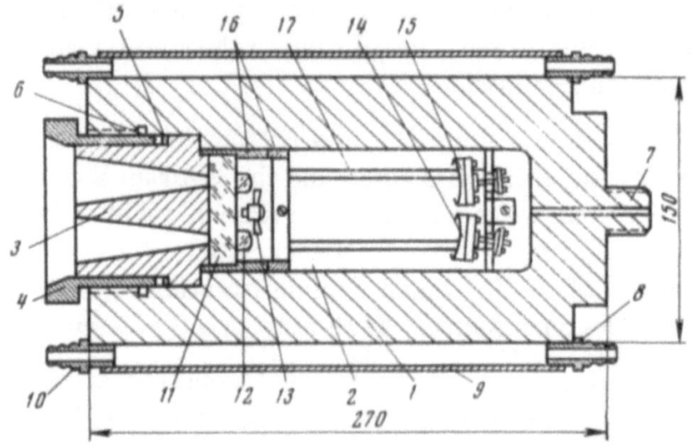

Fig. 35. Cross section of the multiple-pass cuvette of the high-pressure type. 1) Body of cuvette made of stainless steel; 2) cavity inside body containing the mirror system; 3) stopper-obturator with two oval apertures for the entering and leaving beams; 4) pressure nut; 5) sealing ring; 6) steel ring; 7) side tube with thread and capillary for filling the cuvette with the test gas; 8) rings sealed to the body of the cuvette; 9) outer wall of the thermostating jacket of the cuvette; 10) conduit for admitting water to the thermostating jacket from an ultrathermostat; 11) glass entrance window with convergent lenses (12) stuck to it (K_1 and K_2, Fig. 34); 13-15) spherical mirrors (Z_1, Z_2, Z_3, Fig. 34); 16) guide holding the entrance window and creating the initial seal; 17) mirror holder.

with a minimum total error. The minimum spectral width of the monochromator slits was 0.4 cm^{-1}.

At low pressures ($p < 20$ atm) the rotational lines hardly overlapped at all (Fig. 14), and the line widths only slightly exceeded the spectral slit width; in this case the true line widths were determined by ordinary methods, first eliminating the apparatus function by the Rayleigh method.

At high pressures ($p > 20$ atm) the rotational lines broadened and overlapped, and the widths were determined by the Elsasser method [82], the applicability of this being verified by graphical construction for a finite number of lines. The accuracy of measuring the width of the rotational components was estimated as 10%. For a pressure of over 80-100 atm the individual rotational lines merged into an almost continuous background and it became practically impossible to determine the line widths.

The vibrational–rotational band of methane at $\lambda = 1.7\mu$ was studied in most detail, this corresponding to the first overtone of the ν_3 vibration, which consists of the well-resolved components of the P and R branches (Fig. 14). The distance between the rotational components close to the Q branch is about 10 cm^{-1}. We also studied the $\lambda = 2.2\ \mu$ band ($\nu_1 + \nu_3$ vibration). In order to obtain more accurate data for the band at $\lambda = 1.7\ \mu$, we measured the width of the rotational line R_6 (Fig. 14), which corresponds to the intensity maximum in the R branch. The number of rotational lines on the two sides of the R_6 line is quite large, which is important in

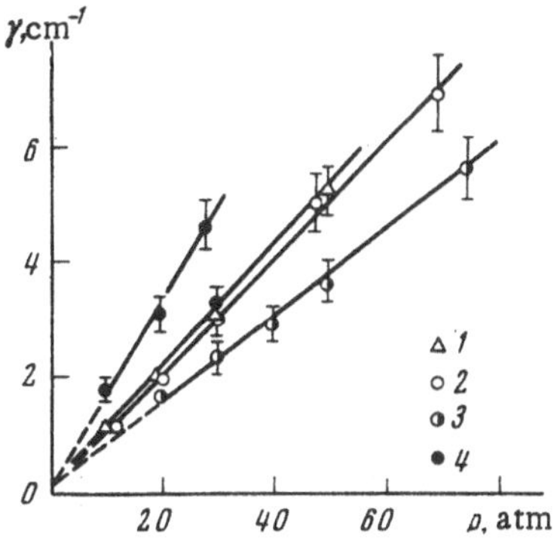

Fig. 36. Width $\gamma = \gamma_{obs} - \gamma_0$ of the rotational line R_6 ($\lambda = 1.7\,\mu$) of methane as a function of the pressure of various gases: 1) N_2; 2) Ar; 3) He; 4) CH_4.

the application of the Elsasser method [82]. We measured the width of this line at various He, Ar, N_2, and CH_4 pressures. In addition to this we measured the width of the R_9, P_3, P_6, and P_9 lines at various Ar pressures. For the $\lambda = 2.2\,\mu$ band we measured the width of the R_1 line at various pressures of He, Ar, and N_2.

The results of our measurements of the line widths γ are presented in Fig. 36, where $\gamma = \gamma_{obs} - \gamma_0$ is the line width due to the collision of methane molecules with molecules of another nature, γ_{obs} is the observed line width, and γ_0 is the line width due to the collisions of the absorbing molecules with each other and the intrinsic width; we determined γ_0 experimentally from the absorption spectrum of pure methane at the pressures used in our study ($p < 1$ atm). Under our experimental conditions the pressure of the absorbing gas remained constant and $\gamma_0 \leq 0.3$ cm^{-1}. It follows from Fig. 36 that the width of the rotational line R_6 of methane depends linearly on pressure, as in the case of other gases.

The effectiveness of the broadening depends on the nature of the extraneous gas and increases in the sequence He, Ar, N_2, CH_4; Ar and N_2 cause almost the same amount of broadening, while the intrinsic broadening is much greater than that due to molecules of another nature. Analogous results were obtained for the R line of the $\lambda = 2.2\,\mu$ band of methane.

We measured the contour of individual rotational lines for various nitrogen pressures. By way of example, Fig. 37 shows the contours of the absorption coefficient of the R_6 line for various nitrogen pressures. On the contour of the line at p = 30 atm we have marked points corresponding to the dispersion contour. We see that the contour of individual rotational com-

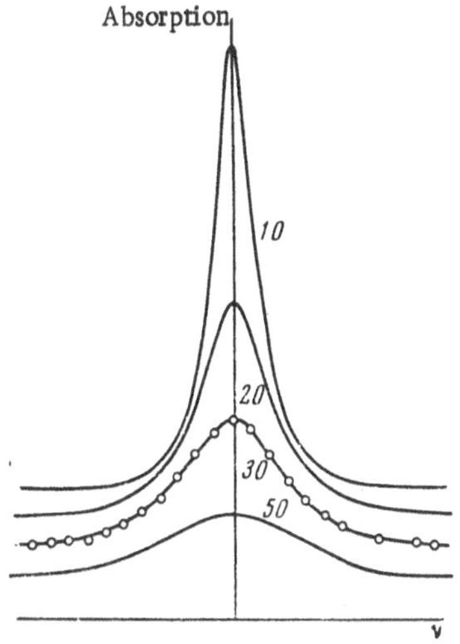

Fig. 37. Width and shape of absorption-index contour of the R_6 line of methane as functions of nitrogen pressure. Circles indicate theoretical points of the dispersion contour. Numbers on curves indicate nitrogen pressure in atmospheres.

ponents of the absorption coefficient lie close to the dispersion distribution; this indicates that under our experimental conditions the impact mechanism was operative in the broadening of the rotational lines, so that we may calculate the optical collision cross sections σ between the methane molecule and molecules of other natures. In accordance with the impact theory [17]

$$\sigma = \frac{\pi c k T}{\bar{v}} \frac{\gamma}{p},$$

where

$$\bar{v} = \sqrt{\frac{8RT}{\pi}\left(\frac{1}{M_{CH_4}} + \frac{1}{M_f}\right)};$$

we calculated the values of σ for all the mixtures considered. The results of the calculation are presented in Table 10 (in units of 10^{-16} cm^2); for comparison the quantities in brackets indicate the gas-kinetic collision cross sections. In the same methane band (λ = 1.7 μ) we studied the width of the rotational lines as a function of the number J. We measured the widths of the lines P_9, R_9, P_6, and P_3 for argon pressures of 20, 30, and 40 atm. (For large values of the number J the lines were double, which prevented us from obtaining reliable results.) The measurements showed that the broadening of lines with small J was greater than that of lines with large; the broadening of lines differing only in the sign of J was almost identical. The results of these measurements are presented in Fig. 38; they agree closely with the experimental data of other authors [83].

Apart from the pressure dependence of the widths of various rotational lines of the components of the P and R branches, we also studied the pressure dependence of the widths of the rotational components of the Q branch. It was experimentally very difficult to measure these components of the Q branch, since, in the case of methane, they are only resolved at low pressures, while at high pressures of the extraneous gas the fine structure of the Q branch vanishes altogether, so that, in this case, we cannot employ the usual method of determining the line widths in the absorption spectrum. We determined the width of the rotational components of the Q branch indirectly by measuring the width of the resultant contour of the absorption band of the Q branch and then proceeding by calculation, taking the positions and relative intensities of the components as known.

TABLE 10. Optical Collision Cross Sections of the Methane Molecule with Various Gases

Extraneous gas	Band λ = 1.7 μ						Band λ = = 2.7 μ, R_1
	R_6	R_9	P_3	P_6	P_9	component of Q-branch	
He	21 (17)					6 (17)	18 (17)
Ar	50 (28)	48	65	48	48	35 (28)	44 (28)
N_2	50 (30)					50 (30)	60 (30)
CH_4	70 (15)					57 (15)	

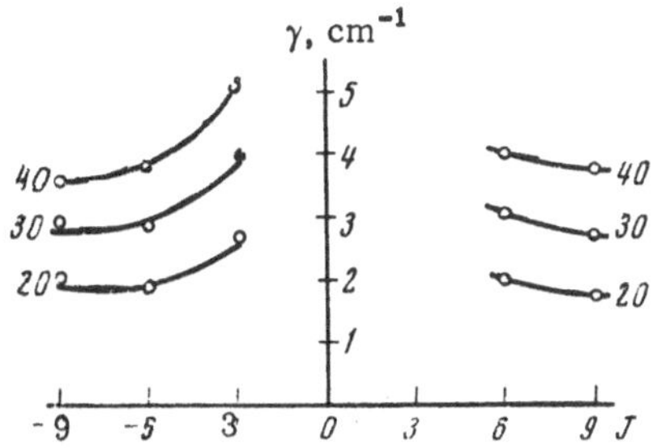

Fig. 38. Width of the rotational lines of methane with various J values as a function of argon pressure (numbers on the curves).

According to [84], the Q branch has a structure arising from the interaction of vibration and rotation, the position of the rotational components of the Q branch being determined by the relation $\nu = \nu_0 + \alpha J(J + 1)$, while the intensity distribution of the components is given by the expression

$$I_\nu = I_0 \sqrt{1 + \frac{4\nu}{\alpha}} e^{-\frac{B\nu}{\alpha kT}},$$

where α is the interaction constant for the vibration and rotation, J is the rotational quantum number, and B is the constant of rotation. The position of the components of the Q branch of methane at $\lambda = 1.7\ \mu$, i.e., the constants ν_0 and α in the expression for ν, were determined experimentally from the absorption spectrum at a low methane pressure.

For this purpose we recorded the spectrum using a through-type cuvette of variable length with a greater light transmission. The side walls of this cuvette were made of a metal bellows. The cuvette enabled us to vary the length of the optical path from 60 to 120 μ and the methane pressure from 0.5 to 2 atm. Using a cuvette of this type, the width of the apparatus function was 0.4 cm^{-1}. Under these conditions the rotational components of the Q branch in the spectrum are fairly well resolved (Fig. 39) and their frequencies may be reliably measured. We found that

$$\nu = [6006.1 + 0.067 J(J+1)]\ cm^{-1}.$$

At low pressures, for which the fine structure components of the Q branch were partly resolved, we determined their widths by graphical analysis of the contour of the Q branch. At high pressures the structure was poorly expressed, and we plotted a theoretical curve for the relationship between the total width of the Q branch γ_Q and the width of its components (Fig. 40). By constructing this curve we found the contours of the Q branch in which the widths of the components were all assumed equal, the intensity distribution with respect to J was taken in accordance with theory, and the frequencies of the lines in accordance with our measure-

Fig. 39. Contour of the Q branch of methane ($\lambda = 1.7\ \mu$) at low pressure (p < 1 atm). Spectral slit width 0.4 cm^{-1}. Vertical segments indicate theoretical intensity distribution of the components of the Q branch with different J.

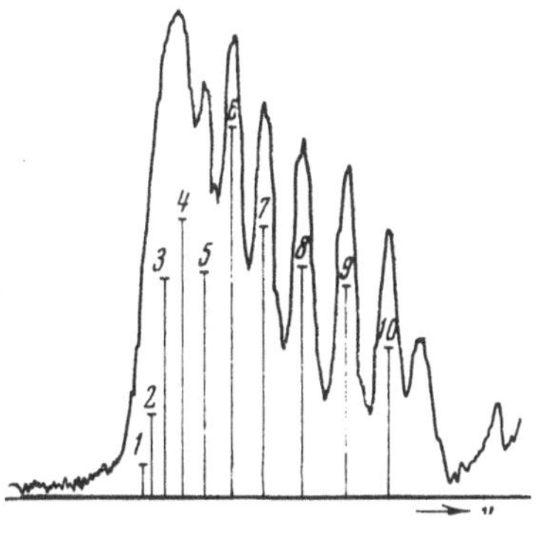

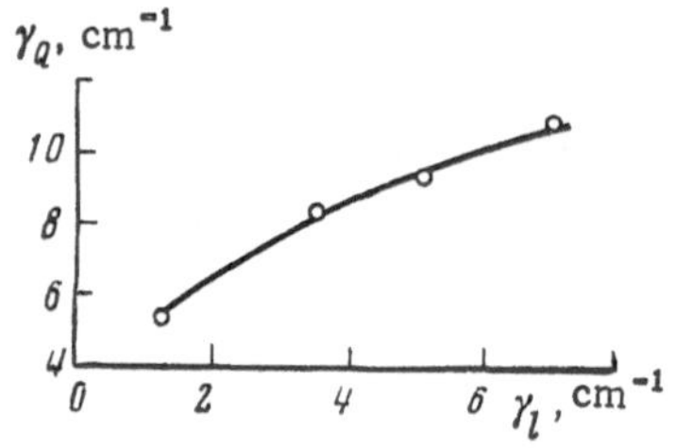

Fig. 40. Dependence of the total width of the Q-branch contour γ_Q on the width of the constituent components γ.

ments at low pressure (see above). By measuring the resultant width γ_Q of the Q branch at different pressures and using the curve of Fig. 40, we were able to determine the width γ of the components of the Q branch. Since the shape of the Q-branch contour underwent substantial changes with varying pressure (Figs. 39 and 41), the accuracy of our determination of the widths of individual components was naturally much lower than it was in the case of the lines of the P and R branches, amounting to some 20%.

The measured widths of the Q-branch components are shown as functions of the pressures of various gases in Fig. 42, which also shows the pressure dependence of the R_6 line for comparison.

We also calculated the optical collision cross sections for the components of the Q branch on the assumption of the impact broadening mechanism. The results are presented in Table 10.

It follows from Fig. 42 and Table 10 that the broadening of the rotational lines of the P and R branches and of the components of the Q branch are quite different, the difference depending on the nature of the perturbing gas. Thus, in the case of a CH_4 + He mixture, the cross section for the R_6 line is more than three times that of the components of the Q branch. In a CH_4 + Ar mixture this difference is less sharp, and in a CH_4 + N_2 mixture the cross sections of the R_6 line and the components of the Q branch are almost identical. In the case of the

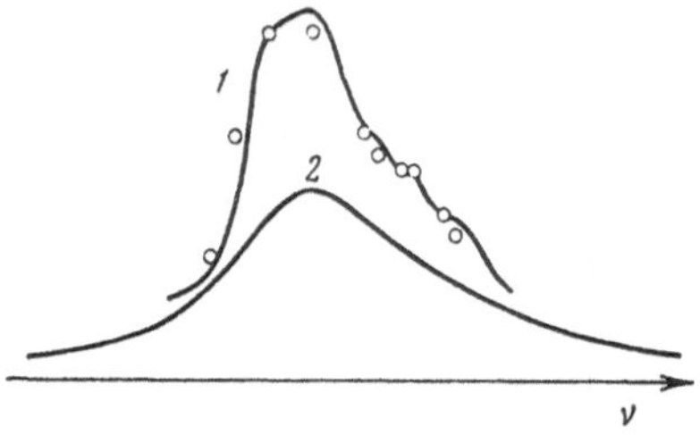

Fig. 41. Contours of the Q branch of methane for nitrogen pressures of 12 atm (1) and 50 atm (2).

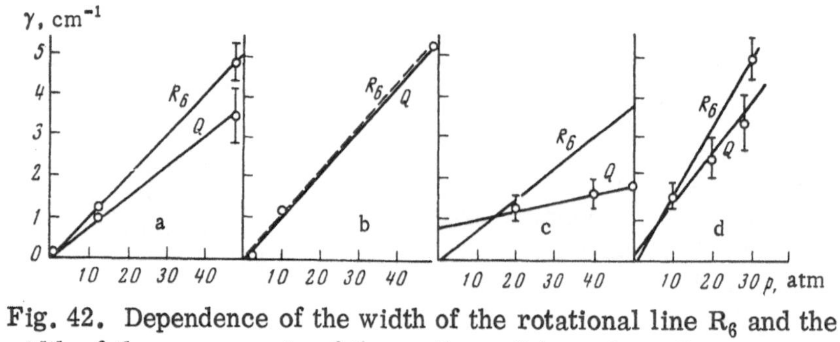

Fig. 42. Dependence of the width of the rotational line R_6 and the width of the components of the methane Q branch on the pressure of various gases: a) Ar; b) N_2; c) He; d) CH_4.

CH_4 + He mixture the optical cross section of the components of the Q branch is almost three times less than the gas-kinetic cross section, whereas, in the other mixtures, the optical cross section is greater than the gas-kinetic cross section. The greatest difference between the optical and gas-kinetic cross sections occurs for the intrinsic broadening of methane. It should be noted that in the case of the intrinsic broadening of the methane lines the optical collision cross sections for the R_6 lines and the components of the Q branches are not very different from one another (70 and 57), whereas, in the case of the intrinsic broadening of the Raman lines of nitrogen [72], oxygen [79], and CO_2 [77], the optical cross sections for the components of the Q branch are much smaller than they are for the Raman lines of the rotational spectrum. In the case of the components of the Q branch of methane, however, in the Raman spectrum [80] the broadening of the Q branch is quite different for polarized and unpolarized lines.

We also see from Table 10 that the collision cross sections depend slightly on the number J; the greatest section is that of the P_3 line. However, the cross sections of lines differing only in the sign of J (P_6 and R_6, P_9 and R_9) practically coincide. The dependence of the collision cross sections on the nature of the extraneous gas for the line R_l of the $\lambda = 2.2\ \mu$ band of methane is practically the same as for the R_6 line of the $\lambda = 1.7\ \mu$ band, although the line width of the latter was measured to a lower accuracy.

The observed differences in the broadening of the rotational lines of the P and R branches ($\Delta J = 1$) and the components of the Q branch ($\Delta J = 0$) of methane indicate different broadening mechanisms when the methane molecules collide with various other types of molecules. On the basis of the foregoing data it is still difficult to provide any satisfactory explanation for the observed experimental facts relating to the broadening of the different lines in the molecular spectra or to indicate the specific mechanisms responsible for the broadening. However, there is no doubt that a study of the broadening of the rotational lines of the P, R, S, and O branches and the components of the Q branches of various rotational–vibrational bands, and also the lines of the purely rotational spectrum, will help in explanation of the mechanism underlying the intermolecular interaction of complex molecules.

§ 2. Dependence of the Parameters of the Infrared Absorption Bands on Viscosity

One of the methods of studying intermolecular interactions is that of analyzing the effect of the rotational motion of the molecules of a liquid on the line width of the vibrational spectra of the molecules. Many papers have been devoted to this subject; these have considered the width and intensity of the Raman [85-90] and infrared absorption [91] lines on the viscosity. It was shown in [91, 92] that the rotational thermal motion of the molecules might lead to an additional line broadening in the molecular spectra of liquids, the extent of this broadening $\Delta\omega$ being related to the mean reorientation time of the molecules in the liquid τ_0 by the expression $\Delta\omega = 2/\tau_0$. For the Brownian motion of a liquid particle the time τ_0 may in turn be derived from $\tau_0 = \eta a^3/kT$, where η is the viscosity of the liquid, a is the radius of a particle, k is Boltzmann's constant, and T is the absolute temperature. Thus, by measuring $\Delta\omega$ we may estimate the value of τ_0 and also the extent of the potential barrier which the molecule has to overcome in changing its orientation.

In all the papers just mentioned, the viscosity of the test substance was varied by changing the temperature. It was assumed that, since the viscosity depended much more strongly on temperature than would correspond to a linear law, the change in the actual term T could be neglected in the expression for τ_0 and it might be considered that the value of $\Delta\omega$ (and hence τ_0) was simply determined by the viscosity η.

Earlier [93] we attempted to study the dependence of the parameters of the infrared absorption bands, namely, the band width δ, the absorption coefficient at the maximum k_m, and the integrated absorption coefficient k_∞, on the viscosity of a solution at constant temperature. For this purpose the test substance was dissolved in a solvent, the molecules of which were capable of polymerizing, and the band width of the dissolved substance was studied in relation to the viscosity of the solution, which varied substantially in the course of polymerization. Thus, in these experiments, the viscosity of the solution was varied by virtue of a change in the potential barrier [94] due to the polymerization of the molecules whereas, in [85-91], the viscosity was varied by changing the velocity of the thermal motion of the molecules (by changing the temperature).

The concentration of the dissolved substance was taken so small that it was reasonable to consider the molecules of the dissolved substance as being solely surrounded by molecules of the solvent. Under these conditions the broadening effect is determined simply by the interaction between the molecules of the dissolved substance and the solvent molecules.

The choice of subjects for study was determined, on the one hand, by the fact that the absorption band of the dissolved substance under consideration had to lie in the "transparency window" of the absorption spectrum of the solvent, and, on the other hand, by the fact that the band in question had to correspond to a type of molecular vibration which would be modulated by the rotational motion of the molecule. We accordingly recorded the absorption spectra of a large number of different polymerizing substances (solvents) as well as the spectra of the substances to be introduced into the polymer.

As a result of these investigations, we selected styrene as solvent; this has a fairly wide transparency window in the spectral range 2800-2000 cm^{-1} and polymerizes quite easily, its viscosity varying over a very wide range.

As dissolved substance we studied compounds the molecules of which contained S−H and C−D groups, since the absorption bands of these groups lay in the transparency window of the absorption spectrum of styrene. We studied the absorption spectra of thiophenol, deuterized methyl alcohol, deuterochloroform, deuterobenzene, and others. The most convenient from the practical point of view was fully deuterized benzene C_6D_6, and most of the investigations were applied to this substance. We studied the parameters of the absorption band of the C−D group with a frequency of ν = 2290 cm^{-1}, corresponding to the principal completely symmetric plane vibration of symmetry E_{1u}, as a function of the viscosity of the polystyrene. According to [81], the selection rules specify that the vibration of symmetry E_{1u} should have two nonzero components of dipole moment (M_x and M_y) lying in the plane of the molecule; hence, the random modulation of the dipole moment by the Brownian rotational motion should broaden the absorption band of this vibration.

Deuterobenzene was also a particularly convenient subject for study because it had a relatively high boiling point (about 80°C), so that the temperature of the solution could be raised slightly in order to accelerate polymerization without changing the concentration of the dissolved substance. The temperature of the solution was only raised to accelerate polymerization; the absorption spectra were always measured at 20°C.

The viscosity of the solutions was measured by means of ordinary viscosity meters (viscometers) of the Ostwald type with capillaries of different radii. The accuracy of the relative viscosity measurements was 1.5-2% or better. However, the absolute accuracy of the viscosity measurements was considerably lower in view of the particular way in which the experiment was conducted. In order to keep the conditions of measuring the absorption spectra constant, the working absorption cuvette, filled with a solution of deuterobenzene in styrene, was hermetically sealed for the whole duration of the experiment, the viscosity of the solution in the cuvette varying meanwhile over a very wide range.

Naturally, under these conditions, the viscosity of the solution in the working cuvette could not be measured directly. In addition to the working cuvette, demountable control cuvettes were therefore also prepared and filled with the same solution. The solution in these cuvettes was used for measuring the viscosity. Under these conditions of operation it was essential that the viscosity of the solutions in the working and control cuvettes should always be the same. However, preliminary experiments showed that the rate of polymerization of the solutions was affected by the thickness of the layer of solution, its concentration, and the material and even the shape of the cuvette. Hence, in order to ensure identical conditions of polymerization in the working and control cuvettes, all the cuvettes were made from the same material (rock salt) with the same shape and thickness (0.5 mm), and the polymerization conditions were maintained identical.

For control purposes we made some direct measurements of the viscosity in the working and control cuvettes; these showed that the viscosities in the two cuvettes differed by no more than 8-10%, subject to the foregoing conditions being observed. These figures relate to relatively low viscosities; for larger values the scatter becomes less severe. We may therefore consider that the accuracy of the viscosity measurements in the working cuvette was no worse than 10%.

The absorption spectra were recorded by means of a double-beam infrared spectrophotometer with an IKS-11 monochromator, using an LiF prism. The spectral slit width in the working region was about 10 cm^{-1}, this being about one-third of the width of the C−D absorption band under consideration. The apparatus slit-width function was eliminated graphically by the Rayleigh method. The rate of scanning the spectrum was chosen so as to minimize the systematic error associated with the inertia of the recording system. The temperature of the solutions was kept constant, while the spectra were being recorded (room temperature ≈20°C). The viscosity of the solution varied between 0.8 and 200 cP.

The measurements showed that with increasing degree of polymerization of the styrene, i.e., with increasing viscosity, the width δ of the absorption band ν = 2290 cm^{-1} diminished, while the absorption coefficient at the maximum k_m increased. The integrated absorption coefficient k_∞ remained constant within the limits of experimental error.

The results of the measurements of δ and k_m are shown in Figs. 43 and 44 in relation to the value of $1/\eta$, where η is the measured viscosity of the solution. We see from Fig. 43 that the width of the band δ is a linear function of $1/\eta$ over the range of viscosities in question, and hence the quantity η is proportional to the mean reorientation time τ_0 of the molecules in the solution.

Over the same range of viscosities the absorption coefficient k_m at the maximum is also a linear function of $1/\eta$ (Fig. 44).

Since the quantities δ and k_m were measured to a fairly high accuracy over the whole range of viscosities (as indicated by the constancy of the measured value of k_∞), the scatter of the points in Figs. 43 and 44 is mainly attributable to the considerable errors committed in the viscosity measurements, as implied earlier. However, analysis of the resultant curves by

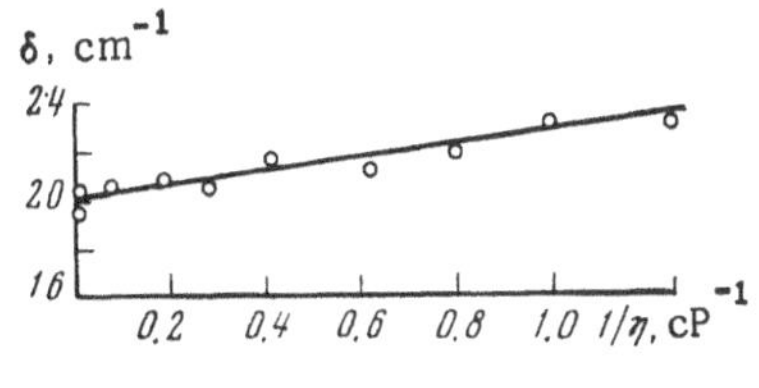

Fig. 43. Dependence of the width of the absorption band δ on $1/\eta$ for the ν = 2290 cm^{-1} band of deuterobenzene.

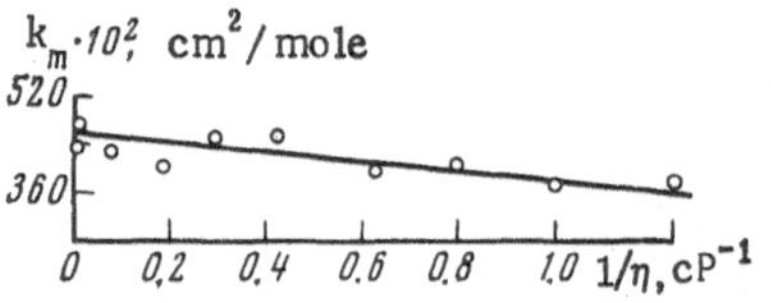

Fig. 44. Dependence of the absorption coefficient k_m at the maximum of the ν = 2290 cm^{-1} band of deuterobenzene on $1/\eta$.

the method of least squares showed that the dependence of δ and k_m on $1/\eta$ was of a linear nature over the whole range of viscosities studied. Whereas in the experiments of [90, 91] the linear relation between δ and $1/\eta$ only held from 0.01 to 10 cP, our own data show that the linear relationship remains valid up to much higher viscosities ($\approx$200 cP). It is interesting to note that on the $\delta = f(1/\eta)$ curve (Fig. 43), the slope of θ = 2.5 is very similar to that of the relationship between the infrared absorption line width and $1/\eta$ obtained in [91] (3.3 for toluene, 2.9 for cyclohexane, and 1.9 for acetonitrile), although in these experiments the viscosity was varied by changing the temperature.

The linear relationship thus found and the assumption that the broadening due to the rotational motion of the molecules fails to occur in the solid state of the material enable us to determine the value of τ_0 for various viscosities of the solution by using the relation $2/\tau_0 = \Delta\omega = \delta(\eta) - \delta_0$, where $\delta(\eta)$ is the measured total width of the absorption band for the specified value of η; δ_0 is the band width in the solid state of the material, which may be determined from the curve by extrapolating $1/\eta$ to zero.

The results so obtained show that on varying the viscosity from 0.8 to 12.5 cP the value of τ_0 changes from $4 \cdot 10^{-12}$ to $2.4 \cdot 10^{-11}$ sec, i.e., τ_0 increases by a factor of six. Our own value of $\tau_0 = 4 \cdot 10^{-12}$ sec for η = 0.8 cP is similar to the value of τ_0 given in [90] for benzene ($\tau_0 = 5 \cdot 10^{-12}$ sec) at room temperature (η = 0.65 cP) using the temperature dependence of the width of the Raman lines. It should also be noted that the results given in [90] were obtained for pure benzene, whereas we studied a solution of deuterobenzene in styrene.

The linear relationship which we observed between the mean time of reorientation of the deuterobenzene molecule in a styrene solution and the viscosity indicates that the conclusion drawn in [90] to the effect that the reorientation potential barrier coincided with the Frenkel viscosity barrier [94] remains valid even for weak solutions, i.e., for the case of interaction between molecules of different natures. We may also consider that the relaxation time of the rotational motion of the molecules in a liquid and the associated process of line broadening in the molecular spectra are determined simply by the viscosity η and not by the manner in which the latter is varied.

It should be noted that the linear dependence of the band width of a dissolved substance on $1/\eta$ obtained in the present experiments may be used (as an indicator) in order to measure the viscosity of polymers over a wide range of viscosities when ordinary methods of measuring viscosity are inapplicable.

In addition to studying the viscosity dependence of the parameters of the C−D absorption band, we also measured the infrared absorption spectra of pure liquid and solid styrene in the range 800-3200 cm^{-1}. Such spectra were studied earlier in [95]; however, the authors confined attention to studying the spectra principally in the region of 800–2000 cm^{-1}. Our own apparatus enabled us to obtain the absorption bands of the principal valence vibrations of the CH_2 and CH_3 groups in the range 2700–3200 cm^{-1} with a higher resolution than that obtained in [95]. The changes taking place in the absorption spectra in the range 800–2000 cm^{-1} agree with the results given in [95]. Figure 45a, b shows our absorption spectra for liquid and solid styrene in the range 2700-3200 cm^{-1}. We see from these figures that the spectra of liquid and solid styrene differ considerably. In the styrene spectrum the intensity of the ν = 2851 and 2923 cm^{-1} absorption bands increased sharply, these bands respectively corresponding to the symmetric

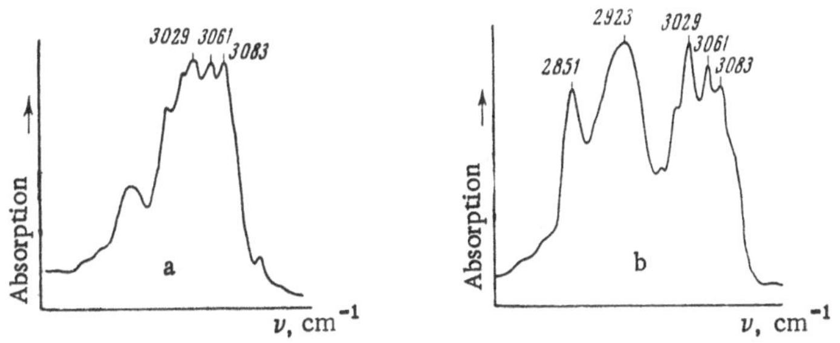

Fig. 45. Absorption spectra of styrene in the range 2700-3200 cm^{-1} (valence vibrations of the CH_2, CH_3 groups). Prisms made of LiF: a) liquid styrene; b) solid, polymerized styrene.

and asymmetric valence vibrations of the CH_2 group. The rise in the intensity of these bands indicates the formation of polyethylene chains in the course of polymerization.

§ 3. Broadening of the Emission Lines of Thallium Atoms by Molecular Hydrogen

In addition to the changes in the molecular spectra, a large number of theoretical and experimental investigations [17, 96-99] have been devoted to changes taking place in atomic spectra (both emission and absorption) under the influence of various physical factors: temperature, pressure, the presence of other kinds of particles, and so on.

This section of the present analysis comprises experimental work of this kind, but differs in method from the earlier investigations.

In the majority of experimental work on the broadening and displacement of emission lines, sources with electrical excitation have been used: electric arcs, sparks, and so forth. The plasma of these sources contains various quantities of foreign particles (molecules, atoms, ions, and electrons), and the observed changes in the parameters of the spectral emission lines are due to interaction with all these particles. Under these conditions it is not always possible to study the broadening and displacement attributable to any one kind of particle in pure form, and in particular to determine the broadening and displacement due to neutral atoms and molecules. The broadening and displacement of the spectral lines due to neutral particles can only be determined from the absorption spectra for lines of the principal series.

In the present investigation we studied the broadening of emission lines due to neutral particles when the atoms were excited during the photo-dissociation of molecules. For this purpose the cuvette containing the molecules of the substance under examination and an extraneous gas is illuminated by radiation from the source of excitation, and the width of the spectral lines of the excited atoms formed by photo-dissociation is measured in relation to the pressure and nature of the extraneous gas. For an appropriate choice of excitation source and sample molecules, photo-dissociation in the cuvette will lead to the formation of excited atoms, no ions or electrons being present, so that the broadening of the emission lines will be solely associated with the interaction of the neutral particles.

In contrast to sources with electrical excitation, the method here proposed enables us to vary the pressure of the extraneous gases over a fairly wide range, as well as varying their nature, by using a variety of atomic and molecular gases. In addition to this, the density of the test working substance in the cuvette may be made fairly low so as to be able to neglect the intrinsic broadening and observe simply the broadening attributable to the extraneous gas.

It should nevertheless be noted that when atoms are excited by photo-dissociation there may be an additional broadening of the lines due to the excess kinetic energy of the free atoms so formed if the energy of the exciting quantum exceeds the minimum energy required to dissociate the molecule and excite the level in question. This effect may in general be of a considerable magnitude, depending on the spectral composition of the radiation from the source of excitation. However, for a specific source and a particular molecule the extent of the extra broadening will be constant, i.e., it will not depend on the pressure of the extraneous gas.

The effect of extraneous gases on the emission spectrum of atoms formed by photo-dissociation has been studied by a number of authors [100-102]. However, in these papers attention was limited to the effect of the extraneous impurities on the intensity of the emission lines (quenching of fluorescence), due to inelastic collisions of the nascent atoms with the impurity atoms or molecules.

The simultaneous study of emission-line broadening and the quenching of fluorescence will provide more comprehensive information as to the interaction of the excited atoms with neutral particles.

In the present investigation [103] we used our own method to study the broadening of the emission lines of thallium atoms at λ = 5350 and 3776 Å formed by the photo-dissociation of TlI molecules due to the action of molecular hydrogen. The λ = 5350 Å line corresponds to a transition from the $7S_{1/2}$ level to the metastable $6P_{3/2}$ and the λ = 3776 Å line to a transition from the $7S_{1/2}$ level to the ground level $6P_{1/2}$.

The experiment was carried out in the following manner. Cylindrical quartz cuvettes with flat sealed end windows were first thoroughly evacuated and then TlI salt was introduced together with dosed amounts of hydrogen, after which the cuvette was again sealed off. In the experiments we used several cuvettes with differing hydrogen pressures, from 0 to 760 mm Hg. The diameter of the cuvettes was 1 cm and the working length 18 cm. The cuvette, together with the photo-dissociation-exciting source (a PRK-2 mercury lamp), was placed in a cylindrical metal jacket with a heater winding wound around it. Since the short-wave boundary [104] of the absorption band of the TlI molecules corresponding to the photo-dissociation of the molecules with the formation of thallium atoms in the excited state $7S_{1/2}$ lies in the region of 2080 Å, the bulbs of the mercury lamps and the cuvettes were made of quartz, which has a high transparency in the region of λ = 2000 Å. With this excitation source, the photo-dissociation of the TlI molecules takes place as a result of the absorption of the mercury lines λ = 2002, 1972, and 1942 Å. For the absorption of these wavelengths by Tl vapor in the cuvette to be appreciable, a molecular concentration [104] of 10^{17} cm^{-3} is required, this being achieved by heating the cuvette to about 460°C.

In order to prevent the condensation of the TlI salt on the end windows of the cuvette, the cuvette holders, fixed to the sidewalls of the jacket, are heated by means of auxiliary heaters to a temperature higher than that of the rest of the cuvette. As a result of this the end windows of the cuvette through which the emission is studied remain transparent.

The fluorescence spectra of the Tl atoms were studied by means of an ISP-28 spectrograph crossed with a Fabry–Perot etalon. The thickness of the spacers employed was varied from 8 to 2 mm, depending on the width of the lines. The contours of the 5350 and 3776 Å lines were determined in the usual way by photographic photometry. It should be noted that the widths of these lines are rather hard to determine in view of the presence of hyperfine structure.*

*These thallium lines also have an isotopic shift amounting [105] to some 0.06 cm^{-1}. Since the relative content of the 205 and 203 thallium isotopes [106] equals 70.5 and 29.5%, this only leads to a slight asymmetry of the observed contour for low hydrogen pressures and causes very little error in the measured width.

The 5350 Å line has two components separated by $\Delta\nu_1 = 0.42\ cm^{-1}$ and the 3776 Å line three components with $\Delta\nu_1 = 0.42\ cm^{-1}$, $\Delta\nu_2 = 0.72\ cm^{-1}$. At low pressures these components were resolved and the individual widths were measured. At high hydrogen pressures p = 450 and 750 mm Hg the components overlapped and the resultant contour had to be analyzed graphically, reducing the accuracy of the measurements.

In order to avoid the reabsorption of the λ = 3776 Å line corresponding to a transition to the ground level, the working length of the cuvette was reduced to 1 cm when studying the effect of hydrogen on the width of this line.

The apparatus function of the etalon was approximately one-thirteenth of the range of dispersion and was eliminated from the resultant experimental values of the widths by normal methods. According to our estimates, the accuracy of the line-width measurements was 15-20%.

The measurements showed that the widths of the λ = 5350 and 3776 Å line components varied from about 0.1 to 0.75 cm^{-1} as the hydrogen pressure varied from 0 to 1 atm. The results are presented in Fig. 46, from which we see that the width γ of both lines may be regarded as a linear function of hydrogen pressure p. The experimentally measured values of the line widths in the absence of hydrogen are: for λ = 5350 Å, $\gamma_0 = 0.06\ cm^{-1}$; for λ = 3776 Å, $\gamma_0 = 0.09\ cm^{-1}$. Calculations of the ordinary Doppler width of these lines at T = 460°C, respectively, give 0.03 and 0.045 cm^{-1}. An estimate of the additional broadening due to the excess kinetic energy of the atoms shows that the extent of this broadening approximately equals the ordinary Doppler width, while the total line width is close to the experimental values.

We also studied the effect of hydrogen on the intensity of the λ = 5350 Å line; we found that over the whole range of hydrogen pressures up to 720 mm Hg there was hardly any quenching, in agreement with the conclusions of [101]. The measurements were made photoelectrically. The mean scatter in the results obtained for different cuvettes was 20%.

Quenching inelastic collisions also, generally speaking, have an effect on the width of the spectral line, reducing the lifetime of the atoms in the excited state. The absence of quenching in the case in question indicates that the observed line broadening is simply due to elastic collisions between the Tl atoms and the H_2 molecules, so that we may use impact (collision) theory in order to calculate the broadening cross section σ due to the elastic collisions of the Tl atoms with the hydrogen molecules from the equation $\pi\gamma = N_{H_2}\langle\bar{v}_{rel}\sigma\rangle$. The ratio γ/N_{H_2} which may be found from the tangent of the angle of inclination in Fig. 46 equalled $9 \cdot 10^{-10}\ cm^3/sec$. If we then calculate the mean relative velocity of the colliding particles

$$\bar{v}_{rel} = \sqrt{\frac{8RT}{\pi}\left(\frac{1}{M_1} + \frac{1}{M_2}\right)},$$

where M_1 and M_2 are the molecular weights of thallium and hydrogen, we may determine σ.

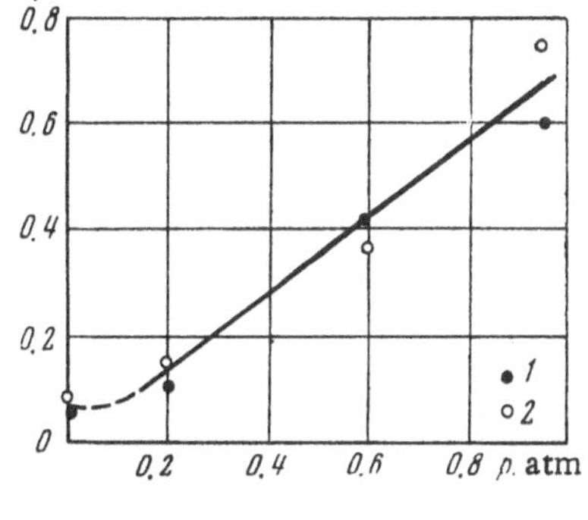

Fig. 46. Dependence of the line width of the thallium 5350 (1) and 3776 Å (2) lines on hydrogen pressure.

For $T = 460°C$, $\overline{v}_{rel} = 2.8 \cdot 10^5$ cm/sec. Hence σ equals $1 \cdot 10^{-14}$ cm^2.

This value differs little from those obtained by analyzing the broadening of the resonance lines of alkali metals in collisions with molecular hydrogen [17].

CHAPTER III

DEVELOPMENT OF SPECTRAL APPARATUS

On the initiative of G. S. Landsberg, the author took part in the design of a number of spectral instruments (in the development of the general construction, the design of the optical system, and the testing of the instrument), which were required for carrying out various kinds of physical investigations in the Optical Laboratory of the Physics Institute of the Academy of Sciences, and in particular for the investigations outlined in Chapters I and II. In this chapter a brief description of these instruments will be presented, together with an indication of their principal characteristics, and also various methods of increasing the dispersion of the spectral instruments.

§ 1. Double-Beam Automatic Infrared Spectrophotometer

The difficulties experienced in obtaining infrared absorption spectra with single-beam spectrometers, such as those used up to the 40's, are already well known. With the appearance of sensitive, low-inertia radiation receivers and the use of electronics for amplifying the signals, it became possible to automate the recording of infrared absorption spectra by using double-beam spectrophotometers [107-109]. In instruments of this kind the demands made on the stability of the source, the amplification factor of the amplifier, and the linearity of the latter are less stringent, and furthermore the instrument directly records the transmission curve of the material under test.

The first double-beam automatic spectrophotometer was made in our laboratory in 1940-1950 and was the first instrument of this type in the Soviet Union [59, 110]. The electronic amplifying unit of the spectrophotometer was developed by M. N. Markov [111]; the first versions of the bolometers were made by A. A. Shubin, and later the bolometers were made by a technology due to M. N. Markov [112].

The spectrophotometer so constructed was used for various experiments in order to establish the optimum working conditions of the amplifying and recording system, to elucidate the role of various factors affecting the accuracy of the recorded curves, to compare the operation of single- and double-beam systems, and so forth. Special attention was paid to the effect of the absorption bands of atmospheric vapor and the absorption bands of the solvent (when studying solutions with a compensation cuvette containing the pure solvent placed in the comparison beam) on the accuracy of the recorded spectrum of the test substance.

It was found that, where the absorption bands of the atmospheric vapor or the solvent occurred, there was a reduction in intensity in both beams, and the absolute value of the signal diminished, causing the motor of the servo system to come into action. This in turn led to a reduction in the sensitivity of the servo system and hence to a distortion of the recorded spectrum, although in the case of exact compensation the actual absorption bands of the atmospheric vapor or solvent failed to appear in the recording.

In order to restore the sensitivity of the servo system in these parts of the spectrum we might increase the intensity of the beams falling on the receiver by increasing the slit width of the monochromators; however, even in this case, the recorded spectrum will be further distorted as a result of the increase in the systematic error due to the effects of the apparatus function of the monochromator. Hence, in general, the presence of the absorption bands of atmospheric vapor or the absorption bands of a solvent additionally distorts the recorded spectrum. Experience showed that these distortions were relatively small if the absorption by the atmospheric vapor or solvent were no greater than 20% in the region of the bands under consideration.

Later the author cooperated with S. G. Rautian in developing the original version of the spectrophotometer into a new universal double-beam auxiliary device (the AIKS-F4), which served to convert any infrared monochromator into an automatic double-beam recording spectrophotometer. Although industry is now producing standard-issue infrared spectrophotometers, our auxiliary device is still of considerable practical interest, since in view of its universality it often proves more convenient for various physical investigations than standard instruments.

The auxiliary device consists of three separate units: the double-beam illuminating system (condenser), the recording device, and the electronic amplifier unit. Figure 47 shows a general view of the auxiliary device with the IKS-11 monochromator, and Fig. 48 shows the optical system of the double-beam condenser. Here M_1, M_2, M_1', and M_2' are concave spherical aluminum mirrors imaging the globar source Q with a magnification of 2 :1 in the plane Q', in which the cuvette C containing the test substance (in the working beam) and the photometric wedge W (in the comparison beam) are situated. The plane mirrors M_3, M_3', and M_4 bring both beams on to the plane of the mirror obturator M_5, which by its rotation alternately directs the working and comparison beams on to the spherical mirror M_6. The obturator mirror is a half disc fixed in an annular guide, its angular frequency being 9 cps.

The axis of the obturator also carries an electromagnetic sensor giving electric pulses for the synchronous detector—amplifier unit. As a result of the introduction of the mirror M_4 the total number of reflections from the mirrors is the same in both beams; this is important

Fig. 47. General form of a universal double-beam auxiliary system, with the monochromator from the IKS-11 spectrometer. 1) Double-beam illuminating system; 2) recording device; 3) preamplifier; 4) amplifier, synchronous detector, and power amplifier.

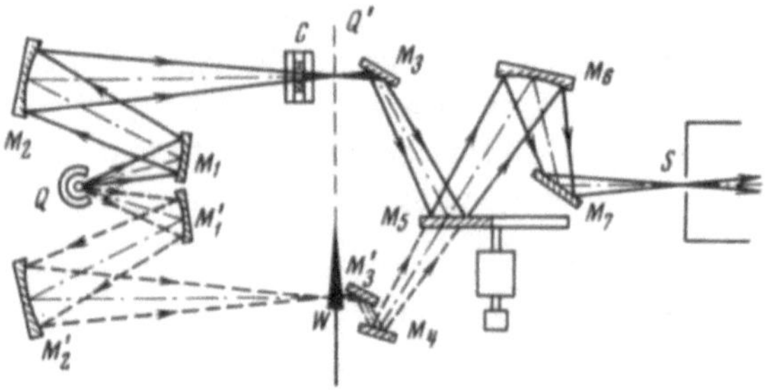

Fig. 48. Optical arrangement of the double-beam illuminating system of the universal auxiliary device.

in order to ensure identity of the working and comparison beams. The last two mirrors (concave M_6 and plane M_7) are common to the two beams; they form a second image of the source, and at the same time one of the photometric wedge on the monochromator slit S (outside the casing of the illuminating system). The focal length of the mirror M_6 and also its disposition are chosen in such a way as to ensure that the image of the source and the photometric wedge on the slit should be greater than or equal to the height of the monochromator slit, while the solid angle of the beam should completely fill the entrance collimator of the monochromator employed. Correspondingly, the dimensions of all the mirrors of the illuminating system (mirrors of rectangular shape) are determined by those of the aperture specified for the beam emerging from the illuminating system and filling the collimator of the monochromator. The photometric wedge W, consisting of eight triangular strips of blackened bronze foil 0.1 mm

Fig. 49. Universal double-beam auxiliary device with a Hilger D-209 prismatic monochromator.

Fig. 50. Universal double-beam auxiliary device with a diffraction monochromator, and general view of a diffraction spectrophotometer of high resolving power.

thick and 100 mm long, fixed to a rigid framework, is mechanically coupled to the motor of the servo system in the recording unit. The radiation source (globar) is kept in a metal casing with water cooling. All the parts of the double-beam illuminating system are situated on a base furnished with adjusting supports and covered with a dust-proof jacket.

The recording unit contains a drum of recording paper (250 × 550 mm), the motor of the servo system, coupled to the photometric wedge and the pen of the automatic recorder, and a synchronous motor with a multiple reducing gear, which rotates the drum. The time for a complete rotation of the drum is 8, 16, 32, 64, or 128 sec. The maximum time for the pen to traverse the whole scale is 8 or 16 sec.

The electronic amplifying unit consists of two parts: 1) the bolometer bridge, preliminary amplifier and bridge generator, enclosed in a thick-walled cylindrical metal sheath, and 2) a separate cabinet with a chassis carrying a narrow-band amplifier, a synchronous detector, a power amplifier, and stabilized rectifiers for feeding the whole circuit. The electrical circuit of the amplifying unit is given in [111]. In order to reduce the induced currents the preamplifier lies close to the bolometer, and the conduits passing from the bolometer to the input stage of the amplifier are enclosed in a metal tube.

The foregoing auxiliary device has been widely used at the Physics Institute of the Academy of Sciences with various monochromators: IKS-11, IKS-6, Hilger D-209 (Fig. 49), a diffraction monochromator [53] of high resolving power (Fig. 50), etc.; as regards the quality of the resultant spectra it is in no way inferior to standard instruments with monochromators of analogous optics.

The auxiliary device has been taken as a base for developing a standard spectrophotometer in the LOMO factory.

* * *

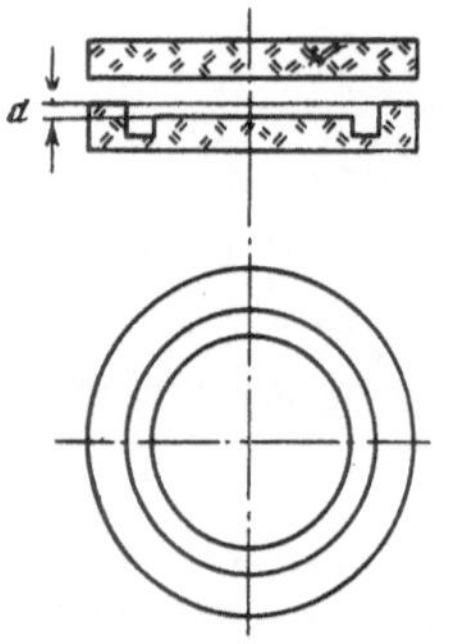

Fig. 51. Arrangement of demountable cuvette of constant thickness; d = thickness of absorbing layer.

For studying the absorption spectra of liquids the author proposed a special construction of a demountable cuvette of constant thickness, which has a number of advantages, particularly when studying liquids of low boiling point. The arrangement of the cuvette is illustrated in Fig. 51. The cuvette consists of two flat window plates 5-6 mm thick, in one of which an annular groove 3-4 mm wide and about 3 mm deep is cut. The central area so formed was polished with a small polisher to a depth d equal to the desired thickness of the absorbing layer. The diameter of the central area was made such as to ensure that, on placing the cuvette in the illuminating system, it should not limit the cross section of the working beam of the spectrophotometer. The plane of the central area had to be parallel to the plane of the outer, unpolished annulus of the plate, the width of which was 5-6 mm. The second plate remained plane parallel. The test liquid is poured into the annular groove of the horizontally supported cuvette so as to cover the central area as well, and the second plate is then put on top. The assembled cuvette is clamped in an ordinary cuvette holder. The annular groove constitutes a ballast space from which capillary forces draw the liquid into the central thin space of the cuvette (even when the amount of liquid in the cuvette has been reduced by evaporation and large air bubbles have formed). Tests showed that the thickness of the working layer of liquid in this type of cuvette is accurately reproduced after repeated assemblies and disassemblies of the cuvette, since it contains no elastic or easily distorted spacers. These cuvettes are therefore suitable for quantitative measurements of absorption spectra.

The technology of making the cuvettes (of CaF_2, NaCl, KBr, and KRS-5) was first developed in the Optical Workshop of the Physics Institute of the Academy of Sciences and is now widely used.

It should be noted that we use cuvettes of this kind as passive-shutter cuvettes, containing translucent solutions, for Q-modulating solid-state lasers. In this case the plates of the cuvettes are made of K-8 glass, and hermetization is achieved by placing the plates in optical contact. If both plates are prepared very carefully, after setting them in optical contact the total deviation of all four surfaces of the central part of the cuvette from the parallel state is no greater than 10". In order to fill the assembled cuvette with the solution, a small radial channel 0.5-1.0 mm wide is made in the outer ring of the cuvette, and this is closed from outside with a cap.

The cuvettes are particularly useful for working with a neodymium glass laser. The solutions commonly used as passive shutters in this case are insufficiently stable and require frequent replacement if an ordinary cuvette is employed. However, in a cuvette with an annular ballast space the worked-out solution in the central working part of the cuvette is continuously renewed with fresh solution from the ballast space. If the ballast space of the cuvette is large enough, the solution within it may be repeatedly used without replacement. Furthermore, the construction envisaged enables us to make cuvettes with a plane-parallel working layer of very small thickness ($\approx$ 1 mm), which in turn reduces the losses associated with the absorption and scattering of the radiation (ordinary and stimulated) in the solvent.

§ 2. A Double-Beam Vacuum Diffraction Spectrophotometer for the Infrared Region

In order to solve a number of spectroscopic problems requiring the use of an instrument of high resolving power, and in particular for solving the problems indicated in Chapters I and II, as well as for developing methods of measuring the true widths and intensities of infrared absorption bands, the author and S. G. Rautian [53], working in the G. S. Landsberg Optical Laboratory of the Physics Institute of the Academy of Sciences, constructed a double-beam spectrophotometer with interchangeable diffraction gratings for the spectral range 1-25 μ (the DAIKS-F1). This instrument was the first Soviet double-beam diffraction infrared spectrophotometer of high resolving power. The general appearance of the spectrophotometer is indicated in Fig. 50. The instrument consists of three main parts: the universal double-beam auxiliary device described in Section 1 of this chapter (only the mirror M_6 being replaced), and preliminary and diffraction monochromators. The optical system of the spectrophotometer as a whole is illustrated in Fig. 52 and the arrangement of the double-beam illuminating system in Fig. 48.

The preliminary monochromator is made on the autocollimation principle and consists of a spherical collimator mirror M_9 (f_9 = 300 mm), plane mirrors M_8 and M_{10}, and interchangeable LiF and KBr right-angled prisms P with refracting angles of 37.5 and 30°, respectively. The reflecting coatings of the prisms (on the plane of the long leg) are of gold; they are produced by thermal sputtering and fixed with a layer of magnesium fluoride. The prisms are fixed to a table, the rotation of this being coupled to the rotation of the diffraction grating.

The principal diffraction monochromator is constructed on the Ebert–Fastie principle [113]. The spherical collimator mirrors M_{11} and M_{12} of rectangular shape have a focal length of $f_{11} = f_{12}$ = 2 m and dimensions of 250 × 300 mm. The mirrors were made in the Optical Workshop of the Physics Institute of the Academy of Sciences, their quality being monitored by the shadow method, which yielded very high-quality mirrors. The aluminum reflecting coatings of the mirrors were deposited by thermal sputtering, care being taken to create aluminum films of as uniform a thickness as possible. Since there was a possibility of the reflecting surface being distorted on fixing the mirrors in the guides, this fixing operation and the setting of the mirrors on the base of the monochromator were carried out under shadow control. All these measures yielded a high-quality focusing system for the principal monochromator, as witnessed by the diffraction pattern clearly observed under the microscope in the image of the narrow entrance slit (zeroth order).

As dispersing elements of the principal monochromator we used interchangeable diffraction echellete gratings G made in the State Optical Institute, having 300 and 200 lines/mm, with an area of 150 × 150 mm, or 50 lines/mm, with an area of 250 × 250 mm. In order to ease interchange of the gratings, each of them is fixed in a separate holder furnished with the necessary adjusting screws, which may be rigidly fixed to the principal rotating table of the monochromator.

The entrance and exit slits S_2 and S_3 of the principal monochromator have a height of 50 mm and, in order to compensate astigmatism, the collimator mirrors are curved in accordance with the Fastie theory [113]; the radius of curvature of the slits in the apparatus under consideration is 125 mm. The radius of curvature of the entrance slit of the preliminary monochromator S_1 (equal to 320 mm) is chosen in such a way that the curvature of the image of slit S_1 formed by the optical system of the preliminary monochromator coincides with the curvature of the slit S_2 in the middle of the working ranges of the prisms employed.

The exit slit of the principal monochromator S_3 is imaged on the bolometer B with a 12-fold reduction by mirrors M_{13} and M_{14}.

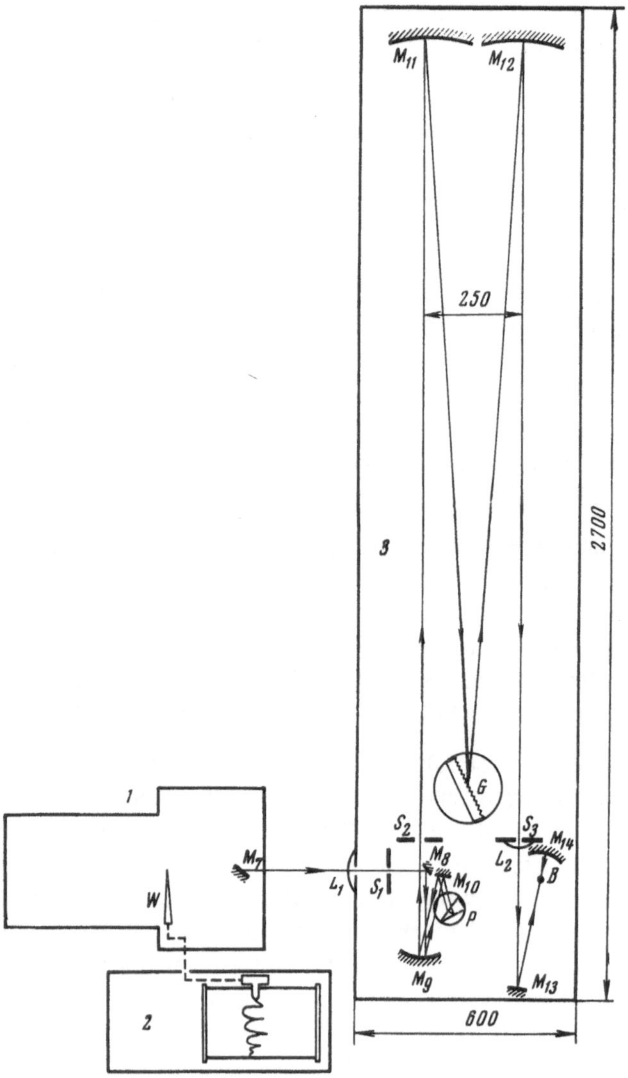

Fig. 52. Optical system of the double-beam infrared diffraction spectrophotometer of high resolving power (DAIKS-F1). The general appearance of the instrument is shown in Fig. 50. 1) Double-beam illuminating system (Fig. 48); 2) automatic recorder; 3) monochromator.

In order to increase the light transmission of the instrument for a specified resolving power we took special precautions to eliminate every possible kind of vignetting by the optical parts of the spectrophotometer, such as usually arise when working with high slits. Since the diffraction grating constitutes the aperture diaphragm in the present arrangement of the monochromator, in order to prevent vignetting by the collimator mirrors M_{11} and M_{12} the height of these (H' = 300 mm) was determined from the condition $H' = H + h$, where H is the height of the grating and h is the height of the slit.

The distance between the prism of the preliminary monochromator and the collimator mirror M_9 is made equal to the focal length of the mirror f_9, as a result of which the mirror M_9 images the prism on the diffraction grating. This condition in turn determines the size of the prism. In addition to this, a converging lens L_1 (f = 600 mm) made of KBr is placed at the entrance slit of the preliminary monochromator S_1; together with the collimator mirror M_9 this images the mirror M_6 of the double-beam illuminating system on the plane of the prism, the distance between the mirror M_6 and the lens L_1 being equal to the focal length of this lens. Finally, at the exit slit of the principal monochromator S_3 is yet another converging lens L_2 (f = 700 mm), also made of KBr; together with mirror M_{12} this images the diffraction grating on the plane of the condenser mirror M_{14}, which focuses the exit slit S_3 on the receiving area of the bolometer.

In view of the use of the converging lenses and the mutual disposition of the optical parts indicated, there is no vignetting in the instrument for any of the pencils of rays passing through the slits S_1, S_2, and S_3. It should also be noted that in the optical system indicated it is quite easy to calculate the dimensions of all the optical elements, taking the dimensions of the diffraction grating as original values.

In this spectrophotometer the diffraction gratings (echelettes) are used not only in the ordinary setting, i.e., with relatively small diffraction angles ($\varphi \approx$ 12-20°) corresponding to "shine" from the wide face of the line, but also with large diffraction angles ($\varphi \approx$ 78-70°) corresponding to "shine" from the small face of the line. We were able to demonstrate (Section 4 of this chapter and [114]) that the latter arrangement of the echelette has a number of advantages in relation to the practical resolving power, the light transmission, the dispersion, the extension of the working range of the spectrum in the long-wave direction, and so on.

The diffraction gratings in their holders are placed on a rotating table with a conical center bearing supported by a thrust ball bearing. The inner axle of the conical bearing is made of quenched steel and the outer sleeve of bronze. The working conical surfaces are carefully polished. When scanning the spectrum the table is rotated by means of a flat lever 500 mm long, rigidly connected to the lower end of the conical axle of the table, and a guide screw with a split nut against which the second, free end of the lever is pressed. In order to ensure uniformity and smoothness of the rotation of the table, the screw and nut are first carefully ground to one another. Our experiments showed that, in order to ensure uniformity of the rotation of the table over the whole working range of angles of rotation (≈ 20°), it was very important to keep a constant pressure on the free end of the lever pressing it to the pin fixed in the nut. Various types of springs failed to achieve this end, and in our instrument the pressure was therefore applied to the lever by means of a Kapron filament passing around a pulley with a freely suspended weight at the other end.

Since the table and conical bearing lie under the vacuum hood, while the guide screw and nut lie outside, a mobile vacuum seal is arranged between the axle of the bearing and the outer part of the lever, using a metal bellows. The guide screw is set in rotation by a synchronous motor through a multiple gear system, which provides eleven velocities of rotation of the diffraction grating, from $5 \cdot 10^{-4}$ to $5 \cdot 10^{-7}$ rad/sec. The rotation of the grating is read along a graduated circle fixed to the axle of the guide screw. A test on the rotation mechanism showed that the rotation of the grating took place uniformly at all velocities without any jumps or seizing.

The radiation receiver is a bismuth bolometer made by M. N. Markov [112]. In order to ensure efficient use of the bolometer, its receiving strip is made curved in accordance with the curvature of the image of the exit slit of the monochromator S_3. The parameters of the bolometer are as follows: threshold of sensitivity $5 \cdot 10^{-11}$ W for an amplifier time constant of 1 sec, conversion factor 30 V/W, time constant 0.02 sec. The electronic amplifying unit of the universal auxiliary device (Section 1 of this chapter, [110,111]) was used with hardly any changes.

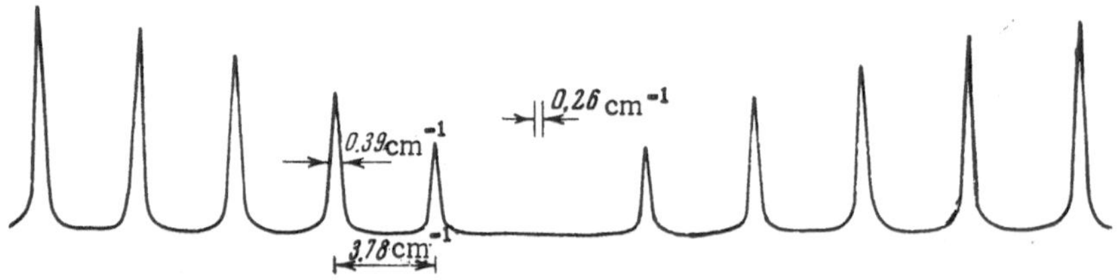

Fig. 53. Absorption spectrum of CO obtained with the diffraction spectrophotometer, placing the echelette (300 lines/mm) at a large diffraction angle (see Section 4). Spectral slit width $\delta\nu = 0.26\,\text{cm}^{-1}$.

Since the rate of scanning the spectrum may vary over quite a wide range, the amplifying system is furnished with a set of time constants: 0.5, 1, 2, 4, 8, and 16 sec.

In the optimum operating mode of the spectrophotometer the spectral slit width in the spectral range 1.5-5 μ is 0.2-0.25 cm^{-1} for a signal-to-noise ratio of the order of 100 and a 1-sec time constant of the amplifying system. By way of illustrating the potentialities of the instrument, Fig. 53 shows the absorption spectrum of CO in the region of 4.56 μ.

No noise appears in the spectrum, so that in this range work may be carried out with a smaller slit width.

As already indicated, our spectrophotometer constitutes a double-beam instrument and hence the absorption bands of the atmospheric vapor are not recorded in the spectrum. However, since the total optical path from source to receiver in the instrument is very long (over 12 m), there is a considerable reduction in the intensity of the beam, which ultimately reduces the practical resolving power (Section 1 of this chapter). Hence, in order to increase the resolving power, the main part of the apparatus (preliminary and principal monochromators and bolometer) are enclosed in a vacuum space. The double-beam illuminating system, however, lies outside the vacuum space (Fig. 50), so that different cuvettes containing the samples for study may readily be interchanged and adjusted.

The preliminary and principal monochromators are assembled on a double-T beam 600 mm high with a shelf width of 300 mm and are covered with a roof welded from sheet iron and strengthened with cross ribs. The roof is fixed to the beam through a rubber spacer. The slit widths and the setting of the prism of the preliminary monochromator are varied by special bellows manipulators without breaking the vacuum in the vacuum space. In the upper wall of the vacuum vessel is a hermetic trap door through which the diffraction gratings are interchanged. The light beam from the double-beam illuminating system is introduced through a KBr window in the side of the beam. The system is evacuated to a pressure of 10^{-1}-10^{-2} mm Hg with a VN-2 pump.

§ 3. Vacuum Spectrophotometer for the Far Infrared Part of the Spectrum

In order to solve a number of physical problems, workers in the Optical Laboratory of the Physics Institute of the Academy of Sciences constructed [115] a high-transmission spectrometer for the spectral range 50-1000 μ. The arrangement of this instrument as a whole is illustrated in Fig. 54. The principal diffraction monochromator is constructed on the Ebert–Fastie principle [113]. The source of radiation Q (a PRK-4 mercury lamp) is imaged by the mirror M_1 on a selective modulator M with sectors made from a KBr crystal. The mirror M_4 images the modulator and hence the source on the entrance slit of the monochromator S_1. The collimator mirrors M_5 and M_6 have a focal length of 1 m and dimensions of 330 × 375 mm.

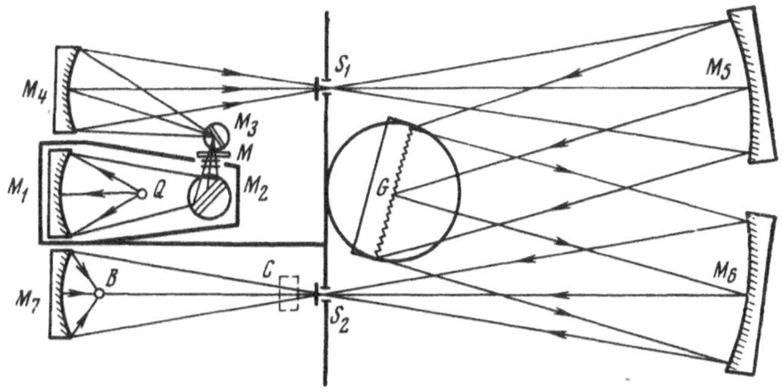

Fig. 54. Optical scheme of the vacuum spectrometer for the far-infrared part of the spectrum.

Here, also, in order to eliminate vignetting, the height of the mirrors, 375 mm, has to equal the sum of the heights of the grating and slit (see Section 2 of this chapter). We use echelettes G 300 × 300 mm in size with constants 0.25, 0.5, 1.0, and 2.0 mm, the last two being made in the workshop of the Physics Institute of the Academy of Sciences. The slits S_1 and S_2, 75 mm in height, are curved in accordance with the Fastie theory [113], and in this instrument have a radius of curvature of 200 mm. The exit slit S_2 is imaged by the spherical mirror M_7 on the receiver B (bismuth bolometer) with a receiving area of 10 × 2 mm made by M. N. Markov. The amplifier employed is a slightly modified version of that described in [111]. Preliminary monochromatization is effected by a combination of absorbing and reflecting filters. For the latter we use mat plane mirrors or echelettes working in the zero order instead of the plane mirrors M_2 and M_3. In place of mirror M_3 we may also use selectively reflecting plates of alkali halide crystals. The cuvette containing the test substance C may be placed either in front of S_1 or after S_2.

The optical part of the spectrometer is mounted on a massive welded frame which runs on rollers into the welded vacuum vessel 600 × 800 × 2000 mm in size. All the mirrors were made in the Optical Workshop of the Physics Institute.

Tests* showed that, as regards resolution actually achieved, the spectrometer was on the same level as the best Soviet and other spectrometers of analogous type. Thus, in the spectral region $\nu = 100\ cm^{-1}$ for an amplifier time constant of $\tau = 37$ sec and a signal-to-noise ratio N = 100, the spectral slit width was $\delta\nu = 0.7\ cm^{-1}$, and for the region of $\nu = 20\ cm^{-1}$ with $\tau = 160$ sec, N = 100, $\delta\nu = 0.4\ cm^{-1}$.

The spectrometer was copied by the Special Designs Office of Instrument Making and manufactured for a number of other institutes of the Academy of Sciences of the USSR.

§ 4. Use of the Echelette at Large Diffraction Angles

The angular dispersion, resolving power, and light transmission are the main characteristics of any spectral instrument, and the possibility of increasing these (particularly in an existing manufactured instrument) is of great practical interest. An analysis of the operation of spectral devices incorporating an echelette diffraction grating shows that in a number of cases these characteristics may be improved in fairly simple ways, e.g., by changing the setting of the grating. This section is devoted to a consideration of such problems.

*The final alignment of the whole spectrometer, the choice of filters, and the tests on the instrument were conducted by I. M. Aref'ev, who later used the instrument in his work on the low-frequency absorption spectra of various substances.

1. First let us briefly consider the main characteristics of various diffraction gratings.

The expressions for the angular dispersion $d\varphi/d\lambda$ and the theoretical resolving power $R = \lambda/\delta\lambda$ of diffraction gratings of any type – echelette, echelle, and Michelson echelon (see below) – may be expressed in the form (see, for example, [116])

$$\frac{d\varphi}{d\lambda}=\frac{\sin\psi+\sin\varphi}{\lambda\cos\varphi};$$

$$R=\frac{A_0}{\lambda}(\sin\psi+\sin\varphi), \quad (1)$$

where ψ and φ are the angles of incidence and diffraction, reckoned from the normal to the plane of the grating, λ is the wavelength, and A_0 is the length of the ruled part of the grating in a direction perpendicular to the lines.

It follows from Eq. (1) that for any grating (independently of the period of the grating and the shape of the line) both quantities increase with increasing diffraction angle φ. The dependence of $d\varphi/d\lambda$ and R on the diffraction angle φ is particularly obvious when using diffraction gratings in instruments based on the now widely employed autocollimation principle or similar instruments, in which $\varphi \approx \psi$ and both angles lie on the same side of the normal to the grating. In this case, Eq. (1) transforms into

$$\frac{d\varphi}{d\lambda}=\frac{2\tan\varphi}{\lambda}; \quad R=\frac{2A_0\sin\varphi}{\lambda}. \quad (2)$$

However, an increase in $d\varphi/d\lambda$ and R due to an increase in the diffraction angle often leads (other conditions being equal) to a fall in the intensity of the spectrum, which limits the increase in φ.

It is well known that the resultant energy distribution in the spectra of different orders of a diffraction grating is determined by the product of two functions. One of these, having a diffraction distribution $I' \approx [(\sin u)/u]^2$, is determined by diffraction at an individual line of the grating and depends on the shape of the line; the other function, not depending on the shape of the line, is due to the interference of rays from all the lines of the grating and has the form $I'' \approx (\sin nv/\sin v)^2$ (where n is the total number of lines in the lattice), with sharp principal maxima of various orders but all of the same intensity. The position of the principal maxima of the function I'' is determined by the well-known relation $d(\sin\psi + \sin\varphi) = m\lambda$, where d is the grating constant and m is the order of the spectrum.

In the resultant intensity distribution, that particular order of the spectrum, the position of which coincides with the middle of the central maximum of the function $I' = (\sin u/u)^2$, where $I'_{max} = 1$ for $u = 0$, has the greatest intensity. In the case of the ordinary, nonconcentrating grating, the greatest intensity is that of the zeroth (m = 0), achromatic order of the spectrum, and hence, in this grating, an increase in the angle of diffraction means a transition to higher orders of the spectrum, the intensity of these falling rapidly, i.e., in nonconcentrating gratings an increase in $d\varphi/d\lambda$ and R is always accompanied by a reduction in the intensity of the spectrum.

In the case of an echelette, in which the profile of the line has an asymmetric, triangular form (Fig. 55), the greatest intensity is that of a nonzero order of the spectrum ($m_0 \neq 0$), its value depending on the angle Ω_1 made by the working face of the line with the plane of the grating (Fig. 55). Theory shows [117] that the order of the spectrum m_0 and the diffraction angle φ_m corresponding to the maximum intensity are uniquely related to the wavelength λ_0, the angle of incidence ψ, and the angle Ω_1 by the expressions

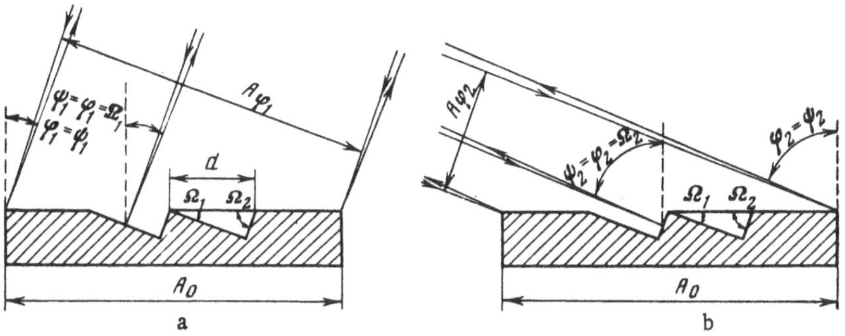

Fig. 55. Arrangement of the echelette when working at the angle of shine. a) From the wide face of the line $\varphi_1 = \psi_1 = \Omega_1$; b) from the narrow face of the line $\varphi_2 = \psi_2 = \Omega_2$, $A_{\varphi_1} < A_{\varphi_2}$; d is the grating constant (for clarity the lines are illustrated greatly enlarged).

$$d(\sin\psi + \sin\varphi_m) = m_0\lambda_0, \quad \psi + \varphi_m = 2\Omega_1 \tag{3}$$

or, for the case of an autocollimation arrangement,

$$2d\sin\Omega_1 = m_0\lambda_0, \quad \psi = \varphi_m = \Omega_1. \tag{4}$$

Condition (4) means that the incident and diffracted rays are perpendicular to the working face of the line, i.e., we have normal specular reflection from the face. Thus, for a particular echelette (d, Ω_1) and a specified wavelength λ_0, the order of the spectrum m_0 and the diffraction angle φ_m in the direction of which the spectral intensity reaches a maximum (the so-called "angle of shine"), are completely defined and, hence, $d\varphi/d\lambda$ and R also have quite specific values. An attempt to increase $d\varphi/d\lambda$ and R for a specified echelette (d, Ω_1) by increasing the diffraction angle $\varphi > \varphi_m$ also leads to a fall in the intensity of the spectrum since, in this case, the conditions (3) and (4) are not satisfied. Hence the only way to increase $d\varphi/d\lambda$ and R without reducing the intensity of the spectrum lies in increasing the angle Ω_1 between the working face of the line and the plane of the grating and working in the direction of the angle of shine $\varphi_m \approx \Omega_1$.

Until recently the angle of shine Ω_1 of echelettes could never be made greater than 20° for various technological reasons (for Soviet instruments [118], in most cases, it lay in the range 12-18°), and the working face was the large face of the line, special attention being paid to the quality of this when ruling the grating. Hence, the values of $d\varphi/d\lambda$ and R were relatively low when working with echelettes at the angle of shine. Only recently has a technology been developed for ruling diffraction gratings (called echelles) in which the angle of shine has a greater value (30-65°), and hence echelles have a much greater angular dispersion and resolving power (for the same A_0) than echelettes, without any loss of intensity. This constitutes the main advantage of echelles over echelettes, and hence the replacement of an echelette by an echelle in an existing instrument greatly improves the spectroscopic characteristics.

We note that the Michelson echelon has approximately the same angle Ω_1 as echelles, and hence it has the same advantages (although the technology for making Michelson echelons differs considerably from that of making echelles).

It should nevertheless be noted that the grating constants d in echelles, and still more in the Michelson echelon, are usually much larger (for various technological reasons) than the constants of echelettes designed for working in the same range of wavelengths, and hence the range of dispersion, i.e., that part of the spectrum in which there is a unique relationship between the wavelength and the diffraction angle $\Delta\lambda$, is much smaller for echelles than for echelettes* in accordance with the relation for $\Delta\lambda$:

$$\Delta\lambda = \frac{\lambda}{m} = \frac{\lambda^2}{2d\sin\Omega_1}. \tag{5}$$

2. As already indicated, the profile of the line of echelettes has an asymmetrical triangular shape, and usually in spectral instruments one uses an angle of shine $\varphi_1 = \Omega_1$ corresponding to reflection from the wider face of the line (Fig. 55a). However, according to theory [117], for a grating with this kind of profile, there are two angles of shine $\varphi_1 = \Omega_1$ and $\varphi_2 = \Omega_2$, corresponding to reflection from the large and small faces of the line,† in the direction of which the resultant energy distribution takes a maximum value. We may therefore try changing the orientation of the echelette in a standard instrument so as to use the angle of shine $\varphi_2 = \Omega_2$ corresponding to reflection from the narrow face of the line (Fig. 55b).

Since the profile of the lines of echelettes is usually such that the angle between the wide and narrow face of the line is close to (slightly greater than) 90°, while Ω_2 is much greater than Ω_1 ($\Omega_2 = 90° - \Omega_1$ and for $\Omega_1 = 20°$, $\Omega_2 = 70°$), when working in shine from the second, narrow face $\varphi_2 = \Omega_2$ the values of $d\varphi/d\lambda$ and R will be much greater (in the ratio $\tan\Omega_2/\tan\Omega_1$ and $\sin\Omega_2/\sin\Omega_1$) than when working in shine from the wide, normal face. Thus, for angles $\Omega_1 = 20°$ and $\Omega_2 = 70°$, the angular dispersion increased by almost eight times and the theoretical resolving power by about three times.

It should be noted that working at the angle of shine from the second, small face of an echelette essentially corresponds to the normal working of an echelle. However, from the point of view of the light transmission of the spectral instrument as a whole, the use of the second angle of shine $\varphi_2 = \Omega_2$ at first glance seems distinctly unfavorable. The point is that, in diffraction spectral instruments, the grating usually forms the aperture diaphragm, i.e., it determines the cross section $\sigma = A_\varphi H$ of the beams in the instrument and hence its geometrical transmission, while the width of the beam for the normal arrangement of the grating A_{φ_1} (Fig. 55a) approximately equals the height of the grating H ($A_{\varphi_1} = A_0\cos\varphi_1 \approx H$). Thus, on using the angle of shine from the small face of the line, the cross section of the beam is $\sigma_2 = A_{\varphi_2}H$ and hence the geometrical transmission will be much smaller than in the case of shine from the wide face, since $A_{\varphi_2} = A_0\cos\varphi_2$ is smaller than A_{φ_1} (for $\Omega_2 = 70°$ and $\Omega_1 = 20°$ the beam cross section diminishes by about a factor of three).

However, a more detailed analysis of the working of the spectral instrument shows that the second mode of orienting the grating may, in a number of cases, present some considerable advantages [114], even in relation to the transmission of the instrument as a whole, despite the reduction in the cross section of the beam ($A_{\varphi_2} < A_{\varphi_1}$).

This is the case, for example, when studying a continuous spectrum by means of a monochromator (particularly when studying absorption spectra) if the resolving power of the instrument is limited by random measuring errors. Here the slit widths are much larger than the normal $S > S_{10} = f\,\frac{\lambda}{A_\varphi}$, and the flux separated out by the monochromator is determined [120]

*Echelles designed for working in the visible part of the spectrum have a constant of $d \approx 0.1$-0.003 mm and, correspondingly, the work is carried out in fairly high orders of the spectrum, m = 200-7 for a dispersion range of $\Delta\lambda$ = 25-700 Å. In the Michelson echelon, correspondingly, $d \approx$ 5-10 mm, $m \approx 10^4$, $\Delta\lambda \approx 0.2$ Å. At the same time echelettes (1200-600 lines/mm) are used in the first and second orders of the spectrum, and $\Delta\lambda \approx$ 5000-2500 Å.

†Later, S. G. Rautian showed [119] that apart from these two main directions of shine there are other directions corresponding to double reflections from both faces within the line.

by the expression

$$\Phi = \varepsilon B \frac{hH}{f} (\delta\lambda)^2 k_\varphi A_\varphi \frac{d\varphi}{d\lambda}. \quad (6)$$

Here B is the brightness of the source of radiation referred to unit range of wavelengths; H and A_φ are the height and width of the cross section of the diffracted pencil of rays; h is the height of the slit; $f_1 = f_2 = f$ is the focal length of the collimator; ε is the transmission coefficient of the instrument; $\delta\lambda = \frac{S}{f}\frac{d\lambda}{d\varphi}$ is the spectral, and S the geometrical slit width. Equation (6) differs from the corresponding equation in [120] simply in respect of the explicit introduction of the factor k_φ defining the energy distribution with respect to various orders of the diffraction pattern (as mentioned earlier, although for exact shine $k_\varphi = 1$). Let us compare the fluxes for the two angles of shine. These two cases are distinguished simply by the diffraction angle φ, and of the characteristics of the instrument entering into Eq. (6) the angle φ only affects k_φ, $d\varphi/d\lambda$ and A_φ. Since $A_\varphi = A_0 \cos\varphi$, on substituting this expression into Φ, using $d\varphi/d\lambda$ from Eq. (1) and denoting all the factors independent of φ as $C_1 = \varepsilon B \frac{hH}{f}\frac{A_0}{\lambda}$ we obtain the following for Φ:

$$\Phi = C_1 (\delta\lambda)^2 k_\varphi (\sin\psi + \sin\varphi) \quad (7)$$

or, for an autocollimation arrangement (or the analogous Ebert–Fastie or Pfund–Hardy schemes) in which $\varphi \approx \psi$,

$$\Phi = 2C_1 (\delta\lambda)^2 k_\varphi \sin\varphi. \quad (8)$$

If we suppose that, on working at the angle of shine from the wide and narrow faces of the line,

$$k_{\varphi_1} = k_{\varphi_2}, \quad (9)$$

it then follows from (8) that, for a specified spectral slit width $(\delta\lambda)_1 = (\delta\lambda)_2$, i.e., for the same practical resolving power a transition from one arrangement to the other leads to a flux increase* of $\Phi_2/\Phi_1 = \sin\varphi_2/\sin\varphi_1$, while for a specified value of the flux $\Phi_1 = \Phi_2$ or $(\delta\lambda)_1^2 \sin\varphi_1 = (\delta\lambda)_2^2 \sin\varphi_2$ (i.e., a constant signal-to-noise ratio) there will be a reduction in the spectral slit width

$$\frac{(\delta\lambda)_2}{(\delta\lambda)_1} = \sqrt{\frac{\sin\varphi_1}{\sin\varphi_2}} = \frac{(\delta\nu)_2}{(\delta\nu)_1}. \quad (10)$$

This is associated with the fact that the angular dispersion increases faster with increasing φ (as $\tan\varphi$) than the beam cross section diminishes (as $\cos\varphi$).

These conclusions only apply if Eq. (9) is satisfied; this in fact holds if the lines of the grating have a triangular shape with ideally plane and mutually perpendicular faces, and if $\varphi = \psi = \Omega$. In real echelette gratings, as already discussed, the wider face of the line is usually of better quality than the short "nonworking" face, and the faces of the line are not strictly perpendicular to one another. Hence, the final solution of the question as to the advantages of the new setting of the echelette has to be solved experimentally; it is quite possible that the gain will differ for different gratings.

3. In order to verify the theoretical conclusions [53], we made the corresponding measurements in a double-beam diffraction infrared spectrophotometer of the kind described in Section 2 of this chapter (the DAIKS-F1).

---

*If, however, in one particular instrument we place a new grating of greater length A_0' at an angle φ_2 so that $A'_{\varphi_2} \approx H \approx A_{\varphi_1}$, then the gain in flux will be still greater: $\Phi_2/\Phi_1 = \tan\varphi_2/\tan\varphi_1$.

The work was carried out with a State Optical Institute grating No. 2538 with 300 lines per mm and a shine angle of 18°. We recorded the rotational structure of the methane band in the region of 1.7 μ, first with the ordinary orientation of the echelette in the minus first order. The spectral slit width was $(\delta\nu)_1 = 0.46\ \mathrm{cm}^{-1}$. Then we set the echelette in the new orientation (m = 3) and increased the geometrical dimensions of the slits to restore the previous signal-to-noise ratio. The spectral slit width was then $(\delta\nu)_2 = 0.27\ \mathrm{cm}^{-1}$. The ratio $(\delta\nu)_2/(\delta\nu)_1 = 1/1.7$, which is very close to the theoretical value of $1/\sqrt{3}$, obtained from Eq. (10), i.e., the new setting of the grating gave a gain of 1.7 times in resolving power for the same signal-to-noise ratio.

Analogous measurements were carried out with the lines of the mercury spectrum λ = 1.3571 μ (m = 4), 1.3954 μ (m = 4), 1.5299 μ (m = 3), 1.7114 μ (m = 3); in every case agreement with theory was perfectly satisfactory.

Thus, the foregoing experiments have shown that, from the point of view of increasing the flux for a specified value of $\delta\lambda$ (or reducing $\delta\lambda$ for a specified flux), the new orientation of the echelette is clearly advantageous. These experiments have also confirmed the earlier assumption as to equality of the factors $k_{\varphi_1} = k_{\varphi_2}$.

It should be noted that the gain in flux Φ achieved with the new setting of the echelette for a constant resolving power (or for a constant *spectral* slit width $\delta\lambda$) is associated with a corresponding increase in the *geometrical* slit width S of the monochromator in the new orientation. It follows in fact from the condition $(\delta\lambda)_1 = (\delta\lambda)_2$ or $\frac{S_1}{f}\frac{1}{(d\varphi/d\lambda)_1} = \frac{S_2}{f}\frac{1}{(d\varphi/d\lambda)_2}$ that

$$\frac{S_2}{S_1} = \frac{(d\varphi/d\lambda)_2}{(d\varphi/d\lambda)_1} = \frac{\tan\varphi_2}{\tan\varphi_1}.$$

At the same time the expression for the flux Φ given in (6) may be written in the form

$$\Phi = c_2(\delta\lambda)^2 A_\varphi k_\varphi\left(\frac{d\varphi}{d\lambda}\right) = c_2\delta\lambda A_\varphi k_\varphi\left(\delta\lambda\frac{d\varphi}{d\lambda}\right) = c_2 A_\varphi k_\varphi \delta\lambda\frac{S}{f}, \tag{11}$$

where $c_2 = \varepsilon B(hH/f)$, i.e., for a constant $\delta\lambda$ the flux is proportional to the slit width S. Allowing for the change in the slit widths $S_2/S_1 = \tan\varphi_2/\tan\varphi_1$ and the beam cross sections $A_{\varphi_2}/A_{\varphi_1} = \cos\varphi_2/\cos\varphi_1$ on passing from one orientation to the other, we obtain $\Phi_2/\Phi_1 = \sin\varphi_2/\sin\varphi_1$ from (11) as in the earlier case.

Let us consider some other advantages of the new setting of the echelette.

Another considerable advantage of working at large diffraction angles is the possibility of using a single grating for a wider range of wavelengths. It follows in fact from the basic grating formula $d(\sin\psi + \sin\varphi) = m\lambda$ that for every grating there is a limiting maximum wavelength λ_{max} in a nonzero order of the spectrum of the diffraction pattern* equal to $\lambda_{max} = 2d$ and determined from the condition $\psi = \varphi = 90°$, m = 1. For a grating of 300 lines/mm, for

*We note that on the short-wavelength side there are no such fundamental limitations, since the basic grating formula may be satisfied even for $d \gg \lambda$ for any arbitrarily high order of the spectrum. On the short-wavelength side any limitations are purely practical: the reduction in the reflection coefficient, the increase in the intensity of "ghosts," the small region of dispersion, the worsening of the image quality due to defects in the manufacture of the grating, and so on. In addition to this, for $d \gg \lambda$ we may operate with low orders of the spectrum if $\psi = \varphi$, but the angles have different signs (i.e., the incident and diffracted rays lie on different sides of the normal to the grating): $d(\sin\psi - \sin\varphi) = m\lambda$. This arrangement is, of course, used in glancing-incidence instruments with a bent diffraction grating in the vacuum-ultraviolet and soft x-ray range.

example, $\lambda_{max} = 6.66\ \mu$. In practice, for the ordinary orientation of the echelette and a shine angle of 18°, the working range of wavelengths is 1.6-3 μ (for m = 1), since outside this range the value of k_φ falls sharply. For the new orientation it proved feasible to work up to about 6.5 μ. By way of example, Fig. 53 shows a recording of part of the absorption spectrum of CO close to $\lambda = 4.65\ \mu$ obtained with the same grating of 300 lines/mm but in the new orientation.

The geometrical slit widths are considerably greater in the new orientation than in the conventional one. Hence the tolerances in manufacturing the knife edges of the slits are made considerably less stringent, and (a particularly important matter) the same applies to the accuracy of the rotating mechanism of the grating. Furthermore, in view of the contraction in the beam cross section, the wave aberrations of the collimator mirrors are reduced, and so are the dimensions of the mirrors and the whole instrument.

4. Apart from the foregoing advantages, the new orientation of the echelette may also have some shortcomings. First of all it should be pointed out that, for large diffraction angles φ and high orders m, defects in the ruling of the grating unnoticed at small φ and m may become serious. Naturally these factors play an important part in gratings of different types or gratings cut on different ruling machines, and even for different samples of the same type of grating. We noticed in our own grating in particular that on increasing the diffraction angle the focusing plane of the spectra changed slightly, the displacement extending as much as 6 mm (for a focal length of 2000 mm) in the eleventh order of the green mercury line $\lambda = 5461$ Å. In view of this the exit slit of the principal monochromator had to be readjusted.

Finally, other failings of the new orientation include the reduction in the range of dispersion (5) and the increase in the intensity of the Rowland "ghosts" with increasing order of the spectrum.

We note further that the echelette can only be used at large diffraction angles and over a wide range of wavelengths in the Ebert–Fastie mode [113], in which the slits are curved in accordance with the Fastie principle so as to compensate the curvature of the spectral lines for all wavelengths and all diffraction angles. For other optical arrangements of the monochromators, the curvature of the spectral lines depends on the wavelength and diffraction angle, increasing sharply with increasing diffraction angle as the angular dispersion rises. It should also be noted that in the Ebert–Fastie arrangement the angles of incidence and diffraction differ slightly from one another. Hence the angular magnification $W = \cos\psi/\cos\varphi$ of the grating deviates from unity, the more so the greater the diffraction angle. This leads to two effects: first, the geometrical widths of the entrance and exit slits have to be different, since $S_3 = WS_2$ (Fig. 52); second, for $W \neq 1$ third-order coma appears [121], whereas this is completely eliminated in the so-called Z scheme (such as the Ebert–Fastie scheme of [120]) for W = 1. It is important to observe, however, that at the same time the beam cross section A_{φ_2} diminishes, so that the wave aberration associated with the coma will not increase too rapidly with increasing φ. This clearly explains the fact that, in microscope observations, we detected no asymmetry of the lines for large diffraction angles (60-70°). By way of example we may indicate that for $\varphi \approx 65°$, m = 11 we were able to detect up to eight hyperfine-structure components of the $\lambda = 5461$ Å line of a low-pressure mercury lamp.

Finally, it should be noted that the reduction in the beam cross section makes it less desirable to place the receiver in the beam (unless it is a very small one), since it cuts off a greater proportion of the beam, and it may well be that an extra-axial positioning of the light receiver will be preferable.

5. We have given detailed consideration to the photoelectric recording of a continuous spectrum in the foregoing analysis. However, the number of problems in which the new orientation of the echelette offers specific advantages is not limited to this case. Let us consider, in particular, the case of the photoelectric recording of a bright-line spectrum. For a bright-line spectrum the flux emerging from the monochromator slit is given by the expression [120, 122]

$$\Phi = \varepsilon B_\infty \frac{hH}{f} k_\varphi \delta\lambda A_\varphi \frac{d\varphi}{d\lambda}$$

or, for the case of the autocollimation arrangement,

$$\Phi = 2c_1' k_\varphi \delta\lambda \sin\varphi, \tag{12}$$

which, in contrast to (6) and (8), contains the integrated brightness of the line B_∞; the spectral width $\delta\lambda$ enters to the first power, while $c_1' = \varepsilon B_\infty \frac{H}{f} \frac{A_0}{v}$.

As in the continuous spectrum, so also in the bright-line spectrum a transition from diffraction angle φ_1 to φ_2 with a constant resolving power (constant $\delta\lambda$) leads to an increase in flux in the ratio $\Phi_2/\Phi_1 = \sin\varphi_2/\sin\varphi_1$. Keeping a constant signal-to-noise ratio, i.e., a constant value of the flux Φ or the quantity $(\delta\lambda)_1 \sin\varphi_1 = (\delta\lambda)_2 \sin\varphi_2$, the gain in resolving power $(\delta\lambda)_2/(\delta\lambda)_1 = \sin\varphi_2/\sin\varphi_1$ will be still greater than in the continuous spectrum as represented by Eq. (10). Thus, in respect of a gain in resolving power, the new arrangement of the echelette has still greater advantages for the case of the bright-line spectrum.

6. Finally, we may consider the case of the photographic recording of the spectrum. In photographic recording the photometric quantity determining the degree of film blackening (photometric density) is the exposure

$$E_c = \varepsilon k_\varphi B \frac{HA_\varphi}{f^2} \delta\lambda, \qquad E_l = \varepsilon k_\varphi B_\infty \frac{HA_\varphi}{f^2} \tag{13}$$

for the continuous and bright-line spectra, respectively. Here ε, B, B_∞, H, and A_φ have the same meanings as before in (6) and (12); f is the focal length of the camera objective; and $\delta\lambda$ is the spectral width of the image of the entrance slit of the spectrograph, $\delta\lambda = S\frac{f}{f'} W \frac{d\lambda}{dl}$, where S is the geometrical width of the entrance slit, f' is the focal length of the objective of the entrance collimator, and W is the angular magnification of the echelette.

It follows from (13) that a transition from a small diffraction angle φ_1 to a large angle φ_2 leads (in contrast to the case of photoelectric recording) to a reduction in the exposure for both E_l and E_c (constant $\delta\lambda$ for the case of the continuous spectrum) owing to the reduction in the beam cross section $A_\varphi = A_0 \cos\varphi$. Hence the reorientation of the echelette in an existing instrument, in which we usually have $A_{\varphi_1} \approx H$, will certainly be disadvantageous from the point of view of transmission, since $E_2/E_1 = \cos\varphi_2/\cos\varphi_1 < 1$.

However, the increased angular dispersion may by itself present a distinct interest, since it enables us to increase the accuracy of wavelength measurements. In addition to this, it is well known that the resolving power of a spectrograph may, in a number of cases, be largely determined by the apparatus function of the photosensitive layer. In order to reduce the influence of this factor we must increase the linear dispersion of the instrument; this is usually done by increasing the focal length of the camera objective f. Since $E \sim 1/f^2$, as in Eq. (13), we may well ask whether it would not be better, from the point of view of the exposure, to achieve the same result by increasing the angular dispersion, by changing to the second shine angle. If we keep the linear dispersion constant, $dl/d\lambda = f d\varphi/d\lambda = \text{const}$, Eq. (13) may may be transformed to

$$E_l = c_3 k_\varphi \sin\varphi \tan\varphi,$$

where $c_3 = \varepsilon B \frac{4HA_0}{[(dl/d\lambda)\cdot\lambda]^2}$, from which we see that, in this case, the new orientation of the echelette does in fact offer considerable advantages: $E_2/E_1 = \sin\varphi_2 \tan\varphi_2 / \sin\varphi_1 \tan\varphi_1$. Such a sharp increase in exposure is associated with the fact that, on keeping the linear dispersion $dl/d\lambda$ constant, the increase in the angular dispersion (varying as $\tan\varphi$) which takes place on passing to the large diffraction angle φ_2 enables us to reduce the focal length of the camera objective f very considerably, and this leads to a substantial rise in exposure ($E \sim 1/f^2$), despite the reduction in the beam cross section $A_{\varphi_2} < A_{\varphi_1}$. If we use a new grating of greater length A_0' in the instrument, in such a way that in the new setting it leaves the beam cross section $A'_{\varphi_2} = A_0' \cos\varphi_2 = A_{\varphi_1} \approx H$ unchanged, then the increase in exposure (for $dl/d\lambda$ = const) should be still greater $E_2/E_1 = \tan^2\varphi_2/\tan^2\varphi_1$. Of course, in the latter two cases, the existing instrument will have to be slightly modified so as to replace the camera objective with shorter-focus versions.

It should be noted that the foregoing considerations as to the advantages of setting the echelette to large diffraction angles are also of interest in the design of new high-transmission spectrographs.

Thus, for example, the author used the foregoing method of setting the echelette in designing a high-dispersion spectrograph for studying the emission spectrum of a neodymium glass laser. The spectrograph consists of a standard autocollimation camera of the UF-85 type (f = 1300 mm) and a 200 lines/mm echelette 150 × 150 mm in size, placed on the table at an angle of $\psi \approx \varphi = 75°$. The linear dispersion of this spectrograph in the region of 1 μ is $d\lambda/dl \approx 1.0$ Å/mm, and exceeds the dispersion of all known Soviet diffraction spectrographs (DFS-13, DFS-4, DFS-8, STÉ-1, etc.).

At the same time, the spectrograph is very compact and small in size. We may also note that in using an echelette of the dimensions indicated there is hardly any limitation imposed on the beam cross section by the echelette, and the working section of the beam is determined by the diameter D of the UF-85 camera objective ($A_{\varphi_2} = A_0 \cos\varphi_2 \approx D$), so that there is no additional reduction in the light transmission of the instrument.

7. Our investigations into the properties of practical echelettes in the new orientation, confirming the advantages of the new arrangement, were carried out with fairly "coarse" echelettes of 300, 200, and 50 lines/mm (made in the State Optical Institute), intended for normal working in the infrared part of the spectrum. These echelettes were used to study absorption spectra in the middle and near-infrared parts of the spectrum, and also to study emission spectra in the near-infrared and visible regions. As regards "finer" gratings with 600 and 1200 lines/mm, intended for working in the visible part of the spectrum, the position may be less favorable, since the small face of the lines in these gratings is of poorer quality. Thus there may be a fall in the coefficient k_φ and a worsening of the images of the spectral lines, as well as an increase in the intensity of the "ghosts."

It is therefore essential to make a preliminary experimental test of the practical suitability of the new setting of the echelette in each particular case.

§ 5. Methods of Increasing the Linear Dispersion of Prismatic Spectral Instruments

1. In working with prismatic spectral instruments it is often required to increase the linear dispersion of existing spectral instruments very substantially. As in the case of diffraction instruments, an increase in the linear dispersion chiefly enables us to achieve a greater accuracy of wavelength measurements and measurements of the spectral widths of spectral lines, matters of first-rate importance. However, in the case of prismatic instruments such as spectrographs with photographic recording of the spectrum, in which the rela-

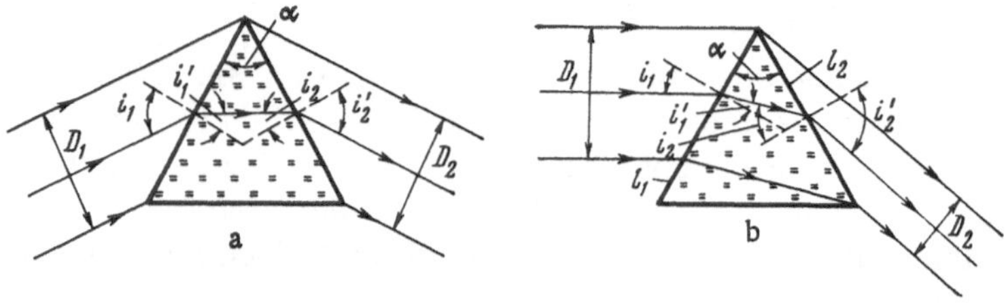

Fig. 56. Course of the rays on refraction in a prism with the angle of incidence (a) equal to the angle of minimum deviation $i_1 = i_{1\,min}$; $D_1 = D_2$, $W = 1$; and (b) smaller than the angle of minimum deviation $i_1 < i_{1\,min}$, $D_2 < D_1$, $W > 1$, $(d\varphi/d\lambda) > (d\varphi/d\lambda)_{i_1 = i_{1\,min}}$. The width of the beam is determined by the exit face of the prism l_2.

tive aperture of the camera objectives is usually great (1 : 4.5 to 1 : 10), the resolving power is often determined by the apparatus function of the photosensitive layer, and an increase in the linear dispersion enables us to increase the practical resolving power very considerably. It is easy to determine the relative aperture of the camera objective for which the resolving power of the instrument as a whole will be determined by the optics of the instrument and not by the photosensitive layer.

If the linear width of the apparatus function of the spectral instrument itself is α_i and its spectral width $\delta\lambda = \alpha_i (d\lambda/dl)$, the resolving power of the instrument may be expressed in general form [123] as*

$$R_i = \frac{\lambda}{\delta\lambda} = \frac{\lambda}{\alpha_i}\frac{dl}{d\lambda}, \tag{14}$$

where $dl/d\lambda = f_2(d\varphi/d\lambda)$ is the linear dispersion and $d\varphi/d\lambda$ is the angular dispersion of the prism.

If the width of the entrance slit of the spectrograph S_1 is almost normal [123], $S_1 \leq S_{10}$ (where $S_{10} = \lambda f_1/D_1$, f_1 is the focal length of the collimator objective and D_1 is the width of the parallel beam entering the prism as in Fig. 56), then the apparatus function (for ideal focusing optics of the instrument) is determined by diffraction in the prism, which constitutes the aperture diaphragm of the instrument. The width of the apparatus function equals $\alpha_i = f_2\lambda/D_2$, where f_2 is the focal length of the camera objective, and D_2 is the width of the beam emerging from the prism (Fig. 56). In this case, substitution of α_i into (14) gives

$$R_i^0 = D_2\frac{d\varphi}{d\lambda}. \tag{15}$$

R_i^0 is the theoretical resolving power constituting the limit for the particular instrument (in the sense indicated above), determined solely by diffraction in the prism, i.e., the diffraction apparatus function.

In the case of wide entrance slits $S_1 \gg S_{10}$ the apparatus function is determined by the slit; its width S_2 equals the width of the image of the entrance slit $S_2 = S_1\frac{f_2}{f_1}W$, where W is the angular magnification of the prism. Thus

*It is well known that, in determining the resolution conditions, it is essential to allow for the noise characteristics of the radiation receiver. Here, for simplicity, we shall consider two monochromatic lines of the same intensity as being resolved if the distance between their maxima equals the width of the apparatus function.

$$R_i = \frac{\lambda}{S_2} \frac{dl}{d\lambda}. \tag{16}$$

In an analogous way we may define the resolving power of the photosensitive layer

$$R_{ph} = \frac{\lambda}{\alpha_{ph}} \frac{dl}{d\lambda}, \tag{17}$$

where α_{ph} is the width of the apparatus function of the photosensitive layer. We may approximately estimate α_{ph} as $\alpha_{ph} = 1/N$, where N is the number of lines per mm in the photographic image of an equidistant black and white grating in which the lines can still just be distinguished.

The resolving power of the spectrograph as a whole (including the photosensitive layer) will be determined by the spectral instrument and not the layer if R_i^0 [Eq. (15)] is greater than R_{ph} [Eq. (17)] or α_{ph} is smaller than α_i.

From the two inequalities $R_{ph} > R_i^0$ or $\alpha_{ph} < \alpha_i$ we may readily derive the limiting condition imposed on the relative aperture of the camera objective: $D_2/f_2 < \lambda/\alpha_{ph}$; only on satisfying this will the resolving power be determined by the spectral instrument. For the average types of photographic emulsion (N = 50-100 lines/mm, i.e., α_i = 0.02-0.01 mm) and various λ we may easily estimate the value of D_2/f_2 for which the resolving power of the spectral instrument will be equal to that of the photosensitive layer: $D_2/f_2 = \lambda/\alpha_{ph}$. The results are presented in Table 11.

Since the majority of prismatic spectrographs are principally intended for studying low-intensity sources of radiation or, in the case of applied analytical problems, for obtaining photographs of spectra with relative short exposure times, they usually incorporate high-transmission camera objectives, since the exposure of the spectral lines E_l (for a slit apparatus function) is determined solely by the solid angle of the beam emerging from the camera objective

$$E_l = \varepsilon B_\infty \frac{HD_2}{f_2^2}, \tag{18}$$

where H is the height of the prism. Usually, in such instruments, D_2/f_2 = 1 :4.5-1 :10; i.e., it is considerably greater (particularly for short wavelengths) than the limiting values of D_2/f_2 given in Table 11. Hence, in these instruments, the theoretical resolving power (15) is not realized, the actually achieved resolving power of the instrument R_{ph} being much smaller than the theoretical R_i^0 of the particular instrument for a relative aperture of D_2/f_2.

We easily deduce from (15) and (17) that $R_{ph}/R_i^0 = \lambda/\alpha_{ph} \cdot f_2/D_2$. By way of example we may quote an estimate of this ratio: for $D_2/f_2 = 1/10$ with α_{ph} = 0.02 mm and λ = 5000 Å, $R_{ph}/R_i^0 = 1/4$; for $D_2/f_2 = 1/4.5$ with α_{ph} = 0.02 mm and λ = 5000 Å, $R_{ph}/R_i^0 = 1/8$.

For shorter wavelengths this ratio is still smaller. Thus high-transmission spectrographs usually have considerable unused reserves of resolving power.

TABLE 11. Values of D_2/f_2 for Which $R_{ph} = R_i^0$ for Various Parts of the Spectrum

λ, Å	D_2/f_2		λ, Å	D_2/f_2	
	α_{ph}=0.01	α_{ph}=0.02		α_{ph}=0.01	α_{ph}=0.02
7000	1:15	1:29	4000	1:25	1:50
6000	1:17	1:33	3000	1:34	1:67
5000	1:20	1:40	2000	1:50	1:100

We note, furthermore, that in cases in which the resolving power of the spectral instrument is determined by the photosensitive layer (photographic film) it is inappropriate to work with a slit close to the normal $S_1 \approx S_{10} = f_1\lambda/D_1$, since the width of the photographic image of the spectral lines is greater than the optical image of the slit. Hence the geometrical width of the entrance slit S_1 may be increased to a value S'_{10} such that the size of its image $S'_{20} = S'_{10}\frac{f_2}{f_1}W$ may be equal to the width of the apparatus function of the photosensitive layer α_{ph}, or, in other words, such that the resolving power of the photosensitive layer R_{ph} may be equal to that of the spectral instrument R_i with a slit apparatus function. From the equations $R_i \simeq R_{ph}$ or $\alpha_{ph} = S'_{20}$ we may derive an expression of S'_{10}:

$$S'_{10} = \alpha_{ph}\frac{f_1}{f_2}\frac{1}{W}. \tag{19}$$

This may be called the generalized normal slit width.

In increasing the slit from S_{10} to S'_{10} we lose hardly any resolving power, but at the same time we may make a considerable gain in the brightness of the spectral lines, since an increase in the slit leads to a rise in the amount of flux passing into the monochromator ($\Phi \sim S_1$), yet the width of the image determined by the photosensitive layer remains practically unchanged. Hence, for slit widths $S_{10} < S_1 < S'_{10}$, the brightness of the spectral lines E_l is proportional to S_1, and only for $S_1 > S'_{10}$ does it cease depending on the slit size and become defined by Eq. (18). For different instruments the difference between S'_{10} and S_{10} may be considerable, and it will depend on the relation between R_i^0 and R_{ph}, since $S'_{10}/S_{10} = R_i^0/R_{ph} = \alpha_{ph}/\lambda \cdot D_2/f_2$. For the examples given above, S'_{10} is four to ten times greater than S_{10}. Hence, by working with slits close to S'_{10} we may achieve a considerable gain in light transmission in such instruments without any loss of resolving power.

For a particular instrument and a particular photographic emulsion the value of S'_{10} may readily be determined experimentally from Fig. 57, which relates the width of the photographic image S_{ph} of a narrow spectral line to the width of the entrance slit S_1. The slit width S'_{10} in this diagram corresponds to the bend on the curve at which the width of the photographic image $S_{ph} \approx \alpha_{ph} \approx$ const passes into a linear relationship $S_{ph} \approx S_2 = S_1\frac{f_2}{f_1}W$. We note further that, for $R_{ph} > R_i^0$, $S'_{10} < S_{10}$; however, working with slits $S_1 < S_{10}$ is undesirable, as we then gain nothing in resolving power but lose in brightness.

All that we have said regarding the limitations imposed on the relative aperture of a camera objective for which the theoretical resolving power is realized and the choice of slit widths relates not only to prismatic spectrographs but also to those of the diffraction type, since relations (14), (16)-(18) and others are valid for these as well. In particular, the expression for the theoretical resolving power of the grating (1) may be brought to the form $R = A d\varphi/d\lambda$, which coincides with (15), since A corresponds to D_2 for a prism.

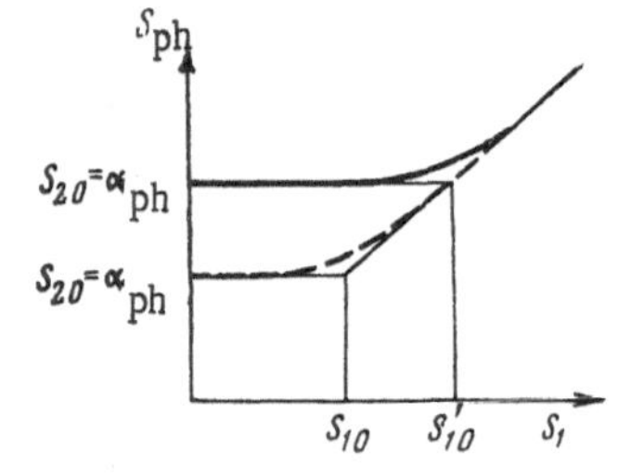

Fig. 57. Dependence of the width of the photographic image S_{ph} of a monochromatic spectral line on the width of the entrance slit S_1. Solid curve $\alpha_{ph} > \alpha_i$, broken curve $\alpha_{ph} \leq \alpha_i$.

2. If in a prismatic spectrograph the theoretical resolving power is not realized because $R_{ph} < R_i^0$, the only way to increase the practical resolving power and realize the R_i^0 of the particular instrument according to (17) lies in increasing the linear dispersion $dl/d\lambda = f_2 d\varphi/d\lambda$ by increasing the focal length of the camera objective f_2. Since α_{ph} = const, increasing f_2 will lead to a rise in $dl/d\lambda$ and hence R_{ph}, while the value of R_i^0 will remain constant. In this way, by increasing f_2, we may change the sign of the inequality $R_{ph} < R_i^0$ to $R_{ph} > R_i^0$.

However, an increase in $dl/d\lambda$ achieved by increasing the angular dispersion, for example by increasing the number of prisms or changing one prism for another with a material of higher dispersion $dn/d\lambda$, will not lead to any change in the sign of the inequality between R_{ph} and R_i^0, since both quantities (17) and (15) are proportional to $d\varphi/d\lambda$. At the same time, according to (15) and (17), an increase in the angular dispersion leads to an increase in the absolute value of the practical resolving power R_{ph}, which is quite significant, although even by itself an increase in the linear dispersion is extremely important in practice, as mentioned earlier. However, both these methods (increasing f_2 or $d\varphi/d\lambda$) involve a considerable modification to the manufactured spectral instrument, particularly if the linear dispersion has to be greatly increased.

However, our detailed analysis of the operation of spectral prisms has shown that the linear dispersion and practical resolving power may be considerably increased without seriously disturbing the standard instrument, but only by changing the dispositions of the prisms, i.e., taking them out of the minimum deviation condition [124].

It is well known that in the majority of existing prismatic spectral instruments the prisms are set in the position of minimum deviation (in which the beam suffers the least deviation in passing through the prism system). We shall not discuss the reasons for this arrangement of the prisms, but simply note that from the point of view of angular dispersion it is certainly not the best.

The general theory of the diffraction of a beam in a prism [125, 126] shows that the angular dispersion of a prism in the position of minimum deviation is relatively low, and it may be considerably increased if the angle of incidence of the beam on the entrance face of the prism i_1 (Fig. 56) is made smaller than that corresponding to the minimum deviation i_{1min}. We remember that setting the prism in the minimum-deviation position corresponds to a symmetrical path of the ray through the prism when $i_1 = i_2'$, $i_2 = i_1'$ (Fig. 56a), $D_1 = D_2$, $W = 1$.

The expression for the angular dispersion of the prism takes the form [125, 126]

$$\frac{d\varphi}{d\lambda} = \frac{\sin\alpha}{\cos i_1' \cos i_2'} \frac{dn}{d\lambda}, \tag{20}$$

where α is the refracting angle and $dn/d\lambda$ is the dispersion of the prism material.

Figure 58 shows the angular dispersion as a function of the angle of incidence calculated for a prism* made of S-11 glass with a refracting angle of $\alpha = 60°$ and $n_G = 1.673$.

The calculation shows that, if the angle of incidence is reduced by 12° relative to that at the minimum-deviation condition, the angular dispersion is doubled. On further reducing the angle and bringing it close to the limiting angle i_{1lim} (corresponding to the glancing emergence of the ray from the prism, $i_2' = 90°$) the angular dispersion rises sharply and may become very considerable. However, taking the prism out of the position of minimum deviation is also accompanied by a number of disadvantageous effects which may complicate the use of the prism in this setting. A detailed analysis of these effects shows that taking the prism out of the position of minimum deviation is nevertheless frequently of considerable practical value.

*All subsequent calculations will relate to the same prism.

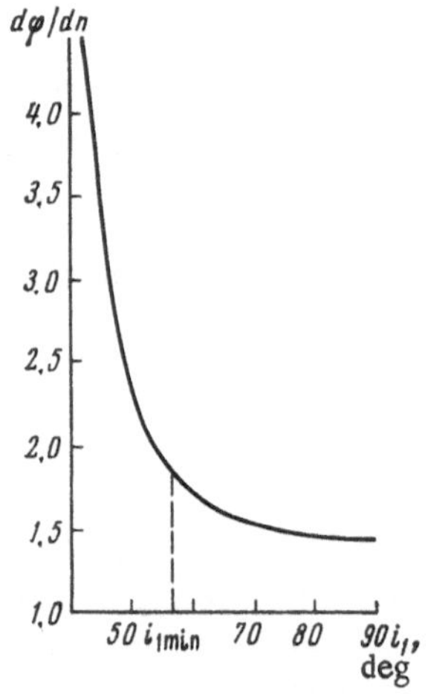

Fig. 58. Dependence of the angular dispersion of a prism $d\varphi/dn = \sin\alpha/\cos i_1'\cos i_2'$ on the angle of incidence i_1 on its front face ($\alpha = 60°$, $i_{1lim} = 41°10'$, $i_{1min} = 56°46'$, $n = 1.673$).

Let us briefly consider the phenomena associated with taking the prism out of the minimum-deviation position.

Angular Magnification of the Prism

If two parallel pencils (of the same wavelength λ) with an angle of di_1 between them fall on a prism, then after refraction in the prism the angle between them di_2' will, in general, not be equal to di_1. Calculation shows that $di_2' = W di_1$, where W is the angular magnification of the prism

$$W = \frac{\cos i_1 \cos i_2}{\cos i_1' \cos i_2'} = \frac{D_1}{D_2}. \quad (21)$$

Here, D_1 and D_2 are the widths of the parallel pencils (Fig. 56). The angular magnification is shown as a function of i_1 in Fig. 59, from which it follows that for $i_1 = 90°$, $W = 0$; for $i_1 = i_{1\,min}$, $W = 1$, and for $i_1 = i_{1\,lim}$ ($\cos i_2' = 0$), $W = \infty$.

In spectrographs, the angular magnification affects the width of the slit image, since $S_2 = S_1 \frac{f_2}{f_1} W$. Hence, on reducing the angle of incidence $i_1 < i_{1\,min}$, when W increases, the increase in the angular dispersion will be accompanied by a broadening of the optical image of

Fig. 59. Dependence of the angular magnification of a prism on the angle of incidence i_1.

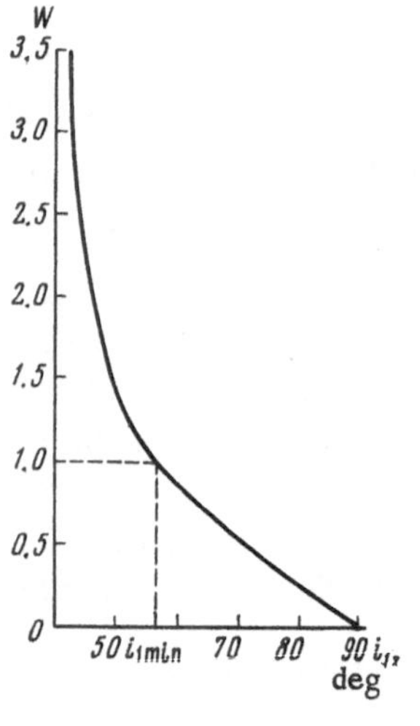

the spectral lines. Furthermore, if $S_2 > \alpha_{ph}$, the increase in the width of the image will lead to a reduction in the illumination of the spectral lines. If, however, $S_2 < \alpha_{ph}$, i.e., $R_{ph} < R_i$ (and we are particularly interested in this case), then the increase in the width of the optical image of the spectral lines associated with the increase in the angular magnification W will, within certain limits (while the condition $S_2 < \alpha_{ph}$ is satisfied) not lead to any reduction in illumination since, under these conditions, the width of the photographic image will be mainly determined by the apparatus function of the photosensitive layer.

Astigmatism of the Prism

If a homocentric converging or diverging pencil of monochromatic light falls on the prism, then, after refraction, it becomes astigmatic. Theory shows that at the minimum-deviation position the astigmatism is small (though not zero), but increases sharply when the prism is taken out of this position. When a parallel pencil falls on the prism, however, there is no astigmatism, and after refraction the beam remains parallel whatever the position of the prism. However, when using a lens collimator, which always has a certain residual chromatic aberration, the slit can only be sited at infinity for a narrow range of wavelengths and, hence (as experience shows) on moving away from the position of minimum deviation we can only obtain a stigmatic image of the spectral lines for a short range of the spectrum, of the order of 100 Å, although, in many cases, this is quite sufficient. The extent of the spectral range giving a stigmatic image of the lines depends, of course, on the quality of the collimator objective and the range of wavelengths.

Losses from Reflection

Since the angles of incidence of the rays on the faces of the prism are fairly large, so are the losses in Fresnel reflection during refraction. For the prism which we have been considering, when set in the position of minimum deviation ($i_{1\,min} = 56°46'$), the reflection losses amount to about 20%. On taking the prism out of the position of minimum deviation the reflection losses increase (owing to the increase in the angle of incidence on either the first or the second face), and hence the transmission coefficient in Eq. (18) diminishes. However, calculation shows that for slight deviations from the minimum-deviation position the transmission coefficient ε changes very little. Thus, on reducing the angle of incidence i_1 by 10° relative to $i_{1\,min}$ the transmission coefficient falls by only 10%, whereas the angular dispersion increases by a factor of 1.9.

Resolving Power of the Prism

The foregoing expression (15) for the theoretical resolving power $R_i^0 = D_2 d\varphi / d\lambda$ may be transformed by substituting $d\varphi / d\lambda$ from (20) and $D_2 = D_1/W$ from (21):

$$R_i^0 = D_2 \frac{\sin \alpha}{\cos i_1' \cos i_2'} \frac{dn}{d\lambda} \quad \text{or} \quad R_i^0 = D_1 \frac{\sin \alpha}{\cos i_1 \cos i_2} \frac{dn}{d\lambda}\,. \tag{22}$$

Analysis of these expressions with due allowance for the finite size of the prisms shows [127] that the resolving power of a prism of finite dimensions has a maximum when the prism is placed in the position of minimum deviation and diminishes on moving away from this position, whether i_1 becomes greater or less than $i_{1\,min}$. Analysis also shows that the main reason for the fall in R_i^0 is the reduction in the width of the beams D_1 and D_2 due to their limitation by the entrance or exit face of the prism and the change in the angular magnification. For $i_1 < i_{1\,min}$ the width of the beam is limited by the exit face l_2 (Fig. 56b) and for $i_1 > i_{1\,min}$ by the entrance face l_1. In particular, the reduction in the value of $R_i^0 = D_2 d\varphi / d\lambda$ for $i_1 < i_{1\,min}$ indicates that the beam cross section D_2 falls more rapidly than the angular dispersion $d\varphi / d\lambda$ rises. Figure 60 shows the dependence of $R_i^* = \frac{R_i^0}{l} \frac{d\lambda}{dn}$ on the angle of incidence i_1 obtained by calcula-

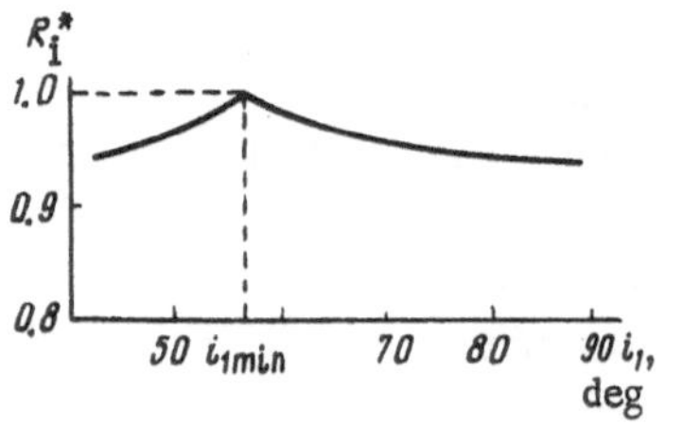

Fig. 60. Dependence of the theoretical resolving power of the prism R_i^* on the angle of incidence i_1.

tion from Eq. (22). Since the absolute value of R_i^0 depends on the dimensions of the prism and the value of $dn/d\lambda$, in the figure we simply present the value of R_i^*, which only depends on the angles. For $i_1 < i_{1\,min}$, $R_i^* = \frac{R_i^0}{l_2}\frac{d\lambda}{dn} = \frac{\sin\alpha}{\cos i_1'}$, where R_i^0 is the quantity in (22) transformed to the form $R_i^0 = l_2 \frac{\sin\alpha}{\cos i_1'}\frac{dn}{d\lambda}$, $l_2 = \frac{D_2}{\cos i_2'} = \text{const}$ is the length of the exit face of the prism (Fig. 56b). Analogously, for

$$i_1 > i_{1\,min} \quad R_i^* = \frac{R_i^0}{l_1}\frac{d\lambda}{dn} = \frac{\sin\alpha}{\cos i_2}, \quad l_1 = \frac{D_1}{\cos i_1} = \text{const}$$

is the length of the entrance face (Fig. 56b). It is important for us here that the fall in the theoretical resolving power on taking the prism out of the minimum-deviation position takes place quite slowly by comparison with the rise in the angular dispersion (Fig. 58). If for the initial setting of the prism at minimum deviation $R_{ph} < R_i^0$ (i.e., the resolution is determined by the photosensitive layer), on reducing the angle of incidence $i_1 < i_{1min}$ the quantity $d\varphi/d\lambda$, and hence $R_{ph} = \frac{\lambda}{d_{ph}} f_2 \frac{d\varphi}{d\lambda}$ of Eq. (4), will increase rapidly (Fig. 59), while the quantity $R_i^0 = D_2 \frac{d\varphi}{d\lambda}$ of Eq. (15) (Fig. 60) will slowly diminish, and we may therefore come to a situation in which $R_{ph} \geq R_i^0$, i.e., the practical resolving power will be determined not by the photosensitive layer but by the instrument. If the initial "reserve" in the theoretical resolving power of the instrument were fairly large on setting the prism at minimum deviation (as in the example under consideration), then on reaching the state $R_{ph} = R_i^0$ the resultant practical resolving power may be considerably greater than the original, despite the slight fall in the theoretical resolving power. As indicated earlier, the brightness of the spectral lines will hardly diminish at all, despite a certain increase in the width of the optical image.

Experiments carried out by the author with a number of instruments support these conclusions.

3. Thus, taking the prism out of the position of minimum deviation enables us to increase the linear dispersion and practical resolving power of high-transmission spectrographs. If the angle of incidence on the prism is slightly reduced $i_{1\,min} - i_1 \leq 10°$, the transmission of the instrument, and hence the brightness of the lines, will also change very little. However, there is no point in increasing the angular dispersion of one prism very greatly by taking it considerably away from the minimum-deviation position, as this greatly reduces the transmission coefficient on account of reflection losses at the exit face of the prism, at which the exit angle approaches 90° and the reflection coefficient approaches unity ($\varepsilon \to 0$).

In practice it is more reasonable to use the foregoing method of increasing the linear dispersion and practical resolving power by taking not a single prism but a system of several prisms out of the minimum-deviation position. We note that in the majority of modern prismatic instruments the dispersion system usually consists of three prisms with a refracting angle α close to 60°, so that the following methods of dispersion may be applicable to a large number of existing instruments.*

The angular dispersion of a system of prisms is in general not an additive quantity. For three prisms in air the expression for the angular dispersion may be expressed in the form

$$\left(\frac{d\varphi}{d\lambda}\right)_{1,2,3}=\left(\frac{d\varphi}{d\lambda}\right)_1 W_2W_3+\left(\frac{d\varphi}{d\lambda}\right)_2 W_3+\left(\frac{d\varphi}{d\lambda}\right)_3, \tag{23}$$

where $(d\varphi/d\lambda)_j$ is the angular dispersion of an individual prism, the expression for which is given in Eq. (20), and W_j is the angular magnification (21). Thus, for the whole system,

$$W_{1,2,3}=W_1W_2W_3. \tag{24}$$

In the particular case in which all three prisms are set in the position of minimum deviation, $W_{1,2,3} = 1$, $W_1 = W_2 = W_3 = 1$, and the dispersion of the system equals the sum of the dispersions of the individual prisms

$$\left(\frac{d\varphi}{d\lambda}\right)_{1,2,3}=\left(\frac{d\varphi}{d\lambda}\right)_{1m}+\left(\frac{d\varphi}{d\lambda}\right)_{2m}+\left(\frac{d\varphi}{d\lambda}\right)_{3m},$$

where $\left(\frac{d\varphi}{d\lambda}\right)_{jm}$ is the angular dispersion of a prism in the position of minimum deviation.

We note that, in contrast to the angular dispersion, the theoretical resolving power of the system of prisms is additive for any setting of the prisms, and the expression for this in the case of three prisms may be expressed in the form

$$R_{1,2,3}=R_1+R_2+R_3=D_6\left(\frac{d\varphi}{d\lambda}\right)_{1,2,3}, \tag{25}$$

where D_6 is the width of the beam emerging from the refracting face of the last prisms, $(d\varphi/d\lambda)_{1,2,3}$ is the dispersion of the whole system defined in (23).

Analysis of Eq. (23) shows that, if all the prisms are taken out of the minimum-deviation position in such a way that the angles of incidence on the entrance faces of all the prisms are smaller than the angle of incidence at minimum deviation, then the angular dispersion of each prism will increase (Fig. 58): $(d\varphi/d\lambda)_j > (d\varphi/d\lambda)_{jm}$. Furthermore, the angular magnification of each prism will be greater than unity: $W_j > 1$ (Fig. 59). As a result of this, the angular dispersion of the system will increase considerably, even if each prism is only taken very slightly away from the minimum-deviation position. Calculation shows that the resultant theoretical resolving power (25) will then only fall slightly, since each term in the sum (25) only diminishes a very little if the prisms are taken a short way away from minimum deviation. The reflection losses increase very slightly. However, the angular magnification of the whole system (24) increases, and this leads to a slight additional broadening of the optical images of the spectral lines. Thus, if the resolving power of the spectrometer as a whole is determined by the photosensitive layer with the system set at minimum deviation and the linear dispersion is inadequate for the purpose in hand, both the linear dispersion and the practical resolving power may be considerably increased by moving each of the prisms slightly away from minimum deviation, without any serious loss of transmission.

We tested this method of increasing the dispersion in a three-prism spectrograph (α = 60°, n = 1.7) with a relative aperture of 1 : 6 and a camera objective focal length of f_2 = 550

*A system of three prisms with a refracting angle of $\alpha \approx 60°$ and n ≈ 1.6-1.7 set at minimum deviation is optimal in respect of the angular dispersion and reflection losses, as analysis confirms.

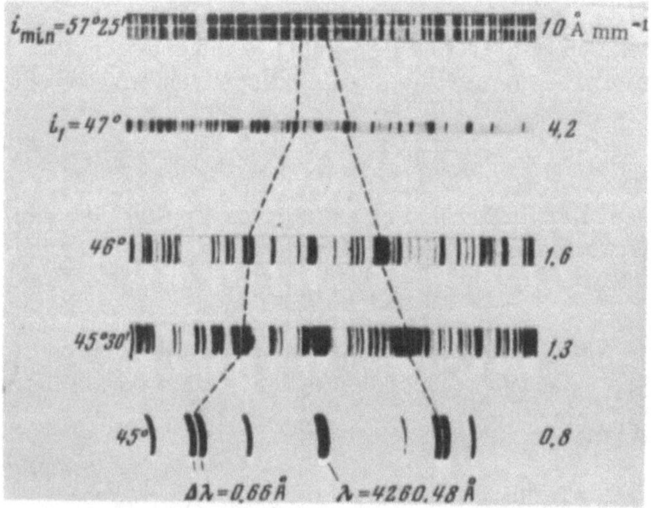

Fig. 61. Spectra of iron in the region of λ = 4260 Å obtained on taking the prisms out of the minimum-deviation position to various extents. Angular magnification of the system $W_{1,2,3} > 1$.

mm. On setting the prisms at minimum deviation, the angle of incidence on the face of the prisms was $i_{1\,min} = 57°25'$, and the reciprocal of the linear dispersion in the region of λ = 4200 Å was $(d\lambda/dl)_m$ = 10 Å/mm. On reducing the angle of incidence on the prisms to i_1 = 45°, the linear dispersion increased by a factor of 12 and the reciprocal of this parameter became 0.8 Å/mm (Fig. 61). Thereupon a number of closely situated lines not resolved in the spectrum obtained with the prisms in the minimum-deviation position became clearly resolved in the new setting. The quality of the image of the spectral lines was entirely satisfactory.

Having a system of prisms also enables us to arrange these in such a way that the angular magnification of the whole system $W_{1,2,3}$ equals unity while the angular dispersion of the system exceeds that obtained with each prism set at minimum deviation. For the case of three identical prisms $\alpha_1 = \alpha_2 = \alpha_3$, $dn_1/d\lambda = dn_2/d\lambda = dn_3/d\lambda$, this may be done if the middle prism is set at minimum deviation ($W_2 = 1$) and the first and third are arranged so that $W_{1,2,3} = W_1$, $W_3 = 1$, or $W_1 = 1/W_3$. This may be satisfied if the angle of incidence on the face of the first prism i_1 (with $i_1 > i_{1\,min}$!) is made equal to the angle of emergence of the ray from the third prism, i.e., if the passage of the ray through the whole system is symmetric, $D_1 = D_6$. In this case the expression for the angular dispersion may be converted to the form

$$\left(\frac{d\varphi}{d\lambda}\right)_{1,2,3} = 2\left(\frac{d\varphi}{d\lambda}\right)_3 + \left(\frac{d\varphi}{d\lambda}\right)_{2m} W_3. \tag{26}$$

For this setting ($i_1 > i_{1\,min}$) the angle of emergence of the ray from the last prism will be greater than the angle at minimum deviation, while the angle of incidence on the third prism will be smaller than the angle at minimum deviation. Thus, in accordance with Fig. 58,

$\left(\frac{d\varphi}{d\lambda}\right)_3 > \left(\frac{d\varphi}{d\lambda}\right)_{3m}$ and $W_3 > 1$. As a result of this, the angular dispersion of the system $\left(\frac{d\varphi}{d\lambda}\right)_{1,2,3}$ will be greater than its dispersion with each prism set at minimum deviation, i.e., $\left(\frac{d\varphi}{d\lambda}\right)_{1,2,3} >$

$\left(\frac{d\varphi}{d\lambda}\right)_{1m}+\left(\frac{d\varphi}{d\lambda}\right)_{2m}+\left(\frac{d\varphi}{d\lambda}\right)_{3m}$. Such an arrangement, with W = 1, may be achieved with any number of prisms greater than unity, in particular in autocollimation spectrographs (of the KSA-1 type) and in spectrographs with a system of constant-deviation prisms (of the ISP-51 type) with two trihedral prisms and one Abbe prism, which remains in the position of minimum deviation. The advantage of the latter system (in particular in the ISP-51) is the fact that on taking the trihedral prisms out of the position of minimum deviation, both for $W_{1,2,3} = 1$ and for $W_{1,2,3} > 1$ (but $\left(\frac{d\varphi}{d\lambda}\right)_{2m}$), the angle between the axes of the collimator and camera stays equal to 90°. In the case of a spectrograph with three ordinary 60° prisms, however, for any mode of withdrawing the prisms from the minimum-deviation position, the angle between the collimator and camera has to be changed.

In the ISP-51 instrument we were able to increase the dispersion by a factor of five (both for $W_{1,2,3} > 1$ and for $W_{1,2,3} = 1$), although any further increase in the dispersion was usually limited by the quality of the glass composing the prisms, e.g., the presence of cords (flaws) and other inhomogeneities, the influence of which increases on moving away from minimum deviation and worsens the quality of the image of the spectral lines.

We note that, in the case of a spectrograph with an autocollimation prism, the angular dispersion of the prism may be increased by rotating it in such a way that the angle of incidence on the entrance face of the prism may be greater than the angle at minimum deviation ($i_1 > i_{1\,min}$!).

The foregoing methods of increasing the linear dispersion are applicable to prisms made of various kinds of glass and other isotropic materials. However, in the case of prisms made of crystalline quartz (Cornu prisma) these methods failed. On taking such a prism out of the minimum-deviation position, the ray within the prism does not pass along the optic axis, so that the spectral lines become doubled as a result of birefringence.

4. In the foregoing discussion we have considered methods of increasing the linear dispersion and practical resolving power of high-transmission spectrographs when the theoretical resolving power of the instrument at minimum deviation is not realized, i.e., $\alpha_{ph} > \alpha_i$ or $S_1 < S'_{10}$. However, in practice, we often require to increase the linear dispersion, even in cases in which the work is being carried out with wide slits $S_1 > S'_{10}$ (for example, in spectroanalytical work), i.e., when the apparatus function is the slit function $S_2 = S_1 \frac{f_2}{f_1} W$. In these cases any increase in the linear dispersion achieved by taking the prisms away from minimum deviation always leads to a reduction in the brightness of the spectral lines, not only on account of the reduction in the transmission coefficient ε in (18) associated with the increasing reflection losses, but also on account of an increase in the width of the image S_2 [or, what amounts to the same thing, a reduction in the width D_2 in (18)] because of the increase in the angular magnification. This applies not only to one prism and the system of prisms considered above with $W_{1,2,3} > 1$, but also to the arrangement of the prisms with $W_{1,2,3} = 1$. In the latter case, although there is no additional broadening of the image of the slit S_2, nevertheless the width of the beam $D_2 = D_1$ diminishes as a result of its limitation by the entrance face of the prism l_1, since $D_1 = l_1 \cos i_1$, and for $i_1 > i_{1min}$, $D_1 < D_{1m}$ [124].

The advantage of the arrangement with $W_{1,2,3} = 1$ is simply the fact that an increase in the linear dispersion here leads to an increase (within certain limits) in the practical resolving power (16), since S_2 remains constant while $d\varphi/d\lambda$ increases [although R_1^0 will diminish in accordance with (15), as already indicated]. For the arrangement with $W_{1,2,3} > 1$, the practical resolving power (16) will diminish in view of the fact that S_2 rises more rapidly than $d\varphi/d\lambda$ (for one S-11 prism with $\alpha = 60°$ this has been demonstrated by direct calculation [124]).

5. Apart from the photographic method of recording the spectrum, we may also consider the photoelectric method.

In this case the expression for the flux emerging from the monochromator (for $S_1 > S_{10}$ and $f_1 = f_2 = f$) takes the form (Section 4 of this chapter, and [120])

$$\Phi_c = \varepsilon B \frac{hH}{f} (\delta\lambda)^2 D_2 \frac{d\varphi}{d\lambda} \quad \text{and} \quad \Phi_l = \varepsilon B_\infty \frac{hH}{f} \delta\lambda D_2 \frac{d\varphi}{d\lambda} \tag{27}$$

for the continuous and bright-line spectra, respectively. Here, $\delta\lambda = S_2(d\lambda/dl)$ is the spectral slit width determining the practical resolving power, and h is the height of the slits.

Since the quantity $D_2 d\varphi/d\lambda$ entering into the expression for Φ in (27) is the theoretical resolving power R_i^0 (15), and, as already mentioned, diminishes when the prisms are taken out of the minimum-deviation position (Fig. 60), this means that for a specified resolving power ($\delta\lambda$ = const) the flux will also diminish.

6. We have considered cases in which the linear dispersion of manufactured prismatic instruments may be increased without serious modification. It is nevertheless interesting to compare the increase in linear dispersion achieved by taking the prisms away from the minimum-deviation position with the increase secured by increasing the focal length of the camera objective f_2, with $\frac{dl}{d\nu} = f_2 \frac{d\varphi}{d\lambda} = \text{const}$. In this case (for $S_1 > S'_{10}$), the expression for the exposure (18) may be written

$$E_l = \varepsilon B_\infty \frac{HD_2}{f_2^2} = \varepsilon B_\infty \frac{H}{(dl/d\lambda)^2} \left(D_2 \frac{d\varphi}{d\lambda}\right) \frac{d\varphi}{d\lambda}. \tag{28}$$

Here the quantities $D_2(d\varphi/d\lambda) = R_i^0$ and ε become smaller for a slight change in the angle i_1 ($i_1 < i_{1\,min}$), but only slightly, while the angular dispersion $d\varphi/d\lambda$ increases more rapidly, so that the value of the exposure will increase on moving away from minimum deviation. Thus the method of increasing the linear dispersion by moving away from minimum deviation has an advantage as regards light transmission over that of increasing the focal length of the camera objective f_2.

7. Apart from the method of increasing the linear dispersion of manufactured instruments by taking the prisms out of the position of minimum deviation, we also proposed another method of increasing the dispersion of spectral instruments (with any number of prisms), based on the multiple (three to five times) passage of the light beam through the dispersing system. This may be done by having two (or four) plane mirrors with an external reflecting layer fixed at the top and bottom of the camera and collimator objective guides. The mirrors are so oriented that a beam emerging from the collimator objective is reflected from the mirrors, and after passing through the prism three times (or five times with four mirrors) falls into the camera objective.

Figure 62 presents the arrangement for the case of three passages of the ray through the prism; the dispersion of the instrument is thus increased three (or five) times. The theoretical resolving power also increases three (or five) times. If, however, the prisms are appropriately withdrawn from the minimum-deviation position, then the dispersion of the instrument may be substantially increased. The light transmission of the instrument of course diminishes considerably with this method of increasing the dispersion, since only part of the camera objective is operative, and, furthermore, additional losses appear as a result of reflection from the metallic mirrors. The reflection losses at the faces of the prism, however, increase negligibly, since, for prisms with $\alpha \approx 60°$ and $n \approx 1.6$, the angle of incidence at minimum deviation is close to the Brewster angle and hence the reflection losses at the faces of the prism will be no greater than 50%.

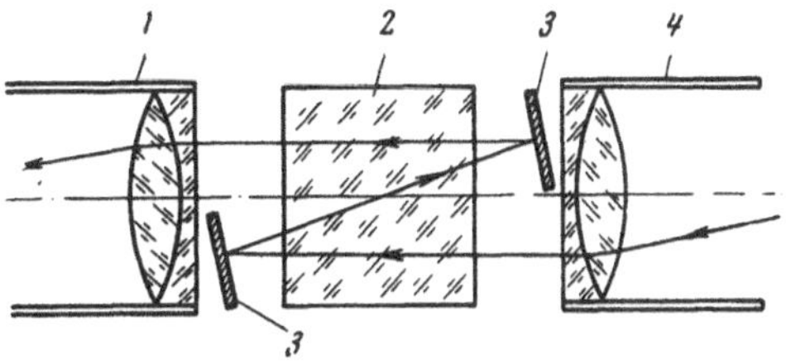

Fig. 62. Arrangement of the mirrors and path of the rays for a threefold passage of the beam through the dispersing system. 1) Camera; 2) prisms; 3) mirrors; 4) collimator.

We tested this method in the ISP-51 instrument, in which we were able to increase the dispersion by five or six times. It should be noted that in this method the light beam passes through the prism outside the principal section, and this leads to a rotation of the image of the slit around the axis. In order to make the lines in the photograph vertical, as before, the collimator slit must be correspondingly rotated around the axis.

The author employed the foregoing methods of increasing linear dispersion when developing a number of spectral-analytical methods of determining trace impurities in various materials, studying Raman spectra, etc.

We also used the principle of taking prisms out of the minimum-deviation position in styloscopes based on the autocollimation system, which were made by the optical workshops of the technological supply unit of the Academy of Sciences of the USSR in 1942-1944.

In setting up the many experiments and discussing the results, the author received inestimable help from G. S. Landsberg, and after his death great assistance was afforded by P. A. Bazhulin.

The author wishes to express his sincere thanks to S. G. Rautian, I. I. Sobel'man, A. A. Shubin, M. N. Markov, M. V. Shishkina, G. N. Zhizhin, V. N. Murzin, I. M. Aref'ev, V. A. Dudkin, T. L. Andreeva, A. I. Maslov, V. N. Sorokin, V. S. Petrov, L. A. Novikova, and all colleagues in the Optical Laboratory who took part in various operations or assisted in the investigations.

LITERATURE CITED

1. K. Angström, Ann. d. Phys., 6:163 (1901).
2. C. Schaefer, Ann. d. Phys., 16:93 (1905).
3. E. V. Bhar, Ann. d. Phys., 29:780 (1909); 33:585 (1910).
4. N. Coggeshall and E. Saier, J. Chem. Phys., 15:65 (1947).
5. R. Richards and H. Thompson, Proc. Roy. Soc., 195:1040 (1948).
6. M. O. Bulanin and V. M. Chulanovskii, Trans. of the Tenth Conf. on Spectroscopy, Vol. 1 [in Russian], Izd. L'vov Univ. (1957), p. 199.
7. H. Welsh, P. Pasher, and A. Dunn, J. Chem. Phys., 19 (1951).
8. M. Crawford, H. Welsh, and I. Locke, Phys. Rev., 75:1607 (1949).
9. B. Vodar, Spectrochem. Acta, 14:213 (1959).
10. I. Ketelaar, Spectrochem. Acta, 14:237 (1959).
11. I. Fahrenford and I. Ketelaar, J. Chem. Phys., 22:1631 (1954).
12. H. Thompson, Spectrochem. Acta, 14:145 (1959).
13. C. Lau, Spectrochem. Acta, 14:181 (1959).
14. L. Bellamy, Spectrochem. Acta, 14:192 (1959).
15. L. J. Bellamy, Infrared Spectra of Complex Molecules, John Wiley, New York (1958).

16. O. Girin and N. Bakhshiev, Usp. Fiz. Nauk, 79:235 (1963).
17. Sh. Chen and M. Takeo, Usp. Fiz. Nauk, 66:391 (1958).
18. I. Kirkwood, J. Chem. Phys., 2:351 (1934).
19. E. Bauer and M. Magat, J. Phys. Rad., 3:319 (1958).
20. N. Bayliss and A. Cole, Austr. J. Chem., 8:26 (1955).
21. M. Josien and N. Fuson, J. Chem. Phys., 22:1169 (1954).
22. P. Maine, L. Daly, and M. Maine, Can. J. Chem., 38:1921 (1960).
23. A. Buckingham, Proc. Roy. Soc., 255:32 (1960).
24. L. Bellamy and R. W. Williams, Proc. Roy. Soc., 255:22 (1960).
25. G. Caldow and H. Thompson, Proc. Roy. Soc., 254:1 (1960).
26. G. S. Landsberg, Izv. Akad. Nauk SSSR, Ser. Fiz., 3:373 (1938).
27. I. Watson and F. Crick, Nature, 175(4356):737 (1953).
28. N. D. Sokolov, Usp. Fiz. Nauk, 57:205 (1955).
29. G. C. Pimentel and A. L. McClellan, The Hydrogen Bond, W. H. Freeman & Co., San Francisco (1960).
30. A. E. Stanevich and I. G. Yaroslavskii, Dokl. Akad. Nauk SSSR, 137:60 (1961).
31. V. I. Malyshev, Candidate's Dissertation [in Russian], Moscow (1940).
32. V. I. Malyshev, Izv. Akad. Nauk SSSR, Ser. Fiz., 4:106 (1940).
33. V. I. Malyshev, Izv. Akad. Nauk SSSR, Ser. Fiz., 5:13 (1941).
34. V. I. Malyshev, Usp. Fiz. Nauk, 63:323 (1957).
35. V. I. Malyshev, Izv. Akad. Nauk SSSR, Ser. Fiz., 9:198 (1945).
36. A. A. Shubin, Izv. Akad. Nauk SSSR, Ser. Fiz., 14:442 (1950).
37. V. Liddel and E. Becker, Spectrochem. Acta, 10:70 (1957).
38. M. Thiel, E. Becker, and G. Pimentel, J. Chem. Phys., 27:95 (1957).
39. E. Becker and V. Liddel, J. Molec. Spectr., 2:1 (1958).
40. V. I. Malyshev and V. N. Murzin, in: Study of Experimental and Theoretical Physics [in Russian], Izd. AN SSSR, Moscow (1959), p. 134.
41. G. S. Landsberg and F. S. Baryshanskaya, Izv. Akad. Nauk SSSR, Ser. Fiz., 10:509 (1946).
42. V. I. Malyshev and M. V. Shishkina, Dokl. Akad. Nauk SSSR, 66:833 (1949).
43. V. I. Malyshev and M. V. Shishkina, Zh. Éksp. Teor. Fiz., 20:297 (1950).
44. G. S. Landsberg and F. S. Baryshanskaya, in: To the Memory of S. I. Vavilov [in Russian], Izd. AN SSSR, Moscow (1958), p. 147.
45. J. Ferry and S. Shulman, J. Am. Chem. Soc., 71:3198 (1949).
46. R. Badger, J. Chem. Phys., 8:288 (1940).
47. G. S. Landsberg and S. A. Ukholin, Dokl. Akad. Nauk SSSR, 16:399 (1937).
48. S. A. Ukholin, Dokl. Akad. Nauk SSSR, 16:403 (1937).
49. V. A. Dudkin, V. I. Malyshev, and S. G. Rautian, Optika i Spektroskopiya, 18:384 (1965).
50. E. Fishman, J. Phys. Chem., 65:2204 (1961).
51. V. I. Dianov-Klokov, Optika i Spektroskopiya, 17:146 (1964).
52. V. I. Malyshev, G. N. Zhizhin, and V. N. Smirnov, Physical Problems of Spectroscopy, Vol. 2 [in Russian], Izd. AN SSSR, Moscow (1963), p. 11.
53. V. I. Malyshev and S. G. Rautian, Izv. Akad. Nauk SSSR, Ser. Fiz., 23:1237 (1959).
54. K. P. Vasilevskii and B. S. Neporent, Optika i Spektroskopiya, 7:572 (1959).
55. D. Rank, W. Birtley, D. Eastman, and T. Wiggins, J. Chem. Phys., 32:296 (1960).
56. W. Benesh and T. Elder, Phys. Rev., 91:308 (1953).
57. K. Shiffrin, Scattering of Light in Cloudy Media [Russian translation], GITTL, Moscow (1951).
58. R. I. Bocharov, Author's Abstract of Candidate's Dissertation [in Russian], Moscow (1955).
59. V. I. Malyshev, M. N. Markov, and A. A. Shubin, Dokl. Akad. Nauk SSSR, 86:273 (1952).
60. V. I. Malyshev, Trans. of the Tenth Conf. on Spectroscopy, Vol. 1 [in Russian], Izd. L'vov Univ. (1957), p. 121.

61. I. Stratton and H. Houghton, Phys. Rev., 38 :159 (1931).
62. J. Johnson and J. Terrell, J. Opt. Soc. Amer., 45 :451 (1955).
63. M. Centeno, J. Opt. Soc. Amer., 31 :244 (1941).
64. G. Jobst, Ann. Phys., 76 :863 (1925); 78 :157 (1925).
65. H. C. van de Hulst, Light Scattering by Small Particles, John Wiley, New York (1957).
66. L. I. Mandelstam and G. S. Landsberg, Phys. Z. Sowietunion, 8 :378 (1935).
67. É. Shpol'skii, Izv. Akad. Nauk SSSR, Ser. Fiz., 11 :401 (1947).
68. H. Primas and H. Grünthard, Helv. Chim. Acta, 37, No. 44 (1954).
69. W. Gordy and P. Martin, J. Chem. Phys., 7 :99 (1939).
70. M. Josein and G. Sourisseau, Bull. Soc. Chim. Fran., 118 :178 (1955).
71. L. Bellamy and H. Hallam, Trans. Faraday Soc., 54 :1120 (1954).
72. R. Adams and I. Katz, J. Molec. Spectr., 1 :306 (1957).
73. A. Gantmakher, M. Vol'kenshtein, and Ya. Syrkin, Zh. Fiz. Khim., 14 :1569 (1940).
74. V. I. Malyshev and I. M. Aref'ev, Physical Problems of Spectroscopy, Vol. 2 [in Russian], Izd. AN SSSR, Moscow (1963), p. 9.
75. I. M. Aref'ev and V. I. Malyshev, Optika i Spektroskopiya, 13 :206 (1962).
76. P. Grange, J. Lascombe, and M. Josein, Spectrochem. Acta, 16 :981 (1960).
77. R. Badger, J. Chem. Phys., 2 :128 (1934); 3 :710 (1935).
78. G. V. Mikhailov, Zh. Éksp. Teor. Fiz., 36 :1368 (1959).
79. G. V. Mikhailov, Zh. Éksp. Teor. Fiz., 37 :1570 (1959).
80. P. A. Bazhulin and Yu. A. Lazarev, Optika i Spektroskopiya, 8 :206 (1960).
81. V. I. Malyshev, S. G. Rautian, and G. N. Zhizhin, Physical Problems of Spectroscopy, Vol. 2 [in Russian], Izd. AN SSSR, Moscow (1963), p. 14.
82. W. Elsasser, Heat Transformed by Infrared Radiation in Atmosphere, London (1942).
83. G. Kortüm and H. Verleger, Proc. Phys. Soc., 63 :462 (1950).
84. H. Herzberg, Vibration and Rotational Spectra of Polyatomic Molecules [Russian translation], IL, Moscow (1949).
85. P. A. Bazhulin and A. V. Rakov, Dokl. Akad. Nauk SSSR, 105 :54 (1955).
86. G. V. Mikhailov, Trans. of the Tenth Conference on Spectroscopy, Vol. 1 [in Russian], Izd. L'vov Univ. (1957), p. 227.
87. N. I. Rezaev, Trans. of the Tenth Conference on Spectroscopy, Vol. 1 [in Russian], Izd. L'vov Univ. (1957), p. 230.
88. A. V. Rakov, Trans. of the Tenth Conference on Spectroscopy, Vol. 1 [in Russian], Izd. L'vov Univ. (1957), p. 229.
89. P. A. Bazhulin and A. I. Sokolovskaya, Trans. of the Tenth Conference on Spectroscopy, Vol. 1 [in Russian], Izd. L'vov Univ. (1957), p. 56.
90. A. V. Rakov, Optika i Spektroskopiya, 7 :202 (1959).
91. A. V. Rakov, Optika i Spektroskopiya, 13 :369 (1962).
92. I. I. Sobel'man, Izv. Akad. Nauk SSSR, Ser. Fiz., 17 :554 (1953).
93. T. L. Andreeva and V. I. Malyshev, Optika i Spektroskopiya, 19 :213 (1965).
94. Ya. I. Frenkel', Kinetic Theory of Liquids [in Russian], Izd. AN SSSR, Moscow-Leningrad (1945).
95. H. Thompson and P. Torkington, Proc. Roy. Soc., 184 :3 (1945).
96. A. Mitchell and M. Zemansky, Resonance Radiation and Excited Atoms [Russian translation], ONTI, Moscow-Leningrad (1937).
97. I. I. Sobel'man, Usp. Fiz. Nauk, 54 :551 (1954).
98. M. A. Mazing, On the Broadening and Shift of Spectral Lines in the Plasma of a Gaseous Discharge, Consultants Bureau, New York (1962).
99. V. F. Kitaeva, Trudy FIAN, 11 :3 (1959).
100. A. N. Terenin, Photochemistry of Salt Vapors [in Russian], GTTI, Leningrad-Moscow (1934).

101. N. A. Prilezhaeva, Zh. Fiz. Khim., 5:1239 (1934).
102. H. Hanson, J. Chem. Phys., 23:1391 (1955).
103. V. A. Dudkin, T. L. Andreeva, V. I. Malyshev, and V. N. Sorokin, Optika i Spektroskopiya, 19:177 (1965).
104. I. M. Frank, Trudy GOI, No. 87, 3 (1933).
105. D. A. Jackson, Z. Phys., 75:223 (1932).
106. G. W. C. Kaye and T. H. Laby, Tables of Physical and Chemical Constants, John Wiley, New York (1959).
107. E. Lehrer, Z. Techn. Phys., 23:169 (1942).
108. N. Wright and L. Herscher, J. Opt. Soc. Amer., 37:211 (1947).
109. J. White and M. Liston, J. Opt. Soc. Amer., 40:29 (1950).
110. V. I. Malyshev, M. N. Markov, and A. A. Shubin, Izv. Akad. Nauk SSSR, Ser. Fiz., 17:654 (1953).
111. M. N. Markov, Zh. Tekh. Fiz., 24:1867 (1954).
112. M. N. Markov, Trans. of the Tenth Conference on Spectroscopy, Vol. 1 [in Russian], Izd. L'vov Univ. (1957), p. 403; Dokl. Akad. Nauk SSSR, 108:428 (1956).
113. W. Fastie, J. Opt. Soc. Amer., 42:647 (1952).
114. V. I. Malyshev and S. G. Rautian, Optika i Spektroskopiya, 6:550 (1959).
115. I. M. Aref'ev, V. I. Malyshev, and S. G. Rautian, Trans. of the Fifteenth Conf. on Spectroscopy, Vol. 2 [in Russian], VINITI, Moscow (1965), p. 650.
116. G. Harrison, J. Opt. Soc. Amer., 39:522 (1949).
117. H. Rowland, Phil. Mag., 35:397 (1893).
118. F. M. Gerasimov, I. Tel'tevskii, S. Nespelov, and V. Sergeev, Trans. of the Tenth Conference on Spectroscopy, Vol. 1 [in Russian], Izv. L'vov Univ. (1957), p. 394.
119. S. G. Rautian, Optika i Spektroskopiya, 1:564 (1959).
120. A. S. Toporets, Monochromators [in Russian], GTII, Moscow (1952).
121. I. V. Peisakhson, Optika i Spektroskopiya, 5:671 (1958).
122. P. Jacquinot, J. Opt. Soc. Amer., 44:761 (1954).
123. I. M. Nagibina and V. K. Prokof'ev, Spectral Instruments and Technique of Spectroscopy [in Russian], Mashinostroenie, Leningrad (1967).
124. V. I. Malyshev, Izv. Akad. Nauk SSSR, Ser. Fiz., 14:746 (1950).
125. H. Kayser, Handbuch der Spectroscopie, Leipzig (1900).
126. S. Czapski and O. Eppenstein, Grundzüge der Theorie der Optischen Instrumente, Leipzig (1924).
127. A. Hammer, Spectrochim. Acta, 2:365 (1944).